# ADVANCES IN EXPERIMENTAL MEDICINE AND BIOLOGY

## Volume 764

For further volumes:
http://www.springer.com/series/5584

Nigel Curtis · Adam Finn
Andrew J. Pollard

Editors

# Hot Topics in Infection and Immunity in Children IX

 Springer

*Editors*
Nigel Curtis
Professor of Paediatric Infectious
Diseases
The University of Melbourne;
Infectious Diseases Unit
The Royal Children's Hospital
Melbourne;
Infectious Diseases & Microbiology
Research Group
Murdoch Children's Research Institute
Parkville, Australia
nigel.curtis@rch.org.au

Andrew J. Pollard
Professor of Paediatric Infection &
Immunity
Director, Oxford Vaccine Group
University of Oxford
Oxford, UK
andrew.pollard@paediatrics.ox.ac.uk

Adam Finn
David Baum Professor of Paediatrics
University of Bristol
Bristol, UK
adam.finn@bristol.ac.uk

ISSN 0065-2598
ISBN 978-1-4614-4725-2      ISBN 978-1-4614-4726-9 (eBook)
DOI 10.1007/978-1-4614-4726-9
Springer New York Dordrecht Heidelberg London

# Preface

The ninth 'Infection and Immunity in Children' (IIC) course was held in July 2011 at St Catherine's College, Oxford, UK. This book, the ninth in the series, comprises chapters based on presentations made at that course. The series provides succinct and readable updates on virtually every topic of relevance to Paediatric Infectious Diseases (PID).

The tenth edition is currently in preparation following the 2012 course, which once again had a programme delivered by renowned top-class speakers. The programme for the eleventh course, planned for 26–28th June 2013, is being finalised as this book goes to press.

PID has emerged as a powerful discipline for the improvement of child health in Europe and worldwide over the last quarter of a century. The European Society for Paediatric Infectious Diseases (ESPID) now holds the largest annual conference in PID in the world and sponsors a wide range of educational activities for trainees and specialists. Among them is the Oxford IIC course, which, with other ESPID-sponsored activities, is an integral part of the two-year Oxford Diploma in PID. This course began in 2008 and continues to enrol trainees from all over Europe, being the first recognised academic qualification of any kind in PID outside North America.

The future of PID promises to be eventful, challenging and fascinating, as new infections emerge, old infections evolve and new diagnostic techniques, treatments and vaccines become available. There is much that is new to learn about and we hope this book will provide a further useful contribution to the materials available to trainees and practitioners in our important and rapidly developing field.

Melbourne, Australia                                          Nigel Curtis
Bristol, UK                                                      Adam Finn
Oxford, UK                                              Andrew J. Pollard

# Acknowledgments

We thank all the speakers from the 2011 'Infection and Immunity in Children' (IIC) course who gave their time both to attending the meeting and to contributing presentations, discussion and chapters for this book. We also thank their co-authors, many of whom are trainees who have also attended the course on one or more occasions.

Since its early days, the course has been organised by Sue Sheaf. Its continuing and growing success is in large part due to her efforts, which include all the necessary liaison with the faculty, the handling of applications and bursaries for all the delegates, working with the college with respect to all practical arrangements for the course, the audiovisual team, the social programme and, of course, cajoling us, the organisers, into doing our part, to name but a few. Without Sue, the Oxford IIC course would not be what it is and we cannot thank her enough.

Another difficult task is the administration of the production of successive editions of this book. As anyone who has ever been involved in putting together a publication of this kind will know, this is a major undertaking requiring a skilled and dedicated team of professionals. Pamela Morison is that team. Careful copyediting and correction of the materials provided by the authors, editors and publishers is essential to ensuring a high-quality final product and Pam is the person who does this work. She also has to persuade all these parties to deliver the materials in a timely fashion, which Pam does with the right mix of persuasiveness and insistence. For Pam's good-natured approach, diplomatic skills and all her hard work we thank her and gratefully share with her the credit for this book's production.

We thank the European Society for Paediatric Infectious Diseases (ESPID) for consistent support and financial assistance for this and previous courses and also for providing bursaries, which have paid the costs of many young ESPID members' attendance. We also acknowledge the recognition given to the course by the UK Royal College of Paediatrics and Child Health (RCPCH).

Finally, we are grateful to the staff of St Catherine's College, Oxford, UK where the course was held and to several pharmaceutical industry sponsors who generously offered unrestricted educational grants towards the budget for the meeting.

# Contents

# Contributors

**Inge Ahout** Department of Pediatrics, Radboud University Medical Centre, Nijmegen, The Netherlands
e-mail: I.Ahout@cukz.umcn.nl

**Mona A. Al-Dabbagh** Division of Infectious and Immunological Diseases, Department of Pediatrics, BC Children's Hospital, Vancouver, Canada; King Abdulaziz Medical City, Jeddah, Saudi Arabia
e-mail: dabbaghM@ngha.med.sa

**Mary E. Appleton** Pediatric Infectious Diseases, Dalhousie University, Canadian Center for Vaccinology, IWK Health Center, Halifax, Canada
e-mail: mary.appleton@iwk.nshealth.ca

**Shai Ashkenazi** Professor of Pediatrics, Sackler Faculty of Medicine, Schneider Children's Medical Center, Petach Tikva, Israel
e-mail: ashai@post.tau.ac.il

**Maria E. Bottazzi** Sabin Vaccine Institute and Texas Children's Center for Vaccine Development, Departments of Pediatrics and Molecular Virology & Microbiology, National School of Tropical Medicine, Baylor College of Medicine, Houston, USA
e-mail: Bottazzi@bcm.edu

**Andrew Cant** Institute of Cellular Medicine, Newcastle University, Newcastle upon Tyne, UK
e-mail: andrew.cant@nuth.nhs.uk

**Jonathan R. Carapetis** Telethon Institute for Child Health Research, University of Western Australia, Perth, Australia.
e-mail: jcarapetis@ichr.uwa.edu.au

**Stuart C. Clarke** Academic Unit of Clinical & Experimental Sciences, Faculty of Medicine, University of Southampton and Health Protection Agency, Southampton, UK
e-mail: S.C.Clarke@soton.ac.uk

**Jonathan C. Craig** School of Public Health, University of Sydney, Centre for Kidney Research, Children's Hospital at Westmead, Sydney, Australia
e-mail: Jonathan.craig@sydney.edu.au

**Nigel Curtis** Department of Paediatrics, The University of Melbourne; Infectious Diseases Unit, The Royal Children's Hospital Melbourne; Infectious Diseases & Microbiology Research Group, Murdoch Children's Research Institute, Parkville, Australia
e-mail: nigel.curtis@rch.org.au

**Toni Darville** Division of Pediatric Infectious Diseases, College of Medicine, University of Pittsburgh Medical Center, Pittsburgh, PA, USA
e-mail: darvilletonil@chp.edu

**Simon Dobson** Division of Infectious and Immunological Diseases, Department of Pediatrics, BC Children's Hospital, Vancouver, Canada
e-mail: sdobson@cw.bc.ca

**Eric Dumonteil** Laboratorio de Parasitologia, Centro de Investigaciones Regionales "Dr Hideyo Noguchi", Universidad Autonoma de Yucatan, Merida, Yucatan, Mexico
e-mail: edumonte@tulane.edu

**Ahmed Fahal** University of Khartoum, Mycetoma Research Centre, Khartoum, Sudan
e-mail: ahfahal@hotmail.com

**Saul N. Faust** Academic Unit of Clinical & Experimental Sciences, Faculty of Medicine, University of Southampton, University Hospital Southampton NHS Foundation Trust, NIHR Wellcome Trust Clinical Research Facility, Southampton, UK
e-mail: s.faust@soton.ac.uk

**Gerben Ferwerda** Department of Pediatrics, Radboud University Medical Centre, Nijmegen, The Netherlands
e-mail: J.Ferwerda@cukz.umcn.nl

**Adam Finn** Paediatrics, University of Bristol, Bristol Royal Hospital for Children, Bristol, UK.
e-mail: adam.finn@bristol.ac.uk

**Steve M. Green** Health Protection Agency, Southampton, UK
e-mail: S.Green@soton.ac.uk

**Ronald de Groot** Department of Pediatrics, Radboud University Medical Centre, Nijmegen, The Netherlands
e-mail: R.deGroot@cukz.umcn.nl

**Gabrielle M. Haeusler** Infectious Diseases Unit, The Royal Children's Hospital Melbourne, Parkville, Australia
e-mail: gabrielle.haeusler@rch.org.au

**Michael J. Heffernan** Sabin Vaccine Institute and Texas Children's Center for Vaccine Development, Departments of Pediatrics and Molecular Virology & Microbiology, National School of Tropical Medicine, Baylor College of Medicine, Houston, USA
e-mail: mjheffer@bcm.edu

**Stefanie Henriet** Department of Pediatric Infectious Diseases and Immunology, Nijmegen Institute for Infection, Inflammation and Immunity, Radboud University Nijmegen Medical Centre, Nijmegen, The Netherlands
e-mail: S.Henriet@cukz.umcn.nl

**Steven M. Holland** Laboratory of Clinical Infectious Diseases, National Institute of Allergy and Infectious Diseases, National Institutes of Health, Bethesda, USA
e-mail: SHOLLAND@niaid.nih.gov

**Peter J. Hotez** Sabin Vaccine Institute and Texas Children's Center for Vaccine Development, Departments of Pediatrics and Molecular Virology & Microbiology, National School of Tropical Medicine, Baylor College of Medicine, Houston, USA
e-mail: hotez@bcm.edu

**David Isaacs** Department of Infectious Disease and Microbiology, Children's Hospital at Westmead, Westmead, Australia
e-mail: davidi@chw.edu.au

**Dominic Kelly** Children's Hospital, Oxford, UK
e-mail: dominic.kelly@paediatrics.ox.ac.uk

**Gilat Livni** Sackler Faculty of Medicine, Department of Pediatrics A, Schneider Children's Medical Center of Israel, Petach Tikva, Israel
e-mail: gilat@orange.net.il

**Noni E. MacDonald** Pediatric Infectious Diseases, Dalhousie University, Canadian Center for Vaccinology, IWK Health Center, Halifax, Canada
e-mail: noni.macdonald@dal.ca

**Tim Mailman** Dalhousie University, IWK Health Center, Halifax, Canada
e-mail: tim.mailman@iwk.nshealth.ca

**Kathryn Maitland** KEMRI-Wellcome Trust Programme, Kilifi, Kenya; Department of Paediatrics, Imperial College, London, UK
e-mail: kathryn.maitland@gmail.com

**Oliver Morris** Academic Unit of Clinical & Experimental Sciences, Faculty of Medicine, University of Southampton, University Hospital Southampton NHS Foundation Trust, NIHR Wellcome Trust Clinical Research Facility, Southampton, UK
e-mail: omorris@doctors.org.uk

**James P. Nataro** Department of Pediatrics, University of Virginia School of Medicine, University of Virginia Children's Hospital, Charlottesville, USA
e-mail: JPN2R@hscmail.mcc.virginia.edu

**Peter Olupot-Olupot** Mbale Regional Referral Hospital, Mbale, Uganda

**Ann Pallett** University Hospital Southampton NHS Foundation Trust, Southampton, UK
e-mail: Ann.Pallett@uhs.nhs.uk

**Andrew D. Pearson** Department of Advanced Computational Biology, University of Maryland, USA
e-mail: drandrewpearson@googlemail.com

**Andrew J. Pollard** Oxford Vaccine Group, University of Oxford; Children's Hospital, Oxford, UK
e-mail: andrew.pollard@paediatrics.ox.ac.uk

**Athimalaipet V. Ramanan** Bristol Royal Hospital for Children, Bristol; Royal National Hospital for Rheumatic Diseases, Bath, UK
e-mail: Athimalaipet.Ramanan@UHBristol.nhs.uk

**Marion R. Roderick** Bristol Royal Hospital for Children, Bristol, UK
e-mail: rodericks1000@hotmail.com

**Paul Roderick** Academic Unit of Primary Care and Population Sciences, Faculty of Medicine, University of Southampton, UK
e-mail: pjr@soton.ac.uk

**Jennifer C. Smith** Pediatric Infectious Diseases, Dalhousie University, Canadian Center for Vaccinology, IWK Health Center, Halifax, Canada
e-mail: jaurora777@yahoo.ca

**Catharina Svanborg** Department of Microbiology, Immunology and Glycobiology, Institute of Laboratory Medicine, Lund University, Lund, Sweden
e-mail: Catharina.Svanborg@med.lu.se

**Marc Tebruegge** Academic Unit of Clinical & Experimental Sciences, Faculty of Medicine, University of Southampton, University Hospital Southampton NHS Foundation Trust, Southampton, UK
e-mail: M.Tebruegge@soton.ac.uk

**Andrew Tuck** Academic Unit of Clinical & Experimental Sciences, Faculty of Medicine, University of Southampton and Health Protection Agency, Southampton, UK
e-mail: A.C.Tuck@soton.ac.uk

**Alex van Belkum** bioMérieux, Microbiology Unit, La Balme-Les-Grottes, France
e-mail: alex.vanbelkum@biomerieux.com

**Wendy W. J. van de Sande** Medical Microbiology and Infectious Diseases, Erasmus MC, Rotterdam, The Netherlands
e-mail: w.vandesande@erasmusmc.nl

**Paul E. Verweij** Department of Medical Microbiology, Nijmegen Institute for Infection, Inflammation and Immunity, Radboud University Nijmegen Medical Centre, Nijmegen, The Netherlands
e-mail: P.Verweij@mmb.umcn.nl

**Adilia Warris** Department of Pediatric Infectious Diseases and Immunology, Nijmegen Institute for Infection, Inflammation and Immunity, Radboud University Nijmegen Medical Centre, Nijmegen, The Netherlands
e-mail: A.Warris@cukz.umcn.nl

**Gabrielle J. Williams** School of Public Health, University of Sydney, Centre for Kidney Research, Children's Hospital at Westmead, Sydney, Australia
e-mail: gabrielle.williams1@health.nsw.gov.au

**Kenneth M. Zangwill** Pediatric Infectious Diseases, David Geffen School of Medicine at UCLA, Harbor-UCLA Medical Center, Los Angeles, USA
e-mail: kzangwill@uclacvr.labiomed.org

Speakers at the ninth 'Infection and Immunity in Children' (IIC) course held in July 2011 at St Catherine's College, Oxford, UK

# Innovation for the 'Bottom 100 Million': Eliminating Neglected Tropical Diseases in the Americas

Peter J. Hotez, Eric Dumonteil, Michael J. Heffernan and Maria E. Bottazzi

**Abstract**

An estimated 100 million people in the Latin American and Caribbean (LAC) region live on less than US$2 per day, while another 46 million people in the US live below that nation's poverty line. Almost all of the 'bottom 100 million' people suffer from at least one neglected tropical disease (NTD), including one-half of the poorest people in the region infected with hookworms, 10 % with Chagas disease, and up to 1–2 % with dengue, schistosomiasis, and/or leishmaniasis. In the US, NTDs such as Chagas disease, cysticercosis, toxocariasis, and trichomoniasis are also common among poor populations. These NTDs trap the poorest people in the region in poverty, because of their impact on maternal and child health, and occupational productivity. Through mass drug administration (MDA), several NTDs are on the verge of elimination in the Americas, including lymphatic filariasis, onchocerciasis, trachoma, and possibly leprosy. In addition, schistosomiasis may soon be eliminated in the Caribbean. However, for other NTDs including hookworm infection, Chagas disease, dengue, schistosomiasis, and leishmaniasis, a new generation of 'anti-poverty vaccines' will be required. Several vaccines for dengue are under development by multinational pharmaceutical companies, whereas others are being pursued through non-profit product development partnerships (PDPs), in collaboration with developing country manufacturers in Brazil and Mexico. The Sabin Vaccine Institute PDP is developing a primarily preventive bivalent recombinant human hookworm vaccine,

P. J. Hotez (✉) · M. J. Heffernan · M. E. Bottazzi
Sabin Vaccine Institute and Texas Children's Center for
Vaccine Development, Departments of Pediatrics and
Molecular Virology & Microbiology, National School
of Tropical Medicine, Baylor College of Medicine,
Houston, USA
e-mail: hotez@bcm.edu

M. J. Heffernan
e-mail: mjheffer@bcm.edu

M. E. Bottazzi
e-mail: Bottazzi@bcm.edu

E. Dumonteil
Laboratorio de Parasitologia, Centro de Investigaciones
Regionales "Dr Hideyo Noguchi", Universidad
Autonoma de Yucatan, Merida, Yucatan, Mexico
e-mail: edumonte@tulane.edu

N. Curtis et al. (eds.), *Hot Topics in Infection and Immunity in Children IX*,
Advances in Experimental Medicine and Biology 764, DOI 10.1007/978-1-4614-4726-9_1,
© Springer Science+Business Media New York 2013

**Table 1.1** Poverty in the Americas. (Data from [1, 4])

| | Region | |
|---|---|---|
| | Latin America & Caribbean | United States |
| Population in 2010 | 578 million | 306 million |
| Living on less than $1 per day in 2005 | 8.2 % | |
| Estimated number of people living on less than $1 per day | 48 million | |
| Living on less than $2 per day in 2005 | 17.1 % | |
| Estimated number of people living on less than $2 per day | 99 million | |
| Living below the poverty line in 2010 | | 15.1 % |
| Number of people living below the poverty line | | 46 million |

**Table 1.2** Countries in the Americas with the lowest Human Development Indices. (Data from [3])

| Country | 2008 HDI rank | Comparator countries with similar HDI rank |
|---|---|---|
| Haiti | 148 | Sudan, Kenya |
| Guatemala | 121 | South Africa, Gabon |
| Nicaragua | 120 | South Africa, Gabon |
| Honduras | 117 | Mongolia, Kyrgyzstan |
| Bolivia | 111 | Egypt, Indonesia |
| Guyana | 110 | Egypt, Indonesia |
| El Salvador | 101 | Algeria, Cape Verde, Vietnam |
| Paraguay | 98 | Sri Lanka, Iran |

which is about to enter phase 1 clinical testing in Brazil, as well as a new therapeutic Chagas disease vaccine in collaboration with several Mexican institutions. The Chagas disease vaccine would be administered to seropositive patients to delay or prevent the onset of Chagasic cardiomyopathy (secondary prevention). Together, MDA and the development of new anti-poverty vaccines afford an opportunity to implement effective control and elimination strategies for the major NTDs in the Americas.

## 1.1 Introduction: Poverty in the Americas

Today, almost one-fifth of the 578 million people who live in the Latin American and Caribbean (LAC) region live in severe poverty. As shown in Table 1.1, an estimated 50 and 100 million people live on less than US$1.25 per day (the World Bank poverty level) and US$2 per day, respectively [1]. These populations represent the LAC region's bottom 100 million. Poverty has an uneven distribution in the LAC region, such that the bottom 100 million cluster in defined geographical areas of Mexico and Central America, South America, and in areas with large indigenous populations. For example,

almost one-half of rural households in Mexico are considered poor, most of them in southern Mexico; one-half of the population of Central America also lives in poverty [2]. Four of the eight LAC countries with the lowest human development indices (HDIs) are found in Central America, including Guatemala, Nicaragua, Honduras, and El Salvador (Table 1.2) [3], where poverty disproportionately affects indigenous populations [2]. Among South American populations, there are an estimated 20 million poor in Brazil, particularly in the northeastern part of the country, and especially among women and children [2]. Some of the poorest people in LAC live in indigenous communities and mountainous areas of Bolivia, Peru, and Ecuador and in the Chaco of Bolivia and Paraguay [2, 3]. More than one-third of impoverished LAC populations live in the Andean region [2]. The Caribbean countries of Haiti and Guyana also stand out for their low HDIs [3]. Overall, the HDIs of the poorest countries in the LAC region are similar to many poor African and Asian countries (Table 1.2).

Poverty in the Americas is not restricted to the LAC region [4–6]. With the recent recession in the US the percentage of people living below the poverty line has increased to its highest level in more than 50 years [4]. It was recently determined that the US poverty rate has increased to 15 %, with 46 million people living below the poverty line [4]. Poverty in the US also has an uneven distribution, with the highest rates along the Gulf Coast and in South Texas [5, 6]. Poverty

**Table 1.3**  The neglected tropical diseases of the Latin American and Caribbean (LAC) region. (Modified from [9])

| Disease | Estimated DALYs in LAC (million) | Estimated number of cases (million) | Estimated bottom 100 million affected (%) | Countries most affected |
|---|---|---|---|---|
| Hookworm | 0.1–1.9 | 50 | 50 | Brazil, Paraguay, Guatemala, Colombia |
| Ascariasis | 0.1–1.1 | 84 | 84 | Brazil, Mexico, Guatemala, Argentina |
| Trichuriasis | 0.3–1.1 | 100 | 100 | Brazil, Mexico, Colombia, Guatemala |
| Chagas disease | 0.7 | 8–9 | 10 | Bolivia, Mexico, Colombia, Central America |
| Dengue | 0.1 | >0.6 | 1 | Brazil, Caribbean |
| Leishmaniasis | <0.1 | 0.1 | 1 | Brazil, Colombia, Peru, Nicaragua, Bolivia |
| Schistosomiasis | <0.1 | 1.8 | 2 | Brazil, Venezuela, Suriname, Saint Lucia |
| Lymphatic FIlariasis | <0.1 | 0.7 | 1 | Haiti, Brazil, Dominican Republic, Guyana |
| Trachoma | <0.1 | 1.1 | 1 | Brazil, Guatemala |
| Leprosy | <0.1 | <0.1[a] | <1 | Brazil |
| Onchocerciasis | <0.1 | <0.1 | <1 | Guatemala, Mexico, Ecuador, Venezuela, Brazil, Colombia |
| Human rabies from dogs | <0.1 | <100[b] | <1 | Haiti, Bolivia |
| Cysticercosis | ND | 0.4 | <1 | ND |
| Toxoplasmosis | ND | ND | ND | Brazil |

*ND* not determined
[a] New cases
[b] Cases

also occurs in the Canadian and the Alaskan Arctic among Inuit and related populations [7].

## 1.2  Neglected Tropical Diseases in the Americas

Among the LAC region's bottom 100 million, approximately one-half lack adequate food [8]. Equally tragic are the results of our previous analysis showing that almost all the bottom 100 million suffers from one or more neglected tropical disease [9]. Globally, the neglected tropical diseases or 'NTDs' comprise 17 major chronic parasitic and related infections [10–12]. Importantly, they not only adversely affect the health of infected individuals, but also trap people in poverty, because of their long-term impact on child growth and cognitive development, pregnancy outcome, and occupational productivity [11, 13]. Shown in Table 1.3 is a ranking of the leading NTDs in the LAC region by disease burden, measured in disability-adjusted life years.

The leaders are hookworm and other soil-transmitted helminths, Chagas disease (American trypanosomiasis), and dengue, followed by leishmaniasis, schistosomiasis, lymphatic filariasis, trachoma, leprosy, onchocerciasis, and cysticercosis [9]. Almost all of the bottom 100 million are infected with one or more soil-transmitted helminths, including trichuriasis, ascariasis, and hookworm. Approximately 10 % are affected by Chagas disease and 1 % by dengue fever. Precise prevalence data for toxoplasmosis in the LAC region is lacking, however it is believed to be common, especially in Brazil [14]. With the exception of Chagas disease and possibly trachoma, most of the NTDs were introduced into the Americas through the Atlantic slave trade [15], and these conditions still disproportionately affect African-American and indigenous populations [9].

Table 1.3 also shows the disproportionate representation of the most common NTDs in low HDI countries, for example, hookworm in Paraguay and Guatemala, and Chagas disease in Bolivia and in several Central American countries

**Table 1.4** Major neglected infections of poverty in the US. (Modified from [5])

| Disease | Estimated number of cases (million) | Major populations affected |
|---|---|---|
| Toxocariasis | 1–3 | African American |
| Trichomoniasis | 0.9 | African American |
| Chagas disease | Up to 1 | Hispanic American |
| Dengue | 0.1–0.2 | Hispanic American |
| Cysticercosis | 0.04–0.16 | Hispanic American |

[9]. Dengue and dengue haemorrhagic fever (DHF) have become more prevlent throughout the LAC region [16]. Brazil has the largest number of NTDs of any country in the LAC region [9, 17], mostly among impoverished populations in the northeastern part of the country. However, the severe poverty in northeastern Brazil does not affect its HDI sufficiently to rank that country in the bottom tier.

The economic downturn in the US has increased media attention to the plight of the Northern American poor, and with it, increased recognition of a hidden underbelly of neglected infections of poverty [5, 6]. Some NTDs found in the LAC region are also found in South Texas and the Gulf Coast, including Chagas disease, cysticercosis, and dengue (Table 1.4) [5, 6]. Of particular concern is evidence suggesting autochthonous transmission of Chagas disease and dengue in these regions. High prevalence rates of toxocariasis and trichomoniasis also occur among people living in poverty in the US and presumably throughout the Americas [5, 6].

## 1.3 Adverse Effects on Maternal and Child Health, Poverty, and Conflict

Overall, there are three major adverse health and socioeconomic consequences of high prevalence NTDs in the Americas, related to their impact on (1) maternal and child health, (2) poverty, and (3) conflict (Box 1).

**Box 1 Effects of the High Burden NTDs on Maternal and Child Health, Poverty, and Conflict in the LAC Region**

Maternal Child Health
- Hookworm is associated with anemia in children and pregnancy leading to intellectual and cognitive deficits, low neonatal birth weight, and increased maternal morbidity.
- Chagas disease in pregnancy increases the risk of miscarriage and preterm birth and is associated with vertical transmission in 1 in 20 seropositive mothers resulting in neonatal infection and death.
- Dengue and DHF disproportionately affect children and in pregnancy are linked with elevated rates of preterm birth, cesarean delivery, and vertical transmission.

Poverty
- Hookworm is associated with 40 % reduction in future wage earning.
- Chagas disease results in economic losses exceeding US$ 1 billion annually with lifetime costs per patient averaging almost US$ 12,000.
- High dengue economic costs remain under investigation.

Conflict
- Destabilizing effects on agricultural productivity, land use, food security, economic growth.
- Links with guerilla movements and narcotrafficking.

### 1.3.1 Maternal and Child Health

It is not commonly appreciated that the highest burden NTDs in the LAC region, including hookworm and the other soil-transmitted helminths, Chagas disease, and dengue exert special adverse consequences for the health of children and women.

Hookworms cause chronic intestinal blood loss leading to iron deficiency anemia and protein malnutrition. Hookworm anemia is more pronounced in women and children because of their unique iron metabolisms [18]. Throughout the LAC region *Necator americanus* is the predominant species causing hookworm disease and anemia [9, 18]. In children this infection is associated with deficiencies in physical growth and cognitive development, while in pregnancy it is linked to increased maternal morbidity and poor neonatal outcomes [18]. Two recent systematic reviews confirm the association between hookworm infection and anemia in children and pregnant women [19, 20]. Specifically, in the LAC region *N. americanus* infection was recently linked to anemia in young children in rural Minas Gerais State, Brazil [21], and with low neonatal birth weight among pregnant women in Peru [22].

Chagas disease has also emerged as an important maternal and child health problem, with mother-to-child transmission increasingly recognized as an important route of infection [23]. Pregnancy has been shown to enhance *T. cruzi* parasitemia [24], with vertical transmission and congenital infection (characterized by hepatosplenomegaly, hydrops, and neonatal death) occurring in an estimated 1 in 20 seropositive mothers [23, 25]. In North America alone there are an estimated 40,000 pregnant women with Chagas disease, and several thousand newborns are likely to be infected with *T. cruzi* annually [26]. The numbers throughout Latin America would therefore be expected to be several-fold higher. Aside from congenital infection, Chagas disease during pregnancy increases the risk of miscarriage, preterm birth, and neonatal infection, which may cause infant death or severe sequelae [25].

Dengue and DHF disproportionately affect children in much of the developing world, including during outbreaks in the LAC region [27, 28]. Additionally, a recent systematic review confirms the harmful effects of dengue during pregnancy, including elevated rates of preterm birth, Cesarean delivery, and vertical transmission [29].

## 1.3.2 Poverty and Conflict

The mechanisms by which the NTDs actually promote or cause poverty have been reviewed previously. These mechanisms include specific effects on maternal and child health, in addition to the health of workers in the tropics, and agricultural productivity [11, 13]. A retrospective analysis of hookworm in the American South revealed that chronic hookworm infection in childhood reduce future wage earning by 40 % [30], postulated to be due to its harmful effects on child cognitive development. Similarly, Chagas disease is associated with US$1.2 billion in economic losses annually, because of its impact on maternal and child health and occupational productivity, as well as the very high costs of treatment, with a lifetime cost averaging almost US$12,000 per patient [31]. Economic losses from dengue in India were recently estimated to be US$27.4 million during a 2006 epidemic [32], however a recent systematic review concluded that the economic literature on dengue is "sparse" and results are conflicting because of the use of inconsistent assumptions [33]. NTDs also promote conflict and warfare through destabilizing effects on agricultural productivity, abandonment of arable lands, food insecurity and decrements in educational attainment and wage earning [34]. In the LAC region an association has been noted between guerilla movements and the trafficking of narcotics with Chagas disease and leishmaniasis [35].

## 1.4 Eliminating NTDs in the Americas

Several high-prevalence NTDs, including lymphatic filariasis, onchocerciasis, and trachoma are being eliminated as public health problems in the LAC region [9]. The term elimination refers to reduction in the incidence of a specific infection to zero or below a threshold that can no longer sustain transmission. Such elimination efforts rely heavily on mass MDA, often with drugs donated by the pharmaceutical industry, including ivermectin for onchocerciasis; diethyl-

**Table 1.5** Latin American and Caribbean countries that have successfully eliminated their neglected tropical diseases as a public health problem. (Modified from [38])

| Disease | Countries |
|---|---|
| Chagas disease | Brazil, Chile, Uruguay |
| Schistosomiasis | Dominican Republic, Guadeloupe, Martinique, Puerto Rico |
| Lymphatic Filariasis | Costa Rica, Suriname, Trinidad and Tobago |
| Trachoma | Mexico |
| Leprosy | All except Brazil |
| Onchocerciasis | Elimination pending in the six affected countries |

carbamazine and albendazole for lymphatic filariasis; and azithromycin for trachoma [36]. Through MDA along with other control measures including integrated vector management, lymphatic filariasis has been eliminated from Costa Rica, Suriname, Trinidad, and Tobago (Table 1.5), with expectations that this disease might also be eliminated from the few remaining endemic countries, including Brazil, Haiti, Dominican Republic, and Guyana by 2020.

Similarly through the Onchocerciasis Elimination Programme of the Americas (OEPA) it should be possible to eliminate onchocerciasis in the six remaining endemic countries, Guatemala, Mexico, Venezuela, Ecuador, Colombia, and Brazil, where approximately 0.52 million people are at risk for infection, while trachoma remains only in Brazil, Guatemala, and in five municipalities of neighboring Chiapas state in Mexico [37–39]. Today, such elimination efforts are being coordinated by the Pan American Health Organization in association with the Interamerican Development Bank and a Global Network for NTDs based at the Sabin Vaccine Institute. It has been estimated that US$128 million will be required to eliminate these three infections in the LAC region by the year 2020 [39], and a special "LAC fund" has been established in order to receive public and private donations for this purpose.

Through MDA with praziquantel, schistosomiasis remains endemic in only four LAC countries, including Brazil, St. Lucia, Suriname, and Venezuela, having been previously eliminated from Martinique, Guadeloupe, Puerto Rico and the Dominican Republic, with an expectation that this disease might be eliminated from the Carib-

bean in the coming years [37, 38]. Because of multi-drug therapy with dapsone, clofazimine, and rifampin, fewer than 34,000 registered cases of leprosy remain in the Americas, with all but about 4,000 cases being in Brazil [40]. Canine rabies has also been largely eliminated. The majority of the cases of canine rabies transmitted to humans occur in poor neighborhoods in Haiti and Bolivia [37].

The highest prevalence NTDs, including hookworm and the other soil-transmitted helminths, Chagas disease, dengue, and leishmaniasis, in addition to schistosomiasis in Brazil, will not be eliminated solely by relying on existing technologies, even though some progress has been achieved with MDA with albendazole/mebendazole and praziquantel (for soil-transmitted helminths and schistosomiasis, respectively) and integrated vector management for Chagas disease, dengue, and leishmaniasis [38]. In the case of hookworm, high rates of mebendazole drug failure and post-treatment reinfection have been reported, and although repeated MDA targeting children may have an impact on reducing the transmission of ascariasis and trichuriasis, it is not expected to have an impact on hookworm infection because of hookworm's unique transmission dynamics among adults [41]. Similarly, integrated vector management has so far been successful for eliminating Chagas disease only in the southern Cone of South America (Brazil, Chile, Uruguay), whereas elsewhere insecticide resistance, reinvasion and recolonization of reduviid vectors after spraying have thwarted control efforts [42]. Therefore, a recent "audacious" call to eliminate all of the 17 NTDs as defined by the WHO will require a new generation of technologies, especially the development of NTD vaccines [38].

## 1.5 New 'Anti-Poverty Vaccines' will be Required

Vaccines to combat the NTD vaccines are also known as the 'anti-poverty vaccines' because of their potential impact on economic development in addition to improving health [13, 43].

**Table 1.6** Vaccines under development for Latin America's neglected tropical diseases. (Modified from [43])

| Disease | Type of vaccine under development | Stage of development | Organization leading vaccine development efforts | Industrial partners |
|---|---|---|---|---|
| Chagas disease | Human therapeutic vaccine | Preclinical | Sabin Vaccine Institute PDP (Chagas Vaccine Initiative) | Birmex and CINVESTAV |
| Cysticercosis | Veterinary vaccine | Animal trials | Universidad Nacional Autonoma de Mexico | Pending |
| Dengue | Human preventive vaccine | Phase 1 and 2 | GSK, Merck & Co., Sanofi-Pasteur, Instituto Butanan | GSK, Merck & Co., Sanofi-Pasteur, Instituto Butanan |
| Foodborne trematode infections | Veterinary vaccine | Animal trials | FIOCRUZ (Oswaldo Cruz Foundation) | Ouro Fino |
| Hookworm infection | Human preventive vaccine | Phase 1 | Sabin Vaccine Institute PDP (Human Hookworm Vaccine Initiative) | FIOCRUZ-Bio-manguinhos; Aeras, Fraunhofer CMB |
| Leishmaniasis | Human preventive, therapeutic, and veterinary | Phase 1 and 2 and animal trials | Infectious Diseases Research Institute (IDRI) | Instituto Butantan |
| Schistosomiasis | Human preventive vaccine | cGMP manufacture* | Sabin Vaccine Institute PDP (Schistosomiasis Vaccine Initiative), FIOCRUZ | Aeras, Instituto Butantan, Ouro Fino |

*cGMP current good manufacturing practices

Table 1.6 lists the major anti-poverty vaccines under development in the Americas, which include new human vaccines against *Trypanosoma cruzi*, dengue virus, hookworm, leishmania species, and schistosoma species, in addition to veterinary vaccines against *Taenia solium* and *Fasciola hepatica* (a food-borne trematode) to prevent transmission to humans [43]. Of these vaccines, only the dengue vaccine is being developed and produced by the multinational vaccine manufacturers, including three vaccines being developed independently by Merck & Co., GlaxoSmithKline, and Sanofi-Pasteur [43, 44]. The other anti-poverty vaccines are largely being developed by non-profit product development partnerships (PDPs) in association with developing country manufacturers [43]. One reason why dengue vaccine development receives investment from multinational pharmaceutical companies is the large potential commercial market for such a product, given that dengue affects people living in wealthier urban centers, whereas the other anti-poverty vaccines target almost exclusively the bottom 100 million in the LAC region, and their counterparts in Africa and Asia [43]. There is some overlap, however, as a Brazilian company Ouro Fino Animal Health is developing a *Fasciola hepatica* vaccine for livestock (http://

www.veterinaryproducts1.com/supplier/ouro-fino-animal-health.html), which could prevent transmission to humans, while the developing country manufacturer, Instituto Butantan is also developing its own dengue vaccine (http://www.butantan.gov.br/home).

Today, efforts to develop and test anti-poverty vaccines targeted for human disease in the Americas are being led by PDPs, in collaboration with Latin American developing country manufacturers [43]. Most of these manufacturers, in turn, are owned and operated by scientific enterprises directly supported by federal and state governments in Latin America, including FIOCRUZ Bio-Manguinhos (through support of the Brazilian Ministry of Health), Instituto Butantan (State of Sao Paulo, Brazil), Birmex (Laboratorios de Biologicos y Reactivos de Mexico, Mexican Ministry of Health), and Cuba's Instituto Finlay [43]. Some examples of PDP-manufacture collaborations are shown in Table 1.6 and include the non-profit Infectious Disease Research Institute (Seattle, Washington), which is working with Instituto Butantan for leishmaniasis vaccine development, while the Sabin Vaccine Institute PDP (Houston, Texas and Washington, DC) is working with FIOCRUZ, (together with the US-based Aeras and Fraunhofer Center for

Molecular Biolotechnology) to develop a human hookworm vaccine, and with CINVESTAV (Centro de Investigacion y de Estudios Avanzados del Instituto Politecnico Nacional (Center for Research and Advanced Studies)) and Birmex for a Chagas disease vaccine. Each of these vaccines is either at the stage of pilot manufacture under cGMP (current good manufacturing practices) or in early clinical development [43].

## 1.6 Hookworm and Chagas Disease Vaccines

Progress in the development of a vaccine against hookworm was reviewed recently [45]. In brief, the human hookworm vaccine is a bivalent vaccine comprised of two recombinant antigens, which are parasitic enzymes involved in blood feeding [45]. The vaccine targets the adult stages of *N. americanus*, the most common hookworm worldwide and almost the exclusive hookworm in the LAC region. One of the antigens is a recombinant *N. americanus* glutathione S-transferase (*Na*-GST-1) expressed in yeast, an enzyme required by the parasite for heme binding and heme detoxification. The other is a recombinant *N. americanus* aspartic protease (*Na*-APR-1) expressed in plants, an enzyme required for hemoglobin degradation. Both recombinant antigens induce high levels of IgG antibody and have demonstrated protective immunity in laboratory animals, with reduced host worm burden and/ or blood loss [45]. The recombinant proteins have completed pilot cGMP manufacture, and *Na*-GST-1 has entered phase 1 clinical testing. Ultimately both antigens would be formulated on alum and possibly combined with a second adjuvant such as a synthetic lipid A [45]. The target product profile of the human hookworm vaccine relies on its use to prevent moderate and heavy *N. americanus* infections in children under the age of 10 years [45]. The vaccine may be incorporated into the Expanded Program on Immunization (EPI) in order to be co-administered with measles and other childhood vaccines to infants (children under the age of one), or it may be co-administered with a single dose of anthelmin-

thic drug in older children already infected with hookworm. The desired efficacy of the human hookworm vaccine is at least 80 % against moderate and heavy hookworm infections for at least 5 years after immunization [45]. The cost-effectiveness of such a vaccine was recently confirmed under a number of different scenarios [46].

Progress on the development of a vaccine for Chagas disease has also been reviewed recently [47]. Unlike the human hookworm vaccine, which is a primarily preventive vaccine, the Chagas vaccine is being proposed as a therapeutic vaccine for the treatment of individuals who have been infected with *T. cruzi* and have seroconverted. In such individuals, who have so-called "indeterminate" status (with no clinical, electrocardiographical or radiological evidence of disease), approximately 20–30 % subsequently develop Chagasic cardiomyopathy [48]. There is an urgent need for new therapeutic approaches for these patients, especially in order to eradicate *T. cruzi* parasites in the myocardium that are responsible for progression to cardiomyopathy and heart failure. A recent meta-analysis of the two currently available drugs benznidazole and nifurtimox concluded that their efficacy in late chronic infection is doubtful and does not result in seroreversion [49, 50], although a larger randomized placebo-controlled study is in progress [51]. Moreover, prolonged treatment courses lasting 2–3 months are required and result in serious side effects in up to one-half of the patients, with 10–20 % discontinuing therapy as a result [52, 53]. The drugs are also contraindicated in pregnancy and are extremely expensive to purchase and administer [54].

The target groups for this therapeutic vaccine are adults, particularly pregnant women to improve birth outcomes and prevent congenital infection, and children over the age of two, in Chagas disease-endemic areas. The vaccine is under development by the Sabin Vaccine Institute PDP and the Texas Children's Center for Vaccine Development in collaboration with the US National Institutes of Health and three Mexican institutions, the Autonomous University of Yucatan (UADY), CINVESTAV, and Birmex (Mexico's public sector vaccine manufacturer).

The bivalent vaccine is comprised of two *T. cruzi* recombinant proteins formulated on alum. One of the antigens is a unique *T. cruzi* 24 kDa antigen (Tc24) and the other belongs to a family of *T. cruzi* surface transialidases (TSA-1). Proof of concept for the protective effect of these antigens is based on experimental immunizations in *T. cruzi*-infected laboratory animals, together with identifiable mechanisms of protective immunity [47, 55–58]. In acutely and chronically-infected mice the combined antigens produced significantly reduced parasitemia and myocardial inflammation compared to controls [58]. Because protection requires the stimulation of CD8$^+$ T cells and production of interferon gamma [58], the vaccine incorporates a second adjuvant comprised of a synthetic lipid A. The requirement for this second adjuvant will be determined pending preclinical studies and early clinical trials.

Ascertaining the feasibility of expression of Tc24 and TSA-1 as soluble recombinant proteins for the purpose of process development, scale-up and current Good Manufacturing Practices (cGMP) manufacture is currently in progress. It is expected that during process development and scale-up, studies to evaluate protein attributes and stability will be established. In addition, formulation studies with alum and other adjuvants will be performed. Ultimately, the ability of these formulations to protect mice from acute and chronic *T. cruzi* infections will be confirmed.

It is anticipated that successfully eliciting Th1 immune responses will be a key to human therapeutic vaccination against Chagas disease [47, 55–58]. Th1-type immune responses are characterized by the generation of CD8$^+$ T cells, which can target intracellular pathogens [55–58]. While several purified recombinant protein vaccines are in clinical development, they are limited by poor immunogenicity and inadequate stimulation of Th1 immunity [59, 60]. Particulate-based systems can increase the delivery of antigens to antigen-presenting dendritic cells, while simultaneously maintaining antigen integrity [60]. Equally important, particulate systems can co-deliver immunopotentiating agents and activate CD8$^+$ cells. Using this strategy, the two recombinant *T. cruzi* antigens under development will be encapsulated in nanoparticles that will also contain an adjuvant molecule such as the TLR3 agonist poly(I:C). It has been previously shown that ovalbumin formulated in this manner elicits antigen-specific CD8$^+$ T cell responses *in vitro* that greatly exceed those produced either by the antigen alone or antigen encapsulated without the TLR3 agonist [61]. We will therefore attempt to simultaneously develop the Chagas disease vaccine as a nanoparticle vaccine using these technologies, and evaluate it in an experimental therapeutic mouse model for *T. cruzi* infection. An alternate delivery system under consideration is a viscous polysaccharide solution, which forms an extracellular antigen depot at the injection site [62]. Still another option is to examine viral vectors and heterologous prime-boost approaches. This approach has been successful for other systems requiring induction of Th1-type immunity [63–65].

## 1.7 Concluding Statement

While estimates indicate that less than US$200 million will be required to eliminate lymphatic filariasis, onchocerciasis, and trachoma from the Americas, there are still the added costs of continuing control efforts with MDA for other NTDs (with costs possibly exceeding US$300 million) [37], in addition to the costs of integrated vector management and the estimated US$1.2 billion required annually for Chagas disease alone, presumably much of which is treatment costs [66] In addition to jumping the scientific hurdles, the challenge of vaccine development includes demonstration of cost effectiveness, and this has now been shown for both human hookworm and Chagas disease vaccines [46, 66]. These two vaccines, in addition to vaccines to prevent leishmaniasis, schistosomiasis, and dengue, represent urgently needed control measures for a full scale elimination effort for all of the major NTDs in the Americas. Such activities in the Western Hemisphere are part of a larger "audacious goal" for elimination of all of the 17 NTDs as the most common infections of the world's poor and legacies of neglect, ignorance, and slavery [38]. Thus, an opportunity is now in hand for the major PDPs, research institutes and developing country manufacturers in the LAC region, together with the major development banks and PAHO to draft an elimination strategy for the NTDs in the Americas.

**Acknowledgements** The development of the human hookworm vaccine is supported by grants to the Sabin Vaccine Institute from the Bill & Melinda Gates Foundation, the Dutch Ministry of Foreign Affairs, and the Brazilian Ministry of Health. The feasibility studies for the development of the Chagas disease vaccine is supported by funds from the Carlos Slim Health Institute (Instituto Carlos Slim de la Salud) and the SouthWest Electronic Energy Medical Research Institute.

# References

1. http://data.worldbank.org/topic/poverty. Accessed 3 Sept 2011
2. http://www.ruralpovertyportal.org/web/guest/region/home/tags/americas. Accessed 3 Sept 2011
3. http://www.eoearth.org/article/Human_Development_Index_for_Latin_America_Caribbean_Nations. Accessed 3 Sept 2011
4. http://www.census.gov/newsroom/releases/archives/income_wealth/cb11-157.html. Accessed 13 Sept 2011
5. Hotez PJ (2008) Neglected infections of poverty in the United States of America. PLoS Negl Trop Dis 2:e256
6. Hotez PJ (2011) America's most distressed areas and their neglected infections: the United States Gulf Coast and the District of Columbia. PLoS Negl Trop Dis 5:e843
7. Hotez PJ (2010) Neglected infections of poverty among the indigenous peoples of the Arctic. PLoS Negl Trop Dis 4:e606
8. http://www.worldhunger.org/articles/Learn/world%20hunger%20facts%202002.htm. Accessed 3 Sept 2011
9. Hotez PJ, Bottazzi ME, Franco-Paredes C, Ault S, Periago MR (2008) The neglected tropical diseases of Latin America and the Caribbean: a review of disease burden and distribution and a roadmap for control and elimination. PLoS Negl Trop Dis 2:e300
10. Hotez PJ, Fenwick A, Kumaresan J, Molyneux DH, Ehrlich Sachs S, Sachs JD, Savioli L (2007) Control of neglected tropical diseases. N Engl J Med 357:1018–1027
11. Hotez PJ, Fenwick A, Savioli L, Molyneux DH (2009) Rescuing the bottom billion through control of neglected tropical diseases. Lancet 373:157–155
12. World Health Organization (2010) Working to overcome the global impact of neglected tropical diseases. First WHO report on neglected tropical diseases. Geneva. http://www.who.int/neglected_diseases/2010report/en/index.html. p 172
13. Hotez PJ, Ferris M (2006) The anti-poverty vaccines. Vaccine 24:5787–5799
14. Furtado JM, Smith JR, Belfort R, Gattey D, Winthrop KL (2011) Toxoplasmosis: a global threat. J Global Infect Dis 3:281–284
15. Lammie PJ, Lindo JF, Secor WE, Vasquez J, Ault SK et al (2007) Eliminating lymphatic filariasis, onchocerciasis and schistosomiasis from the Americas: breaking a historical legacy of slavery. PLoS Negl Trop Dis 1:e71
16. Tapia-Conyer R, Mendez-Galvan JF, Gallardo-Rincon H (2009) The growing burden of dengue in Latin America. J Clin Virol 46(Suppl 2):S3–6
17. Hotez PJ (2008) The giant anteater in the room: Brazil's neglected tropical diseases problem. PLoS Negl Trop Dis 2:e177
18. Hotez PJ, Brooker S, Bethony JM, Bottazzi ME, Loukas A, Xiao SH (2004) Hookworm infection. N Engl J Med 351:799–807
19. Smith JL, Brooker S (2010) Impact of hookworm infection and deworming on anemia in non-pregnant populations: a systematic review. Trop Med Int Health 15:776–795
20. Brooker S, Hotez PJ, Bundy DAP (2008) Hookworm-related anaemia among pregnant women: a systematic review. PLoS Negl Trop Dis 2:e291
21. Brooker S, Jardim-Botelho A, Quinnell RJ, Geiger SM, Caldas IR, Fleming F, Hotez PJ, Correa-Oliveira R, Rodrigues LC, Bethony JM (2007) Age-related changes in hookworm infection, anaemia and iron deficiency in an area of high Necator americanus hookworm transmission in south-eastern Brazil. Trans R Soc Trop Med Hyg 101:146–154
22. Larocque R, Casapia M, Gotuzzo E, MacLean JD, Soto JC et al (2006) A double-blind randomized controlled trial of antenatal mebendazole to educe low birthweight in a hookworm-endemic area of Peru. Trop Med Int Health 11:1485–1495
23. Theiler RN, Rasmussen SA, Treadwell TA, Jamieson DJ (2008) Emerging and zoonotic infections in women. Infect Dis Clin North Am 22:755–772
24. Siriano L da R, Luguetti AO, Avelar JB, Marra NL, de Castro AM (2011) Chagas disease: increased parasitemia during pregnancy detected by hemoculture. Am J Trop Med Hyg 84:569–574
25. Perez-Lopez FR, Chedraui P (2010) Chagas disease in pregnancy: a non-endemic problem in a globalized world. Arch Gynecol Obstet 282:595–599
26. Buekens P, Almendares O, Carlier Y, Dumonteil E, Eberhard M, Gamboa-Leon R, James M, Padilla N, Wesson D, Xiong X (2008) Mother-to-child transmission of Chagas disease in North America:

why don't we do more? Matern Child Health J 12:283–286

27. Giraldo D, Sant'anna C, Perisse AR, March MD, Souza AP, Mendes A, Bonfim M, Hofer CB (2011) Characteristics of children hospitalized with dengue fever in an outbreak in Rio de Janeiro, Brazil. Trans R Soc Trop Med Hyg 105(10):601–603 (Epub 2011 Aug 19)

28. Rodriguez-Barraquer I, Cordeiro MT, Braga C, de Souza WV, Marques ET, Cummings DA (2011) From re-emergence to hyperendemicity: the natural history of the dengue epidemic in Brazil. PLoS Negl Trop Dis 5:e935

29. Pouliot SH, Xiong X, Harville E, Paz-Soldan V, Tomashek KM, Breart G, Buekens P (2010) Maternal dengue and pregnancy outcomes: a systematic review. Obstet Gynecol Surg 65:107–118

30. Bleakley H (2007) Disease and development: evidence from hookworm eradication in the American South. Q J Econ 122:73–117

31. Lee BY, Bacon KM, Connor DL, Willig AM, Bailey RR (2010) The potential economic value of a Trypanosoma cruzi (Chagas disease) vaccine in Latin America. PLoS Negl Trop Dis 4(12):e916

32. Garg P, Nagpal J, Khairnar P, Seneviratne SL (2008) Economic burden of dengue infections in India. Trans R Soc Trop Med Hyg 102:570–577

33. Beatty ME, Beutels P, Meltzer MI, Shepard DS, Hombach J, Hutubessy R, Dessis D, Coudeveille L, Dervaux B, Wichmann O, Margolis HS, Kuritsky JN (2011) Health economics of dengue: a systematic literature review and expert panel's assessment. Am J Trop Med 84:473–488

34. Hotez PJ, Thompson TG (2009) Waging peace through neglected tropical disease control: a US foreign policy for the bottom billion. PLoS Negl Trop Dis 3:e346

35. Beyrer C, Villar JC, Suwanvanichkij V, Singh S, Baral SD et al (2007) Neglected diseases, civil conflicts, and the right to health. Lancet 370:619–627

36. Hotez PJ (2009) Mass drug administration and the integrated control of the world's high prevalence neglected tropical diseases. Clin Pharamcol Therap 85:659–664

37. Schneider MC, Aguilera XP, Barbosa da Silva Jr J, Ault SK, Najera P, Martinez J, Requejo R, Nicholls RS, Yadon Z, Silva JC, Leanes LF, Periago MR (2011) Elimination of neglected diseases in Latin America and the Caribbean: a mapping of selected diseases. PLoS Negl Trop Dis 5:e964

38. Hotez P (2011) Enlarging the "audacious goal": elimination of the world's high prevalence neglected tropical diseases. Vaccine (in press)

39. Bitran R, Martorell B, Escobar L, Munoz R, Glassman A (2009) Controlling and eliminating neglected diseases in Latin America and the Caribbean. Health Aff (Millwood) 28:1707–1719

40. World Health Organization (2011) Leprosy update, 2011. World Epidemiol Rec 36:389–400

41. Hotez PJ, Bethony J, Bottazzi ME, Brooker S, Diemert D, Loukas A (2006) New technologies for the control of human hookworm infection. Trends Parasitol 22:327–331

42. Reithinger R, Tarleton R, Urbina JA, Kitron U, Gurtler RE (2009) Eliminating Chagas disease: challenges and a roadmap. BMJ 338:b1283

43. Hotez P (2011) A handful of 'anti-poverty' vaccines exist for neglected diseases, but the world's poorest billion people need more. Health Aff 30:1080–1087

44. Coller BA, Clements DE (2011) Dengue vaccines: progress and challenges. Curr Opin Immunol 23:391–398

45. Hotez PJ, Bethony JM, Diemert DJ, Pearson M, Loukas A (2010) Developing vaccines to combat hookworminfection and intestinal schistosomiasis. Nat Rev Microbiol 8:814–826

46. Lee BY, Bacon KM, Bailey R, Wiringa AE, Smith KJ (2011) The potential economic value of a hookworm vaccine. Vaccine 29:1201–1210

47. Quijano-Hernandez I, Dumonteil E (2011) Advances and challenges towards a vaccine against Chagas disease. Hum Vaccin (in press)

48. Rassi Jr A, Rassi A, Marin-Neto JA (2010) Chagas disease. Lancet 375:1388–1402

49. Perez-Molina JA, Perez-Ayala A, Moreno S, Fernandez-Gonzalez MC, Zamora J, Lopez-Velez R (2009) Use of benznidazole to treat chronic Chagas' disease: a systematic review with a meta-analysis. J Antimicrob Chemother 64:1139–1147

50. Sarli Issa V, Alcides Bocchi E (2010) Antitrypanosomal agents: treatment or threat? Lancet 376:768–769

51. Marin-Neto JA, Rassi A Jr, Morillo CA et al on behalf of BENEFIT Investigators (2008). Rationale and design of a randomized placebo-controlled trial assessing the effects of etiologic treatment in Chagas cardiomyopathy: the BENznidazole Evaluation for Interrupting Trypanosomiasis (BENEFIT). Am Heart J 156:37–43

52. Pinazo M-J, Munoz J, Posada E, Lopez-Chejade P, Gallego M, Ayala E, del Cacho E, Soy D, Gascon J (2010) Tolerance of benznidazole in treatment of Chagas' disease in Adults. Antmicrob Agents Chemother 54:4896–4899

53. Viotti R, Vigliano C, Lococo B, Alvarez MG, Petti M, Bertocchi G, Armenti A (2009) Side effects of benznidazole as treatment in chronic Chagas disease: fears and realities. Expert Rev AntiInfect Therap 7:157–163

54. Castillo-Riqueime M, Guhl F, Turriago B, Pinto N, Rosas F, Florez Moartinez M, Fox-Rushby J, Davies C, Campbell-Lendrum D (2008) The costs of preventing and treating Chagas disease in Colombia. PLoS Negl Trop Dis 2(11):e336

55. Dumonteil E, Escobedo-Ortegon J, Reyes-Rodriguez N, Arjona-Torres A, Ramirez-Sierra J (2004) Immunotherapy of Trypanosoma cruzi infection with DNA vaccines in mice. Infect Immun 72(1):46–53

56. Zapata-Estrella H, Hummel-Newell C, Sanchez-Burgos G, Escobedo-Ortegon J, Ramirez-Sierra MJ, Arjona-Torres A, Dumonteil E (2006) Control of

Trypanosoma cruzi infection and changes in T-cell populations induced by a therapeutic DNA vaccine in mice. Immunol Lett 103:186–191

57. Sanchez-Burgos G, Mezquite-Vega RG, Escobedo-Ortegon J et al (2007) Comparative evaluation of therapeutic DNA vaccines against Trypanosoma cruzi in mice. FEMS Immunol Med Microbiol 50:333–341

58. Limon-Flores AY, Cervera-Cetina R, Tzec-Arjona JL et al (2010) Effect of a combination DNA vaccine for the prevention and therapy of Trypanosoma cruzi infection in mice: role of $CD4^+$ and $CD8^+$ cells. Vaccine 28(46):7414–7419

59. O'Hagan DT, De Gregorio E (2009) The path to a successful vaccine adjuvant—'the long and winding road'. Drug Discov Today 14:541–551

60. Oyewumi MO, Kumar A, Cui Z (2010) Nano-microparticles as immune adjuvants: correlating particle sizes adn the resultant immune responses. Expert Rev Vaccines 9:1095–1107

61. Heffernan MJ, Kasturi SP, Yang SC, Pulendran B, Murthy N (2009) The stimulation of CD8+ T cells by dendritic cells pulsed with polyketal microparticles containing ion-paired protein antigen and poly(inosinic acid)-poly(cytidylic acid). Biomaterials 30:910–918

62. Heffernan MJ, Zaharoff DA, Fallon JK, Schlom J, Greiner JW (2011) In vivo efficacy of a chitosan/IL-12 adjuvant system for protein-based vaccines. Biomaterials 32:926–932

63. Roller CS, Reyes-Sandoval A, Cottingham MG, Ewer K, Hill AVS (2011) Viral vectors as vaccine platform: deployment in sight. Curr Opin Immunol 23:377–382

64. Cavenaugh JS, Awi D, Mendy M, Hill AVS, Whittle H, McConkey S (2011) Partially randomized, non-blinded trial of DNA and MVA therapeutic vaccines based on hepatitis B virus surface protein for chronic HBV infection. PLoS One 6:e14626

65. Hill AVS, Reyes-Sandoval A, O'Hara G, Ewer K, Lawrie Al, Goodman A, Nicosia A, Folgori A, Colloca S, Coretese R, Gilbert SC, Draper SJ (2010) Prime-boost vectored malaria vaccines. Hum Vaccin 6:78–83

66. Lee BY, Bacon KM, Connor DL, Willig AM, Bailey RR (2010) The potential economic value of a Trypanosoma cruzi (Chagas disease) vaccine in Latin America. PLoS Negl Trop Dis 4(12):e916

# Non-typhoidal Salmonella in Children: Microbiology, Epidemiology and Treatment

2

Gabrielle M. Haeusler and Nigel Curtis

### Abstract

Non-typhoidal *Salmonellae* (NTS) are an important cause of infectious diarrhoea world-wide. In the absence of immune deficiency, gastroenteritis caused by NTS is usually mild, self limiting and rarely requires intervention. NTS are also an important cause of invasive disease, particularly in developing countries, likely secondary to the high prevalence of co-existing malnutrition, malaria and HIV infection. This review provides an overview of the microbiology, epidemiology and pathogenesis of NTS, and compares recommendations for the treatment of NTS gastroenteritis in children.

## 2.1 Introduction

Non-typhoidal *Salmonellae* (NTS) are an important cause of infectious diarrhoea world-wide. In the absence of immune deficiency, gastroenteritis caused by NTS is usually mild, self limiting and rarely requires intervention. NTS are also an important cause of invasive disease, particularly in developing countries, likely secondary to the high prevalence of co-existing malnutrition, malaria and HIV infection. Antibiotic treatment of NTS gastroenteritis has been the subject of a meta-analysis, but questions regarding exactly which patients should be treated and the optimal regimen remain unanswered. This review provides an overview of the microbiology, epidemiology and pathogenesis of NTS, and compares recommendations for the treatment of NTS gastroenteritis in children.

N. Curtis (✉)
Department of Paediatrics, The University of Melbourne;
Infectious Diseases Unit, The Royal Children's Hospital
Melbourne; Infectious Diseases & Microbiology
Research Group, Murdoch Children's Research Institute,
Parkville, Australia
e-mail: nigel.curtis@rch.org.au

G. M. Haeusler
Infectious Diseases Unit, The Royal Children's Hospital
Melbourne, Parkville, Australia
e-mail: gabrielle.haeusler@rch.org.au

## 2.2 Classification and Microbiology

The genus *Salmonella* belongs to the family of Enterobacteraciae. *Salmonella* are separated into two species, *Salmonella enterica* and *Salmonella bongori* (previously classified as subsp. V.), with the former being further classified into six subspecies (I, *S. enterica* subsp. *enterica*; II,

N. Curtis et al. (eds.), *Hot Topics in Infection and Immunity in Children IX*,
Advances in Experimental Medicine and Biology 764, DOI 10.1007/978-1-4614-4726-9_2,
© Springer Science+Business Media New York 2013

**Table 2.1** Number of serotypes in each subspecies. (Based on data from Grimont et al. [1])

| Species | Subspecies | Number of serotypes |
|---|---|---|
| Salmonella enterica | enterica (I) | 1,531 |
| | salamae (II) | 505 |
| | arizonae (IIIa) | 99 |
| | diarizonae (IIIb) | 336 |
| | houtenae (IV) | 73 |
| | indica (VI) | 13 |
| Salmonella bongori (V) | – | 22 |
| Total | | 2,579 |

**Table 2.2** Examples of clinically significant *Salmonella* serotypes and relevant clinical syndromes according to serogroup. (Based on data from Grimont et al. [1])

| Serogroup | Serotype example[a] | Clinical syndrome |
|---|---|---|
| A | Paratyphi A | Enteric fever |
| B | Paratyphi B | Enteric fever |
| | Typhimurium | NTS |
| | Heidelberg | NTS |
| C1 | Paratyphi C | Enteric fever |
| | Choleraesuis | NTS |
| | Virchow | NTS |
| C2 | Newport | NTS |
| D1 | Typhi | Enteric fever |
| | Enteritidis | NTS |
| | Dublin | NTS |

[a] All example serotypes are members of subspecies *Salmonella enterica* subspecies *enterica*

*S. enterica* subsp. *salamae*; IIIa, *S. enterica* subsp. *arizonae*; IIIb, *S. enterica* subsp. *diarizonae*; IV, *S. enterica* subsp. *houtenae*; and VI, *S. enterica* subsp. *indica*) [1]. While an alternative nomenclature describes the genus as a single species, *Salmonella choleraesuis*, the Judicial Commission of the International Committee on the Systematics of the Prokaryotes supports the two-species designation [2].

Salmonellae are motile, Gram negative, facultative anaerobic bacilli which rarely ferment lactose [3]. Within the seven subspecies, more than 2,500 serotypes (or serovars) have been reported (Table 2.1) [1]. Salmonellae are classified according to antigenically diverse surface antigens: polysaccharide O (somatic) antigens, H (flagellar) antigens and Vi (capsular) antigens [3]. Agglutination reactions based on the O-antigen are used by most clinical laboratories to divide Salmonella into serogroups which include, but are not limited to, A, B, C1, C2, D and E (Table 2.2) [1]. Strains in these six serogroups cause most of the NTS infections in humans [4]. Formal serotyping, usually by reference laboratories, is required to differentiate clinically significant serotypes as cross-reactivity occurs. A detailed list of all currently recognised *Salmonella* serotypes is available elsewhere [4].

Salmonella nomenclature and syntax is potentially confusing. An example of a correct *Salmonella* subspecies and serotype designation is *Salmonella enterica* subspecies *enterica* serotype Typhimurium. An accepted abbreviation of this full taxonomic designation is *Salmonella* ser.

Typhimurium (capitalised and not italicised) at the first citation and subsequently *Salmonella* Typhimurium [5].

*Salmonella enterica* subspecies *enterica* contains almost all the serotypes pathogenic for humans [3]. Although many *Salmonella* serotypes exist, they can be broadly categorised as typhoidal or non-typhoidal *Salmonella* (NTS) depending on the clinical syndrome with which they are predominantly associated (Table 2.2). The typhoidal Salmonella include the *S. enterica* subspecies *enterica* serotypes Typhi and Paratyphi A-C and typically cause systemic illness with little or no diarrhoea. The much larger group of NTS primarily induce acute, self-limiting gastroenteritis and is the focus of this review.

## 2.3 Epidemiology

In contrast to *Salmonella* Typhi and *Salmonella* Paratyphi, which are rarely encountered outside endemic countries or in returned travellers, NTS have a worldwide distribution. While the true incidence is unknown, there are an estimated 93.8 million episodes and 155,000 deaths each year attributable to NTS [6]. Data from the World Health Organization (WHO) Global Foodborne Infections Network (GFN) indicate that

*Salmonella* Typhimurium and *Salmonella* ser. Enteritidis account for nearly 80 % of all human isolates reported globally [7].

Unlike *Salmonella* Typhi and *Salmonella* Paratyphi, which have host specificity for humans, NTS can be acquired from both animal and humans. While poultry and eggs remain the most common source of NTS, other animal reservoirs include reptiles, rodents, cats and dogs [8, 9]. A case control study in the United States found that up to 6 % of all sporadic NTS infections are attributable to reptile or amphibian contact [10].

Transmission is predominantly foodborne, although other modes include consumption of contaminated water, contact with infected animals and nosocomial exposure [4, 11]. The incubation for NTS gastroenteritis depends on the host and the inoculum. It is usually 12–36 h, although incubation periods of up to nearly 2 weeks have been reported with certain strains [12–14].

## 2.4  Pathogenesis

Salmonellae are facultative intracellular pathogens that can survive within host macrophages [15]. Unlike typhoidal salmonella, which have the ability to evade the immune system, NTS tend to induce a localised inflammatory response in immunocompetent individuals, provoking a large influx of polymorphonuclear leukoytes to the intestinal lumen [16]. They can also colonise small and large intestinal mucosa thus facilitating prolonged periods of shedding [16].

Host factors predisposing to severe NTS infection include reduced gastric acidity, impaired cell mediated and humoral immunity, and impaired phagocytic function [16–18]. Salmonellae are unable to survive at a gastric pH less than 2.5 [19] and patients with anatomical or functional achlorhydria are at increased risk of developing infection [20]. This is especially relevant to neonates where the combination of relative achlorhydria and frequent milk feeds may contribute to their increased risk of NTS bacteraemia [21].

T-cell immunity is important in controlling *Salmonella* as evidenced by increased suscep-

tibility to invasive NTS in HIV-infection [22, 23] and with corticosteroid use [24]. Children with congenital defects in humoral immunity including X-linked agammaglobulinaemia and common variable immunodeficiency, are also reported to have increased risk of persistent diarrhoea and invasive disease [16]. Impaired phagocytic function seen in chronic granulomatous disease, haemoglobinopathies and malaria similarly increase the risk of invasive NTS infection [16]. In addition, co-infection with *Schistosoma* has been reported to cause prolonged and severe illness due to altered macrophage function and replication and survival of *Salmonella* within the parasite [25].

## 2.5  Clinical Syndromes

Clinical manifestations of NTS can be broadly divided into four groups:
1. Acute gastroenteritis
2. Extra-intestinal infection
3. Non-infectious sequelae
4. Salmonella carriage

### 2.5.1  Acute Gastroenteritis

Non-typhoidal salmonellae usually cause an acute self-limiting gastroenteritis. In contrast to typhoidal salmonella, infection with NTS results in diarrhoea. This is profuse and usually non-bloody. Associated symptoms of fever, abdominal cramping, nausea and vomiting may also occur [4]. Diagnosis is confirmed on stool culture. Fluid and electrolyte disturbances are the most frequent complication of NTS gastroenteritis. Asymptomatic gastrointestinal infection can also occur; however, given the rate of convalescent NTS excretion following acute infection, the true incidence is unknown [26].

### 2.5.2  Extra-intestinal Infection

Overall, NTS bacteraemia is reported in up to 9 % of patients with acute gastroenteritis [27].

Choleraesuis, Heidelberg and Dublin are among some of the serotypes more frequently associated with bacteraemia [28]. The incidence of invasive disease is also modified by factors such as age, region and underlying immune status. Surveillance data from the United States showed the incidence of invasive NTS was 7.8 cases per 100,000 in infants (aged less 1 year) compared to less than 0.8 cases per 100,000 in older children [29]. Significant discordance in the burden of invasive NTS between continents exists with an estimated annual incidence of up to 388 per 100,000 children (aged less 5 years) in Africa [30]. In this continent, the common invasive serotypes are Typhimurium and Enteritidis. The high prevalence of malaria in Africa, and its association with invasive NTS, has been postulated as one reason for this difference [31]. Interestingly, recent studies in the Gambia, Kenya and Tanzania have shown that the marked decline in malaria prevalence has been paralleled by a similar reduction in invasive NTS [32].

Bacteraemia may result in focal NTS infection at any site, including the central nervous system [33, 34]. Risk factors for focal disease are similar to those for bacteraemia [35], although focal infections in non-immunocompromised children are well described [36]. In one study, 7 of 12 (58 %) immunocompromised children developed focal infection compared with 5 of 132 (4 %) non-immunocompromised children [37].

Salmonella bacteraemia usually presents with fever and/or sepsis. Persistent bacteraemia has been documented in afebrile, well-appearing infants with NTS gastroenteritis emphasising the importance of blood cultures in this population [38]. Conversely, bacteraemia has been described in immunosuppressed patients without a history of gastroenteritis [24]. Notably, in the African setting, the majority of invasive NTS cases do not have gastroenteritis.

Salmonella can be cultured from standard blood culture media [3]. Positive urine culture with *Salmonella spp.* may indicate bacteraemia rather than faecal contamination [39]. Similar to *Salmonella* Typhi, culture of bone marrow sam-

ples may increase the diagnostic yield of NTS [30, 40].

### 2.5.3 Non-infectious Sequelae

Non infectious sequelae following infectious gastroenteritis are well described [41]. Reactive arthritis after NTS infection has been reported in up to 29 % of patients [42]. There are conflicting data on the association between antibiotic treatment for acute NTS gastroenteritis and subsequent development of musculoskeletal symptoms, some suggesting a decreased risk and others an increased risk [42].

### 2.5.4 Salmonella Carriage

Salmonella carriage is defined as asymptomatic excretion following acute infection and can be divided into convalescent carriage and chronic carriage.

#### 2.5.4.1 Convalescent Carriage
Convalescent carriage occurs frequently after symptomatic or asymptomatic NTS infection. Fig. 2.1 shows the duration of excretion of NTS in 13 studies according to age [43]. The median duration of NTS excretion in children less than 5 years of age is 7 weeks, with 18 % remaining culture positive at 6 months. Carriage is shorter in older children with a median duration of 3–4 weeks with only 0.3 % remaining culture positive at 6 months. In addition to younger age, factors associated with prolonged duration of excretion include symptomatic infection, treatment with antibiotics and infection with strains other than *Salmonella* Typhimurium (Fig. 2.2) [43]. Episodic excretion is not uncommon.

#### 2.5.4.2 Chronic Carriage
Documented excretion of NTS for more than 1 year is defined as chronic carriage [44]. This occurs in up to 2.6 % of children under 5 years of age and 0.3 % of older children [43].

**Fig. 2.1** Duration of excretion of NTS by age (by permission of Oxford University Press). [43]

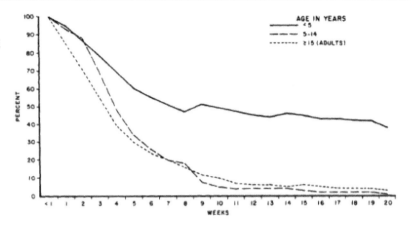

**Fig. 2.2** Duration of excretion of NTS by serotype (by permission of Oxford University Press). [43]

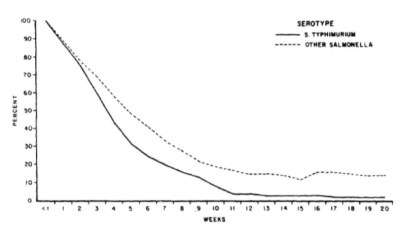

## 2.6  Treatment

Recommendations for the treatment of acute gastroenteritis due to NTS vary (Table 2.3). The rationale for treatment of both acute NTS gastroenteritis and carriage will be the focus of the remainder of this review. Recommendations for the treatment of extra-intestinal complications of NTS are available elsewhere [4].

Antibiotic agents with *in vitro* and *in vivo* activity against NTS include ampicillin/amoxicillin, trimethoprim-sulfamethoxazole, tetracyclines, third generation cephalosporins, macrolides and fluoroquinolones. Aminoglycosides show good *in vitro* activity but poor clinical efficacy and are not recommended [9, 45]. Ciprofloxacin and other fluoroquinolones have previously been restricted in children due an association with arthropathy, first described in young beagles

[46]. However, a systematic review of the safety of ciprofloxacin in children has found that musculoskeletal adverse effects are infrequent (risk 1.6 %, 95 % CI 0.9–2.6 %), generally mild and usually reversible [47].

### 2.6.1  NTS Gastroenteritis

A Cochrane review of trials investigating antibiotic treatment of NTS concluded that there was no evidence to support antibiotic therapy in otherwise healthy children and adults with non-severe diarrhoea [58]. It included 12 randomised controlled trials (RCT) published before 1998, of which only five reported on clinical outcomes that were extractable for meta-analysis. Importantly, almost all trials excluded infants less than 6 weeks and patients with underlying disease and

**Table 2.3** Recommendations for the treatment of acute NTS gastroenteritis. (With duration of treatment and/or specified antibiotic choice when stated in source)

| Source | Age <3 months | Severe infection | Immuno-compro-mised | Haemo-globinopa-thies | Chronic GI disease | Other |
|---|---|---|---|---|---|---|
| **Infectious diseases and paediatric reference books** | | | | | | |
| Textbook of Pediatric Infectious Diseases (Feigin and Cherry) [48] | Yes, 5d | – | Yes, 5d | Yes, 5d | – | – |
| Forfar and Arneil's Textbook of Pediatrics [49] | Yes | – | Yes | – | – | – |
| Evidence-based Pediatric Infectious Diseases (Isaacs) [50] | Yes, 5–7d, ciprofloxacin or azithromycin (empiric) and amoxicillin if susceptible | Yes, 5–7d | Yes, 5–7d | Yes, 5–7d | Yes, 5–7d | Malnourished, 5–7d |
| Principles and Practice of Pediatric Infectious Diseases (Long) [51] | Yes | – | Yes | – | Yes | – |
| Mandell, Douglas and Bennett's Principles and Practice of Infectious Diseases [4] | Yes, 2–3d | – | Yes, 7–14d, fluoroquinolone | – | – | Valvular heart disease; joint disease, 2–3d |
| Nelson Textbook of Pediatrics [52] | Yes, 7–14d | – | Yes, 7–14d | Yes, 7–14d | – | – |
| Oski's Principles and Practice of Pediatrics [53] | Yes | – | Yes | Yes | – | Malnourished |
| Red Book: 2009 Report of the Committee on Infectious Diseases [12] | Yes, as per susceptibilities | – | Yes, as per susceptibilities | Yes, as per susceptibilities | Yes, as per susceptibilities | – |
| Infectious Diseases of the Fetus and Newborn (Remington & Klein) [21] | Yes, 3–5d | – | Yes, 3–5d | – | – | – |
| Rudolph's Pediatrics [54] | Yes | – | Yes | – | – | – |
| The Sanford Guide to Antimicrobial Therapy [45] | Yes (<1y), 7–10d, fluoroquinolone | Yes[b], 7–10d, fluoroquinolone | Yes, 14d, fluoroquinolone | Yes, 7–10d, fluoroquinolone | – | Haemodialysis; vascular aneurysm; prosthetic joints, 7–10d, fluoroquinolone |
| **Other sources** | | | | | | |
| IDSA Guidelines 2001 [55] | Yes (<6m), 3–7d | Yes, 3–7d | Yes, minimum 14d | – | – | Valvular heart disease, 3–7d |
| NICE guidelines: Diarrhoea and Vomiting Caused by Gastroenteritis [56] | Yes (<6m) | – | Yes | – | – | Malnourished |
| UpToDate [57] | Yes, 3–10d | Yes[a], 3–7d | Yes, minimum 1d | Yes, minimum 14d | – | – |

[a] Defined as high fever, severe diarrhoea (>9 stools per day), need for hospitalisation
[b] Defined as hospitalised with fever and severe diarrhoea

severe illness. Five studies were in infants and children ($n = 258$), with only one study including infants less than 4 weeks [59–63]. The review did not identify significant differences in length of illness, diarrhoea or fever between any antibiotic regimen and placebo. Furthermore, although antibiotics were associated with more negative stool cultures during the first week of treatment, clinical relapse was more common and there were more cases of positive cultures at three weeks in the antibiotic-treated group. Adverse drug reactions, including rash, gastrointestinal upset and headache, were also more common in the antibiotic group [58]. Given the exclusion criteria, these results can not be extrapolated to patients at higher risk of severe and invasive infection.

Although the treatment of patients with risk factors for invasive disease seems reasonable, a benefit has not been documented in a RCT. This paucity of data likely explains the discrepancies in treatment recommendations between guidelines and reference books detailed in Table 2.3. Despite this variability, recommendations for the treatment of infants under 3–6 months of age and immunocompromised patients are consistent between all identified references. A 1988 consensus statement for the management of *Salmonella* infections in the first year of life similarly recommends treatment in infants less than 3 months following a blood culture irrespective of the severity of illness [64]. Other conditions where treatment has been recommended, albeit less consistently, include haemaglobinopathies, chronic gastrointestinal disease (for example inflammatory bowel disease), severe infection, malnourished state and vascular or joint disease.

Agents for treatment of NTS can be divided into non-absorbable (colistin and neomycin), absorbable (ampicillin, amoxicillin, tetracycline, macrolide and trimethoprim-sulfamethoxazole) and those with potent intra-cellular activity (fluoroquinolones) [58]. As *Salmonellae* are intracellular pathogens, it is postulated that the latter may be more effective in treatment of this infection and is specifically recommended for treatment of immunocompromised patients in two sources [4, 45]. The Cochrane review included eight trials involving fluroquinolones [58]. Subgroup analysis showed no benefit of treatment

with fluoroquinolones, although negative stool cultures in the first week of treatment were more likely with fluroquinolone treatment compared to other absorbable antibiotics. Consequently, as there are insufficient data to recommend a particular agent, empiric treatment should be based on local susceptibility data and modified according to susceptibility results.

The optimal duration of treatment of NTS gastroenteritis has not been studied. Recommendations vary amongst sources between 3 and 14 days depending on the underlying condition (Table 2.3). Of the trials included in the Cochrane review, duration varied between 1 and 14 days, with 5 days being the most common treatment regimen. The authors commented that the two studies with longer antibiotic treatment duration (10 and 14 days) showed an apparent benefit of antibiotics at 8–21 days [58]. Subsequent to this review, a study reported that 10 days of ofloxacin compared with 5 days resulted in earlier eradication of NTS without an increase in carriage [65]. However, this study was relatively small and did not include a placebo arm.

Other trials have also been published since the last Cochrane review update in 1999 with conflicting results [66, 67]. An updated meta-analysis is awaited [68].

## 2.6.2  Convalescent NTS Carriage

Treatment of convalescent NTS carriage in children has not been subject to RCT but is generally not recommended (Table 2.4).

## 2.6.3  Chronic NTS carriage

There are no RCT investigating the treatment of chronic NTS carriage in healthy children. In adults with chronic carriage, norfloxacin and azithromycin were no better than placebo in eradicating carriage and were associated with higher rates of re-infection and selection of drug-resistant isolates in endemic areas [69]. However, although antibiotics have been shown to prolong NTS excretion, this paradoxical finding may only

**Table 2.4** Recommendations for treatment of convalescent and chronic NTS. (With duration of treatment and/or specified antibiotic choice when stated in source)

| Source | Convalescent excretion | Chronic carriage (>1 year) |
|---|---|---|
| **Infectious diseases and paediatric reference books** | | |
| Textbook of Pediatric Infectious Diseases (Feigin and Cherry) [48] | – | Not recommended |
| Forfar and Arneil's Textbook of Pediatrics [49] | Not recommended | – |
| Evidence-based Pediatric Infectious Diseases (Isaacs) [50] | Not recommended | |
| Principles and Practice of Pediatric Infectious Diseases (Long) [51] | – | – |
| Mandell, Douglas and Bennett's Principles and Practice of Infectious Diseases [4] | 'Not recommended for uncomplicated convalescent excretion' | 'Treatment of persons with asymptomatic carriage of NTS is controversial' |
| Nelson Textbook of Pediatrics [52] | – | – |
| Oski's Principles and Practice of Pediatrics [53] | Not recommended | Not recommended |
| Red Book: 2009 Report of the Committee on Infectious Diseases [12] | – | Not recommended |
| Infectious Diseases of the Fetus and Newborn (Remington & Klein) [21] | Not recommended for infants | – |
| Rudolph's Pediatrics [54] | – | – |
| The Sanford Guide to Antimicrobial Therapy [45] | – | – |
| **Other sources** | | |
| IDSA Guidelines 2001 [55] | – | – |
| NICE Guideline: Diarrhoea and Vomiting Caused by Gastroenteritis [56] | – | – |
| UpToDate [57] | – | Recommended for all patients. Duration: 4–6 weeks; fluoroquinolone |
| | | HIV infection: 'consider suppressive prophylactic therapy if failure to eradicate' |

be relevant following treatment of acute NTS infection. In the Cochrane review, only three trials included asymptomatic patients (7.2 % of all participants) and separate data were not available for subgroup analysis [58].

Treatment of chronic NTS carriage is recommended in one source, based on data for the eradication of chronic *Salmonella* Typi and Paratyphi [57, 70] (Table 2.4). In patients with HIV infection, prolonged ciprofloxacin (minimum 3 months) may also reduce NTS relapse [71] and up to 6 months suppressive treatment is recommended for HIV-infected adolescents and adults with recurrent NTS bacteraemia [72].

## 2.7 NTS and Pregnancy

Pregnancy does not appear to increase the risk of maternal NTS infection. Screening of a large cohort of pregnant women (30,471) at the time of delivery detected NTS in only 60 (0.2 %), of which 17 (28 %) were symptomatic [73]. Of the 60 babies born, only seven (11.7 %) excreted NTS and five (8.3 %) had gastroenteritis. Transplacental infection can occur in the setting of maternal NTS bacteraemia and is often lethal to the foetus [74]. Transmission of NTS through breast milk has also been described, including three nursery outbreaks due to infected pooled milk [75]. More

**Table 2.5** Infection control recommendations

| Source | Recommendation |
| --- | --- |
| **Infectious diseases and paediatric reference books** | |
| Textbook of Pediatric Infectious Diseases, (Feigin and Cherry) [53] | Hospitalised patient: standard and contact precautions until stool cultures are negative. Surveillance cultures on all neonates in nursery |
| Forfar and Arneil's Textbook of Pediatrics [49] | 'No restriction of activities if stool is normal' |
| Evidence-based Pediatric Infectious Diseases (Isaacs) [50] | – |
| Principles and Practice of Pediatric Infectious Diseases (Long) [51] | Hospitalised patient: standard and contact precautions |
| Mandell, Douglas and Bennett's Principles and Practice of Infectious Diseases [4] | Hospitalised patient: standard and contact precautions |
| | Health care worker: exclusion until asymptomatic and passing formed stool |
| Nelson Textbook of Pediatrics [52] | – |
| Oski's Principles and Practice of Pediatrics [53] | – |
| Red Book: 2009 Report of the Committee on Infectious Diseases [12] | Hospitalised patient: Standard and contact precautions for diapered and incontinent children for duration of illness |
| | Child care: exclusion until diarrhoea resolves. Contacts do not require stool culture. Exclude symptomatic contacts until diarrhoea resolves |
| Infectious Diseases of the Fetus and New-born (Remington & Klein) [21] | Hospitalised neonate: standard and contact precautions. 'When two or more cases are recognized, environmental cultures, cultures of all infants, cohorting, early discharge of infected infants, and thorough cleaning of all possible fomites in the nursery and delivery rooms are recommended. If cases continue to occur, the nursery should be closed to further admissions' |
| Rudolph's Pediatrics [54] | – |
| The Sanford Guide to Antimicrobial Therapy [45] | – |
| **Infection control sources** | |
| CDC: 2007 Guideline for Isolation Precautions [95] | Hospitalised patient: standard and contact precautions for diapered or incontinent persons for duration of illness |
| NHMRC: Australian Guidelines for Prevention and Control of Infection in Healthcare [96] | Hospitalised patient: contact precautions for faecally incontinent patients for duration of illness |
| | Health care worker: exclusion until 24 h after symptoms have resolved |
| **Other sources** | |
| IDSA Guideline 2001 [55] | – |
| NICE guideline: Diarrhoea and Vomiting caused by Gastroenteritis [56] | Childcare/school: exclusion until 48 h after symptoms have resolved |
| UpToDate [57] | Healthcare workers: exclusion until diarrhoea resolves |
| | Food handlers: as above |

recently, outbreaks associated with formula feed have been reported [76].

Treatment of asymptomatic maternal NTS carriage is unlikely to be of benefit, but should be considered for severe infection. There are no guidelines for empiric treatment of neonates with known maternal excretion. Due to the increased risk of severe infection in this group, consideration should be given to screening and treatment, as outlined previously, if NTS is detected in the stool [77].

## 2.8   Antibiotic Resistance

Antibiotic-resistant NTS are associated with increased treatment failure and risk of invasive disease [78]. Worldwide surveillance data has demonstrated an overall increase in antibiotic resistance among NTS, although significant geographical and serotype variability exist [79, 80]. The widespread use of antibiotics in food animals has been implicated in the increasing prevalence of antibiotic resistant NTS [81].

The European Centre for Disease Control (formally Enter-net) and the National Antimicrobial Resistance Monitoring System (NARMS) provide comprehensive surveillance data on rates of NTS drug resistance in Europe and the United States [82, 83]. Data from these studies, as well as reports from middle to low income countries, indicate high rates of ampicillin, amoxicillin and trimethoprim-sulfamethoxazole resistance, particularly among the globally prevalent serotypes *Salmonella* Enerididis and *Salmonella* Typhimurium [82–84]. Of particular concern is the emergence of extended spectrum beta-lactamase (ESBL) genes in NTS [80], as well as reports of carbapenemase-containing NTS isolates [85, 86], both of which confer high level antimicrobial resistance.

Fluoroquinolones and third generation cephalosporins are frequently used for the treatment of NTS that are resistant to conventional antibiotics, although reports of increasing resistance to these second line agents are emerging [80]. Isolates resistant to nalidixic acid are frequently resistant to fluoroquinolones, and this is a recommended screening test for ciprofloxacin resistance [3]. However, although a study found the nalidixic acid disk diffusion method was 100 % sensitive for the detection of reduced ciprofloxacin susceptibility (defined as MIC > 0.125 µg/ml) [87], there have been subsequent reports of isolates that are susceptible to nalidixic acid but exhibit reduced susceptibility to ciprofloxacin [88].

## 2.9   Infection Control

Outbreaks of NTS have been reported in neonatal units, paediatric wards and child care facilities [11, 89]. However, despite the frequency of asymptomatic NTS excretion following acute infection, the risk of transmission from asymptomatic health care workers to patients appears to be small when strict adherence to hand hygiene is observed [90]. Similarly, transmission from asymptomatic food handlers is rare, with one survey identifying only 2 % of food handlers as the source of 566 NTS outbreaks [91].

Methods to control the spread of *Salmonella* include appropriate food preparation, water sanitation and strict hand hygiene. Most resources recommend exclusion of children, health care workers and food handlers from childcare/school or work until 24–48 h after resolution of symptoms (Table 2.5). For symptomatic hospitalised patients, standard and contact precautions are recommended. There are also some non-randomised data that suggest the prophylactic use of antibiotics (trimethoprim-sulfamethoxazole or ciprofloxacin) in addition to strict barrier nursing, may control nosocomial *Salmonella* epidemics [92–94]. Confirmation of clearance of NTS is generally not recommended.

**Acknowledgements**   We wish to thank Steve M. Graham for his helpful comments and suggestions.

## References

1. Grimont PAD, Weill F (2007) Antigenic Formulae of the *Salmonella* Serovars. 9th edn. Institut Pasteur, Paris
2. Tindall BJ, Grimont PAD, Garrity GM, Euzeby JP (2005) Nomenclature and taxonomy of the genus Salmonella. Int J Syst Evol Microbiol 55:521–524
3. Nataro JP, Bopp CA, Fields PI, Kaper JB, Strockbine NA (2011) *Escherichia, Shigella,* and *Salmonella.* In: Versalovic J, Carroll KC, Funke G, Jorgensen JH, Landry ML, Warnock DW (eds.). Manual of Clinical Microbiology, 10th edn. ASM, Washington, DC, p 603–626

4. Pegues DA, Miller SI (2010) *Salmonella* Species, including *Salmonella* Typhi. In: Mandell GL, Bennett JE, Dolin R (eds). Mandell, Douglas and Bennett's Principles and Practices of Infectious Diseases, 7th edn. Elsevier, Philadelphia, p 2887–2903

5. Brenner FW, Villar RG, Angulo FJ, Tauxe R, Swaminathan B (2000) Salmonella nomenclature. J Clin Microbiol 38:2465–2467

6. Majowicz SE, Musto J, Scallan E, Angulo FJ, Kirk M, O'Brien SJ et al (2010) The global burden of nontyphoidal Salmonella gastroenteritis. Clin Infect Dis 50:882–889

7. Vieira A, Jensen AR, Pires SM, Karlsmose S, Wegener HC, Wong DLF (2009) WHO Global Foodborne Infections Network Country Databank—a resource to link human and non-human sources of Salmonella. Int Soc Vet Epidemiol Econ 643:512–517

8. Braden CR (2006) Salmonella enterica serotype Enteritidis and eggs: a national epidemic in the United States. Clin Infect Dis 43:512–517

9. Hohmann EL (2001) Nontyphoidal salmonellosis. Clin Infect Dis 32:263–269

10. Mermin J, Hutwagner L, Vugia D, Shallow S, Daily P, Bender J et al (2004) Reptiles, amphibians, and human Salmonella infection: a population-based, case-control study. Clin Infect Dis 38(Suppl 3):S253–261

11. Weikel CS, Guerrant RL (1985) Nosocomial salmonellosis. Infection Control: IC 6:218–220

12. AAP (2006) Salmonella Infections. In: Pickering LK, Baker CJ, Kimberlin DW, Long SS (eds). Red Book: 2009 Report of the Commitee on Infectious Diseases, 28th edn. American Academy of Pediatrics, Elk Grove Village, p 584–588

13. Blaser MJ, Newman LS (1982) A review of human salmonellosis: I. Infective dose. Rev Infect Dis 4:1096–1106

14. Seals JE, Parrott PL, McGowan JE, Jr., Feldman RA (1983) Nursery salmonellosis: delayed recognition due to unusually long incubation period. Infection Control: IC 4:205–208

15. Garcia-del Portillo F (2001) Salmonella intracellular proliferation: where, when and how? Microbes Infect 3:1305–311

16. Dougan G, John V, Palmer S, Mastroeni P (2011) Immunity to salmonellosis. Immunol Rev 240:196–210

17. Gondwea EN, Molyneuxa ME, Goodallc M, Graham SM, Mastroenig P, Draysonc MT et al (2010) Importance of antibody and complement for oxidative burst and killing of invasive nontyphoidal Salmonella by blood cells in Africans. PNAS 107:3070–3075

18. MacLennan CA, Gondwe EN, Msefula CL, Kingsley RA, Thomson NR, White SA et al (2008) The neglected role of antibody in protection against nontyphoidal salmonella bacteremia in African children. J Clin Invest 118:1553–1562

19. Tennant SM, Hartland EL, Phumoonna T, Lyras D, Rood JI, Robins-Browne RM et al (2008) Influence of gastric acid on susceptibility to infection with ingested bacterial pathogens. Infect Immun 76:639–645

20. Buchin PJ, Andriole VT, Spiro HM (1980) Salmonella infections and hypochlorhydria. J Clin Gastroenterol 2:133–138

21. O'Ryan ML, Nataro JP, Cleary TG (2011) Microorganisms Responsible for Neonatal Diarrhea. In: Remington JS, Klein JO, Wilson CB, Nizet V, Maldonado Y (eds). Remington: Infectious Diseases of the Fetus and Newborn, 7th edn. Elsevier, Philadelphia, p. 359–418

22. Moir S, Fauci AS (2010) Immunology. Salmonella susceptibility. Science 328:439–440 (New York, NY)

23. Celum CL, Chaisson RE, Rutherford GW, Barnhart JL, Echenberg DF (1987) Incidence of salmonellosis in patients with AIDS. J Infect Dis 156:998–1002

24. Ramos JM, Garcia-Corbeira P, Aguado JM, Arjona R, Ales JM, Soriano F (1994) Clinical significance of primary vs. secondary bacteremia due to nontyphoid Salmonella in patients without AIDS. Clin Infect Dis 19:777–780

25. Abruzzi A, Fried B (2011) Coinfection of Schistosoma (Trematoda) with bacteria, protozoa and helminths. Adv Parasitol 77:1–85

26. Jertborn M, Haglind P, Iwarson S, Svennerholm AM (1990) Estimation of symptomatic and asymptomatic Salmonella infections. Scand J Infect Dis 22:451–455

27. Mandal BK, Brennand J (1988) Bacteraemia in salmonellosis: a 15 year retrospective study from a regional infectious diseases unit. BMJ 297:1242–1243 (Clinical research ed)

28. Jones TF, Ingram LA, Cieslak PR, Vugia DJ, Tobin-D'Angelo M, Hurd S et al (2008) Salmonellosis outcomes differ substantially by serotype. J Infect Dis 198:109–114

29. Vugia DJ, Samuel M, Farley MM, Marcus R, Shiferaw B, Shallow S et al (2004) Invasive Salmonella infections in the United States, FoodNet, 1996–1999: incidence, serotype distribution, and outcome. Clin Infect Dis 38(Suppl 3):S149–156

30. Gordon MA (2011) Invasive nontyphoidal Salmonella disease: epidemiology, pathogenesis and diagnosis. Curr Opin Infect Dis 24:484–489

31. Graham SM (2010) Nontyphoidal salmonellosis in Africa. Curr Opin Infect Dis 23:409–414

32. Scott AJ, Berkley JA, Mwangi I, Ochola L, Uyoga S, Macharia A et al (2011) Relation between falciparum malaria and bacteraemia in Kenyan children: a population–based, case-control study and a longitudinal study. Lancet 378:1316–1323

33. Schutze GE, Schutze SE, Kirby RS (1997) Extraintestinal salmonellosis in a children's hospital. Pediatr Infect Dis J 16:482–485

34. Owusu-Ofori A, Scheld WM (2003) Treatment of Salmonella meningitis: two case reports and a review of the literature. Int J Infect Dis 7:53–60

35. Sirinavin S, Jayanetra P, Lolekha S, Layangkul T (1988) Predictors for extraintestinal infection in Salmonella enteritis in Thailand. Pediatr Infect Dis J 7:44–48

36. Galanakis E, Bitsori M, Maraki S, Giannakopoulou C, Samonis G, Tselentis Y (2007) Invasive non-typhoidal salmonellosis in immunocompetent infants and children. Int J Infect Dis 11:36–39

37. Zaidi E, Bachur R, Harper M (1999) Non-typhi Salmonella bacteremia in children. Pediatr Infect Dis J 18:1073–1077

38. Katz BZ, Shapiro ED (1986) Predictors of persistently positive blood cultures in children with "occult" Salmonella bacteremia. Pediatr Infect Dis 5:713–714

39. Ramos JM, Aguado JM, Garcia-Corbeira P, Ales JM, Soriano F (1996) Clinical spectrum of urinary tract infections due on nontyphoidal Salmonella species. Clin Infect Dis 23:388–390

40. Gasem MH, Keuter M, Dolmans WMV, Van Der Ven-Jongekrijg J, Djokomoeljanto R, Van Der Meer JWM (2003) Persistence of Salmonellae in blood and bone marrow: randomized controlled trial comparing ciprofloxacin and chloramphenicol treatments against enteric fever. Antimicrob Agents Chemother 47:1727–1731

41. Ternhag A, Torner A, Svensson A, Ekdahl K, Giesecke J (2008) Short–and long-term effects of bacterial gastrointestinal infections. Emerg Infect Dis 14:143–148

42. Arnedo-Pena A, Beltran-Fabregat J, Vila-Pastor B, Tirado-Balaguer MD, Herrero–Carot C, Bellido-Blasco JB et al (2010) Reactive arthritis and other musculoskeletal sequelae following an outbreak of Salmonella hadar in Castellon, Spain. J Rheumatol 37:1735–1742

43. Buchwald DS, Blaser MJ (1984) A review of human salmonellosis: II. Duration of excretion following infection with nontyphi Salmonella. Rev Infect Dis 6:345–356

44. Corrado ML, DuPont HL, Cooperstock M, Fekety R, Murray DM (1992) Evaluation of new anti-infective drugs for the treatment of chronic carriage of Salmonella. Infectious Diseases Society of America and the Food and Drug Administration. Clin Infect Dis 15(Suppl 1):S259–262

45. Gilbert DN, Moellering RC, Eliopoulis GM, Saag MS (2011) The Sanford Guide to Antimicrobial Therapy, 41st edn

46. Ingham B, Brentnall DW, Dale EA, McFadzean JA (1977) Arthropathy induced by antibacterial fused N-alkyl-4-pyridone-3-carboxylic acids. Toxicol Lett 1:21–26

47. Adefurin A, Sammons H, Jacqz-Aigrain E, Choonara I (2011) Ciprofloxacin safety in paediatrics: a systematic review. Arch Dis Child 96:874–880

48. Pickering LK (2009) Approach to Patients with Gastrointestinal Tract Infections and Food Poisoning. In: Feigin RD, Cherry J, Demmler-Harrison G, Kaplan S (eds). Textbook of Pediatric Infectious Diseases. Elsevier, Philadelphia, p 621–653

49. Graham SM (2008) Salmonellosis. In: McIntosh N, Helms P, Smyth R, Logan S (eds). Forfar and Arneil's Textbook of Pediatrics, 7th edn. Elsevier, Philadelphia, p 1246–1249

50. Isaacs D (2007) Gastrointestinal infections. In: Elliot E, Gilbert R, Moyer V, Pichichero M (eds). Evidence-based Pediatric Infectious Diseases. Blackwell Publishing, United Kingdom, p 74–101

51. Reller ME (2009) Salmonella Species. In: Long SS, Pickering LK, Prober C (eds). Principles and Practice of Pediatric Infectious Diseases, 3rd edn. Elsevier, Philadelphia, p 812–817

52. Bhutta ZA (2011) Salmonella. In: Kliegman RM, Stanton BMD, St Geme J, Schor N, Behrman RE (eds). Nelson Textbook of Pediatrics, 19th edn. Elsevier, Philadelphia, p 948–958

53. Pickering LK (2006) Salmonella Infections. In: McMillan JA, Feigin RD, DeAngelis D, Jones MD (eds). Oski's Pediatrics: Principles and Practice, 4th edn. Lippincott Williams and Wilkins, Philadelphia, p 1112–1116

54. Pavia AT (2011) Salmonella, Shigella, and Escherichia coli. In: Rudolph C, Rudolph A, Lister G, First L, Gershon A (eds). Rudolph's Pediatrics, 22nd edn. McGraw Hill, New York

55. Guerrant RL, Van Gilder T, Steiner TS, Thielman NM, Slutsker L, Tauxe RV et al (2001) Practice guidelines for the management of infectious diarrhea. Clin Infect Dis 32:331–351

56. (2009) Diarrhoea and vomiting caused by gastroenteritis: diagnosis, assessment and management in children younger than 5 years. RCOG, London

57. Hohmann EL (2012) Approach to the patient with nontyphoidal Salmonella in a stool culture. In: Baron EL (ed). UpToDate, Waltham

58. Sirinavin S, Garner P (2000) Antibiotics for treating salmonella gut infections. Cochrane Database Syst Rev (Online):CD001167

59. Garcia de Olarte D, Hugo TS, Nancy AO, Nelson JD, Haltalin KC (1974) Treatment of Diarrhea in Malnourished Infants and Children: A Double-Blind Study Comparing Ampicillin and Placebo. Am J Dis Child 127:379–388

60. Nelson JD, Kusmiesz H, Jackson LH, Woodman E (1980) Treatment of Salmonella gastroenteritis with ampicillin, amoxicillin, or placebo. Pediatrics 65:1125–1130

61. Haltalin KC, Kusmiesz HT, Hinton LV, Nelson JD (1972) Treatment of Acute Diarrhea in Outpatients: Double-Blind Study Comparing Ampicillin and Placebo. Am J Dis Child 124:554–561

62. Kazemi M, Gumpert TG, Marks MI (1973) A controlled trial comparing sulfametboxazole–trimethoprim, ampicillin, and no therapy in the treatment

of salmonella gastroenteritis in children. J Pediatr 83:646–650

63. Macdonald WB, Friday F, McEacharn M (1954) The Effect of Chloramphenicol in Salmonella Enteritis of Infancy. Arch Dis Child 29:238–241

64. Geme JW, 3rd, Hodes HL, Marcy SM, Pickering LK, Rodriguez WJ, McCracken GH, Jr. et al (1988) Consensus: management of Salmonella infection in the first year of life. Pediatr Infect Dis J 7:615–621

65. Voltersvik P, Halstensen A, Langeland N, Digranes A, Peterson LE, Rolstad T et al (2000) Eradication of non-typhoid salmonellae in acute enteritis after therapy with ofloxacin for 5 or 10 days. J Antimicrob Chemother 46:457–459

66. Stoycheva MV, Murdjeva MA (2006) Antimicrobial therapy of salmonelloses–current state and perspectives. Folia Med (Plovdiv) 48:5–10

67. Tsai MH, Huang YC, Lin TY, Huang YL, Kuo CC, Chiu CH (2011) Reappraisal of parenteral antimicrobial therapy for nontyphoidal Salmonella enteric infection in children. Clin Microbiol Infect 17:300–305

68. Onwuezobe IA, Oshun PO (2008) Antbiotics for treating nontyphoidal *Salmonella* diarrhoea. Cochrane Database Syst. Rev

69. Sirinavin S, Thavornnunth J, Sakchainanont B, Bangtrakulnonth A, Chongthawonsatid S, Junumporn S (2003) Norfloxacin and azithromycin for treatment of nontyphoidal salmonella carriers. Clin Infect Dis 37:685–691

70. Zavala Trujillo I, Quiroz C, Gutierrez MA, Arias J, Renteria M (1991) Fluoroquinolones in the treatment of typhoid fever and the carrier state. Eur J Clin Microbiol Infect Dis 10:334–341

71. Nelson MR, Shanson DC, Hawkins DA, Gazzard BG (1992) Salmonella, Campylobacter and Shigella in HIV-seropositive patients. AIDS (London, England) 6:1495–1498

72. Kaplan JE, Benson C, Holmes KH, Brooks JT, Pau A, Masur H et al (2009) Guidelines for prevention and treatment of opportunistic infections in HIV-infected adults and adolescents: recommendations from CDC, the National Institutes of Health, and the HIV Medicine Association of the Infectious Diseases Society of America. MMWR Recomm Rep 58:1–207, (quiz CE1–4)

73. Roberts C, Wilkins EG (1987) Salmonella screening of pregnant women. J Hosp Infect 10:67–72

74. Schloesser RL, Schaefer V, Groll AH (2004) Fatal transplacental infection with non–typhoidal Salmonella. Scand J Infect Dis 36:773–774

75. Cooke FJ, Ginwalla S, Hampton MD, Wain J, Ross-Russell R, Lever A et al (2009) Report of neonatal meningitis due to Salmonella enterica serotype Agona and review of breast milk-associated neonatal Salmonella infections. J Clin Microbiol 47:3045–3049

76. Cahill SM, Wachsmuth K, Costarrica M, Embarek P (2008) Powdered Infant Formula as a Source of Salmonella Infection in Infants. Clin Infect Dis 46:268–273

77. Van Der Klooster JM, Roelofs HJ (1997) Management of Salmonella infections during pregnancy and puerperium. Neth J Med 51:83–86

78. Varma JK, Molbak K, Barrett TJ, Beebe JL, Jones TF, Rabatsky-Ehr T et al (2005) Antimicrobial–resistant nontyphoidal Salmonella is associated with excess bloodstream infections and hospitalizations. J Infect Dis 191:554–561

79. Su L-H, Chiu C-H, Chu C, Ou JT (2004) Antimicrobial resistance in nontyphoid Salmonella serotypes: a global challenge. Clin Infect Dis 39:546–551

80. Parry CM, Threlfall EJ (2008) Antimicrobial resistance in typhoidal and nontyphoidal salmonellae. Curr Opin Infect Dis 21:531–538

81. Angulo FJ, Johnson KR, Tauxe RV, Cohen ML (2000) Origins and consequences of antimicrobial-resistant nontyphoidal Salmonella: implications for the use of fluoroquinolones in food animals. Microb Drug Resist (Larchmont, NY) 6:77–83

82. Crump JA, Medalla FM, Joyce KW, Krueger AL, Hoekstra RM, Whichard JM et al (2011) Antimicrobial resistance among invasive nontyphoidal Salmonella enterica isolates in the United States: National Antimicrobial Resistance Monitoring System, 1996–2007. Antimicrob Agents Chemother 55:1148–1154

83. Meakins S, Fisher IST, Berghold C, Gerner-Smidt P, Tschape H, Cormican M et al (2008) Antimicrobial drug resistance in human nontyphoidal Salmonella isolates in Europe 2000–2004: a report from the Enter-net International Surveillance Network. Microb Drug Resist (Larchmont, NY) 14:31–35

84. Reddy EA, Shaw AV, Crump JA (2010) Community-acquired bloodstream infections in Africa: a systematic review and meta-analysis. Lancet Infect Dis 10:417–432

85. Nordmann P, Poirel L, Mak JK, White PA, McIver CJ, Taylor P (2008) Multidrug-resistant Salmonella strains expressing emerging antibiotic resistance determinants. Clin Infect Dis 46:324–325

86. Miriagou V, Tzouvelekis LS, Rossiter S, Tzelepi E, Angulo FJ, Whichard JM (2003) Imipenem resistance in a Salmonella clinical strain due to plasmid-mediated class A carbapenemase KPC-2. Antimicrob Agents Chemother 47:1297–1300

87. Hakanen A, Kotilainen P, Jalava J, Siitonen A, Huovinen P (1999) Detection of decreased fluoroquinolone susceptibility in Salmonellas and validation of nalidixic acid screening test. J Clin Microbiol 37:3572–3577

88. Hakanen AJ, Lindgren M, Huovinen P, Jalava J, Siitonen A, Kotilainen P (2005) New quinolone resistance phenomenon in Salmonella enterica: nalidixic acid-susceptible isolates with reduced fluoroquinolone susceptibility. J Clin Microbiol 43:5775–5778

89. Chorba TL, Meriwether RA, Jenkins BR, Gunn RA, MacCormack JN (1987) Control of a non-foodborne outbreak of salmonellosis: day care in isolation. Am J Public Health 77:979–981

90. Tauxe RV, Hassan LF, Findeisen KO, Sharrar RG, Blake PA (1988) Salmonellosis in nurses: lack of transmission to patients. J Infect Dis 157:370–373

91. (1987) Food handlers and salmonella food poisoning Lancet 2:606–607

92. Kassis I, Dagan R, Chipman M, Alkan M, Simo A, Gorodischer R (1990) The use of prophylactic furazolidone to control a nosocomial epidemic of multiply resistant Salmonella typhimurium in pediatric wards. Pediatr Infect Dis J 9:551–555

93. Dyson C, Ribeiro CD, Westmoreland D (1995) Large scale use of ciprofloxacin in the control of a Salmonella outbreak in a hospital for the mentally handicapped. J Hosp Infect 29:287–296

94. Linnemann CC, Jr., Cannon CG, Staneck JL, McNeely BL (1985) Prolonged hospital epidemic of salmonellosis: use of trimethoprim-sulfamethoxazole for control. Infection control: IC 6:221–225

95. Siegel JD, Rhinehart E, Jackson M, Chiarello L (2007) Guideline for Isolation Precautions: Preventing Transmission of Infectious Agents in Healthcare Settings. Healthcare Infection Control Practices Advisory Committee (HICPAC)

96. NHMRC (2010) Australian Guidelines for the Prevention and Control of Infection in Healthcare. Commonwealth of Australia 1–262

# Invasive Fungal Infections in Patients with Chronic Granulomatous Disease

3

Stefanie Henriet, Paul E. Verweij, Steven M. Holland and Adilia Warris

## Abstract

Invasive fungal infections are a major threat for chronic granulomatous disease (CGD) patients. The present study provides a comprehensive overview of published invasive fungal infections in the CGD host through an extensive review of epidemiological, clinical, diagnostic and therapeutic data. In addition to the often mild clinical presentation, the currently used diagnostics for invasive aspergillosis have low sensitivity in CGD patients and cannot be easily translated to this non-neutropenic host. *Aspergillus fumigatus* and *A. nidulans* are the most commonly isolated species. *A. nidulans* infections are seldom reported in other immunocompromised patients, indicating a unique interaction between this fungus and the CGD host. The occurrence of mucormycosis is mainly noted in the setting of treatment of inflammatory complications with immunosuppressive drugs. *Candida* infections are infrequently seen and do not cause mucocutaneous disease but do show an age-dependent clinical presentation. The CGD patient is susceptible to a wide range of fungal pathogens, indicating the need to determine the causative fungus, often by invasive diagnostics, to guide optimal and rational treatment. This review summarizes current understanding of invasive fungal infections in patients with CGD and will serve as a starting point to guide optimal treatment strategies and to direct further research aimed at improving outcomes.

A. Warris (✉) · S. Henriet
Department of Pediatric Infectious Diseases and Immunology, Nijmegen Institute for Infection, Inflammation and Immunity, Radboud University Nijmegen Medical Centre, Nijmegen, The Netherlands
e-mail: A.Warris@cukz.umcn.nl

S. Henriet
e-mail: S.Henriet@cukz.umcn.nl

P. E. Verweij
Department of Medical Microbiology, Nijmegen Institute for Infection, Inflammation and Immunity, Radboud University Nijmegen Medical Centre, Nijmegen, The Netherlands
e-mail: P.Verweij@mmb.umcn.nl

S. M. Holland
Laboratory of Clinical Infectious Diseases, National Institute of Allergy and Infectious Diseases, National Institutes of Health, Bethesda, USA
e-mail: SHOLLAND@niaid.nih.gov

N. Curtis et al. (eds.), *Hot Topics in Infection and Immunity in Children IX*, Advances in Experimental Medicine and Biology 764, DOI 10.1007/978-1-4614-4726-9_3, © Springer Science+Business Media New York 2013

**Table 3.1** Epidemiology of invasive fungal infections in CGD: comparison of published registry data

| Geographical region [Reference] | Number of patients | *Aspergillus* infections (%) | Organ involvement (%) | | | | | | AF prophylaxis (%) | Death (*Aspergillus*) (%) |
|---|---|---|---|---|---|---|---|---|---|---|
| | | | Lung | Skin | Liver | Brain | Bone | Septicemia | | |
| Europe [18] | 409 | 26 | 61 | 5 | 3 | 7 | 16 | 2 | 53 | 28 |
| USA [4] | 368 | 33 | 78 | 5 | 2 | 4 | 12 | 0 | NA | 35 |
| UK [1] | 94 | 27 | 85 | NA | 0 | NA | 10 | 0 | 93 | 50 |
| Italy [20] | 60 | 34 | 53 | 0 | 0 | 18 | 24 | 0 | NA | 50 |
| Japan [19] | 23 | 45 | NA | NA | NA | NA | NA | NA | 48 | 67 |
| Spain [172] | 13 | 20 | NA | NA | NA | NA | NA | NA | 77 | 25 |
| Sweden [3] | 21 | 24 | 86 | NA | NA | NA | NA | NA | NA | 0 |

*AF* antifungal, *NA* not available

## 3.1 Introduction

Patients suffering from chronic granulomatous disease (CGD) are well known to be prone to invasive fungal infection, thereby providing a unique opportunity to examine host susceptibility to fungal invasion within the framework of a well defined immune defect. CGD is a rare inherited disorder of the NADPH oxidase in which phagocytes fail to generate the microbicidal reactive oxidant superoxide anion and its metabolites, hydrogen peroxide, hydroxyl anion, and hypohalous acid. Clinically, as a result of the defect in this key innate host defense pathway, CGD patients suffer from recurrent life-threatening bacterial and fungal infections and inflammatory sequelae. The molecular basis of CGD is well understood: CGD is a genetically heterogenous disease caused by mutations in any of the five structural components of NADPH oxidase, including the membrane-bound glycoprotein p91$^{phox (phagocyte oxidase)}$ and p22$^{phox}$ and the cytoplasmatic components p40$^{phox}$, p47$^{phox}$ and p67$^{phox}$. Birth prevalence ranges from 1/450,000 in Sweden, 1/250,000 in the United States and Japan, up to 1/120,000 in the United Kingdom and Ireland [1–4]. Although the majority of CGD patients are diagnosed in early infancy, fungal infection as presenting symptom in adult patients has been reported [5–9]. The suspicion of the presence of CGD as an underlying condition should arise whenever the diagnosis of invasive aspergillosis (IA) is made in the absence of another known risk factor.

*Aspergillus* is one of the most important encountered pathogens [10, 11], and the life-time incidence of IA in children with CGD before the advent of specific antifungal prophylaxis varied between 25 % and 40 % [11, 12]; although since the introduction of azole prophylaxis the incidence seems to decrease [13]. In comparison, the incidence of aspergillosis in children undergoing allogeneic haematopoietic stem cell transplantation (HSCT) or who were treated for hematological malignancy, who typically have a very limited period of susceptibility, lies between 2 % and 6.8 % [13–17].

In more detailed information extracted from the published data of CGD patient registries, the percentage of patients who had at least one infectious episode caused by *Aspergillus* spp. ranges from 26 % in Europe [18] up to 46 % in Japan [19]. More importantly, independent of the epidemiology of fungal infections in the published data, in the cases of pulmonary infection, brain abscesses and osteomyelitis, *Aspergillus* spp. have been the most commonly isolated organisms [1, 4, 18, 20] (Table 3.1). The most common *Aspergillus* spp. affecting CGD patients is *A. fumigatus* [10, 21, 22] followed by *A. nidulans* [23]. Much less attention has been paid to *A. nidulans* as an opportunistic pathogen in humans until recent years, when it was recognized especially in patients with CGD or with histories suggestive of CGD. White et al. [24] suggested in 1988 that CGD patients were at greater risk for *A. nidulans* than other immunocompromised patient populations. Scrutiny of the microbiology database of CGD patients at the National Institute of Health (NIH, Bethesda, MD) over a 10 year-period revealed 6 cases of *A. nidulans,* compared to 17 cases of *A. fumigatus* [23]. In cases of osteomyelitis, *A. nidulans* was isolated in up to 50 % of the cases [23, 25]. Although *A. nidulans* behaves more virulently in CGD patients than *A. fumigatus,* solid information on the relative frequency of IA due to *A. nidulans* in CGD patients is lacking. *Candida* is a less frequently encountered pathogen in CGD and is reported as being responsible for invasive fungal infection in 6 % of all isolated microorganisms among CGD patients registered in the European database, in 10 % in a Swedish CGD population, and up to 14 % in the Italian CGD study [3, 18, 20].

Invasive mucormycosis has not been recognized in patients with CGD as frequently as IA [26, 27]. A review of 929 pediatric and adult cases of mucormycosis revealed 81 % of the individuals having an impaired immune status, but none of these were associated with CGD [28]. In a systematic review of 157 cases of mucormycosis in children, 86 % were associated with risk factors as neutropenia, prematurity, diabetes mellitus and ketoacidosis; but no underlying primary immunodeficiency was noted [29]. However, when CGD patients are subjected to further immune compromising conditions affecting T cell function, such as methotrexate or steroids, typically in the setting of management of their inflammatory complications, mucormycosis has been encountered [30].

Globally, specific data regarding the epidemiology of non-*Aspergillus,* non-*Candida* IFD in CGD patients are scarce or described as "Fungi: not identified". Little detailed information is given on the identification of the causative fungus, and available data are almost exclusively found as individual case reports, making it difficult to assess any epidemiological burden of those fungi within the CGD population.

As a cause of death, fungal infections caused by *Aspergillus* spp. have been responsible for one-third to half of all deaths in CGD patients [2, 4, 20]. Mortality rates due to IFD caused by other fungi are unknown but, since case reports are more common with severe infections, may be overestimated from their true incidence.

Overall, as disease pathology and progression are the results of the complex interaction between the fungal pathogen and the host defense, exploring these host-fungus interfaces of the different fungal pathogens in relation to the host, gives us insight and detailed understanding of this challenging frontline, in order to optimize diagnostic and therapeutic tools.

We reviewed the literature of invasive fungal infections in CGD through topics such as epidemiological data, mycological characterization and identification, clinical manifestations, diagnostic features and in vitro susceptibility data in order to assess these findings within a framework of our current understanding of IFD in CGD.

## 3.2 Methods

### 3.2.1 Search Strategy

A PubMed database search of the English-, French-, and Dutch-language literature from 1970–2010 was performed using keywords "chronic granulomatous disease" and "fungal", "mycoses", "asper-gillosis", "aspergillus", "candidiosis", "candida", "zygomycosis", "dermatophytosis", "scedosporium", "trichosporon", "paecilomyces" and "endemic mycoses". In addition, published data concerning diagnostic features and susceptibility testing were reviewed. All the articles found by this means were systematically reviewed and the references cited in the articles were screened for additional information and cases, or for any duplicate reports of the same patient.

### 3.2.2 Criteria for Inclusion of Case Reports

#### 3.2.2.1 Criteria for Diagnosis of Chronic Granulomatous Disease

The phagocytes of CGD patients lack the ability to generate superoxide and its metabolites due to a defect in the NADPH oxidase complex. As a diagnostic tool multiple tests have been validated in literature: the semi-quantitative nitrobluetetrazolium test (NBT) [31]; the cytochromec test—a quantitative test based on the reduction of c ferricytochrome—and assays using probes of which the chemiluminescent or fluorescent properties are altered by their reaction with reactive oxidants. More recently, a fluorescence assay has been introduced, which uses the conversion of dihydrorhodamine 123 (DHR) to rhodamine 123, in the presence of hydrogen peroxide, being detected by flow cytometry [32, 33]. In the past, patients were diagnosed as having CGD by the occurrence of an appropriate clinicopathologic syndrome as described by Berendes et al. [34] and by the investigation of neutrophil killing defects. The vast majority of the included cases of CGD are confirmed by use of either the NBT test, chemoluminescence, cytochrome c reductase or DHR 123 test. If the patient died before proper diagnostic tests had been performed, inclusion was accepted when the autopsy and histology reports were suggestive of CGD in association with a positive genetic test of a sibling or a first-degree relative. A highly suggestive postmortem analysis in the absence of any genetic argument was not included in this review.

### 3.2.2.2 Documentation of Infection

As there are no definitions to categorize invasive fungal infection in patients with CGD, the principles of the EORTC/MSG criteria for invasive fungal infections [35], which apply to patients with cancer, were followed. Host factors, clinical symptoms and mycological evidence were used to categorize cases as probable or proven invasive fungal infection. Obviously, CGD was the host factor associated with an increased risk for fungal infection. Invasive fungal infection was considered proven in CGD patients with a positive culture from a specimen obtained by an invasive procedure showing histological evidence of growing fungi. Cases were considered probable if radiological features were considered to be consistent with fungal infection by the radiologist and mycological evidence was obtained, including a positive culture or detection of circulating galactomannan or 1,3-beta D glucan. Cases without mycological evidence were not included in this review. In case of a negative culture during the patient's lifetime, inclusion required the demonstration of infection at autopsy by culture of the causative agent from the affected organs.

### 3.2.2.3 Localisation of Infection

Documentation of the primary site of the infection was required. The definitions used were the following. (i) Localized infections were defined by isolation of the fungus from a clinically infected site without positive blood cultures or evidence of dissemination to distant organs; (ii) Disseminated infections were defined by the isolation of the fungal isolate from two or more noncontiguous sites showing clinical or histological evidence of infection, and/or positive blood cultures; (iii) Infections with contiguous spread were defined by the isolation of the fungal isolate from at least one contiguous site showing clinical or histological evidence of infection; (iv) Fungemia indicated the presence of the fungus in the blood.

## 3.3 Results

We identified 116 reported cases fulfilling the inclusion criteria of invasive fungal infections in CGD patients between 1970 and 2010. The demographic and clinical characteristics are summarized in Table 3.2. One hundred twenty-seven fungal species were isolated: 44 (35 %) *A. fumigatus*, 23 (18 %) *A. nidulans*, 5 (4 %) *A. niger*, 1 *A. flavus*, 11 (9 %) species belonging to the phylum *Zygomycota* (mainly *Rhizopus* spp.), 10 (8 %) *Candida* spp., 8 (6 %) *Trichosporon* spp., 6 (5 %) *Paecilomyces* spp., 5 (4 %) *Scedosporium* spp., 4 (3 %) *Penicillium* spp., 2 *Acremonium* spp., 2 *Alternaria* spp., 1 *Inonotus* spp., 1 *Exophiala*, 1 *Chrysosporium* spp., 1 *Fusarium* spp., 1 *Microascus* spp. and 1 *Hansela* spp. (Fig. 3.1). No reports of endemic mycoses in patients suffering from CGD were found following the above described search strategies.

**Table 3.2** Characteristics of invasive fungal infections in 116 CGD patients

| Variable | Cases, n (%) | Mortality, n (%) |
|---|---|---|
| Median age in years (range) | 10 (0–69) | |
| Male/female/unknown | 103/10/3 | 24/4/2 |
| X-linked | 65 (56) | 16 (25) |
| AR | 22 (19) | 6 (27) |
| p47phox | 10 | 2 |
| p67phox | 2 | 1 |
| p22phox | 2 | 0 |
| AR, no details | 8 | 3 |
| Genetics unknown | 29 (25) | 8 (28) |
| Mono-fungal infections | 108 | 26 (24) |
| Mixed-fungal infections | 8 | 4 (50) |
| Antifungal prophylactic treatment | 22 | |
| INF-γ prophylactic treatment | 28 | |
| Lung involvement | 83 | 25 (30) |
| Bone involvement | 47 | 10 (21) |
| Infecting organism (n = 127) | | |
| A. fumigatus | 44 | 13 (30) |
| A. nidulans | 23 | 6 (27) |
| Other Aspergillus spp. | 6 | 3 (50) |
| Zygomycetes | 11 | 6 (55) |
| Paecilomyces spp. | 6 | 0 |
| Scedosporium spp. | 5 | 0 |
| Trichosporon spp. | 8 | 2 (25) |
| Candida spp. | 10 | 4 (40) |
| Miscellaneous group | 14 | 3 (21) |

## 3.4 Fungal Infections in CGD: by Clinical Manifestation

### 3.4.1 Pneumonia

Fungal involvement of the lung can result in two major clinical manifestations: the first is the result of direct infection of the pulmonary tissue or the infection of lung cavities, the second is the ability to trigger an immunological reaction when the fungus is inhaled. The CGD patient can suffer from both clinical identities. However, the clinical course and diagnostic signs and symptoms of invasive fungal pulmonary involvement

differ remarkably from the classical clinical spectrum derived from cancer patients.

Seventy-two percent of the reported fungal infections in CGD had lung-involvement. Some species were more prone to cause pulmonary infection, e.g. Zygomycetes (91 %), A. nidulans (78 %), followed by A. fumigatus (72 %). The majority of the trichosporiosis patients suffered from lung infections as did half of the patients infected with Paecilomyces spp. Among the patients with candidiasis, only two out of ten mentioned pulmonary involvement, which can be explained by the route of infection being the result of disruption of skin or mucosal membrane (endogenous) instead of inhalation of spores (exogenous).

Symptoms include cough, usually non-productive, fever, chest discomfort and progressive dyspnea. However, some species have the propensity to be quite indolent, e.g. the clinical onset of A. nidulans infections was, compared to A. fumigatus, often very mild with nonspecific symptoms. Trichosporon inkin infections have also been described with an asymptomatic or mild presentation. However, the masking of the symptoms in those cases might have been at least partially explained by concomitant immunosuppressive drug use. Hemoptysis, a relatively common symptom of angioinvasive aspergillosis or mucormycosis in the immunocompromised host, has not been reported in CGD patients. In line with this, the typical lung lesions observed in neutropenic patients, i.e., the air crescent and halo signs, have never been reported in CGD cases. Nevertheless, A. fumigatus can hematogenously disseminate to liver, skin, brain. In contrast, A. nidulans tends to spread by contiguity.

A distinct pulmonary fungal infection syndrome in CGD is acute fulminant fungal pneumonia, "mulch pneumonitis", a medical emergency highly associated with and probably pathognomonic for CGD. High-level exposure to aerosolized fungi, such as that which occurs during spreading of mulch, lawn care, gardening, haying or digging in soil, leads to rather stereotypical signs and symptoms of pneumonia with hypoxia and diffuse infiltrates on chest radiographs. This condition can deteriorate rapidly

**Fig. 3.1** Distribution (%) of 127 fungal isolates out of 116 cases of invasive fungal disease in CGD reported between 1970–2010

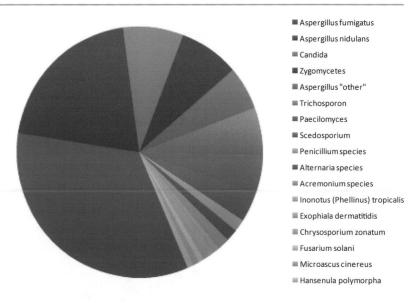

- Aspergillus fumigatus
- Aspergillus nidulans
- Candida
- Zygomycetes
- Aspergillus "other"
- Trichosporon
- Paecilomyces
- Scedosporium
- Penicillium species
- Alternaria species
- Acremonium species
- Inonotus (Phellinus) tropicalis
- Exophiala dermatitidis
- Chrysosporium zonatum
- Fusarium solani
- Microascus cinereus
- Hansenula polymorpha

leading to acute hypoxemic respiratory failure, mechanical ventilation, and death. Since the first report in 1986 [36], 11 cases have been described [36–40]. Mean age was 23 years, ranging from 7–64 years, 4 out of 11 patients were previously well and mulch pneumonitis was the initial presentation leading to the diagnosis of CGD. In more than half of these cases multiple organisms were isolated, including *Aspergillus* spp. in association with members of the phylum Zygomyceta and/or *Penicillium* spp. *A. fumigatus* and *A. nidulans* were the only causative agents isolated in all cases of mulch pneumonitis. The mortality rate was high (8/11, 73 %) despite aggressive antifungal treatment, mechanical ventilation, extra-corporal membrane oxygenation and pulse corticosteroids (82 %) for the acute respiratory distress syndrome. However, it is obvious that with early recognition and prompt institution of antifungals to control the infection and steroids to regulate inflammation that response rates will improve. More recently developed immune regulatory agents might be of interest to further improve outcome. The pathological role of the agents of mucormycosis in this syndrome is unclear, and treatment directed at the *Aspergillus* component alone is probably adequate.

Aspergilloma and chronic necrotizing pulmonary aspergillosis have not been reported in CGD patients. Allergic bronchopulmonary asper-

gillosis (ABPA) in the CGD patient is rare, but has been reported [41]. However, it seems much more likely that what was described as ABPA in a CGD patient, acute respiratory failure after heavily mulching, should be more appropriately classified as "mulch pneumonitis".

### 3.4.2   Osteomyelitis

Bone-involvement was reported in almost half of the cases (41 %) of IFD in CGD, being the second most commonly encountered organ. *A. fumigatus* and *A. nidulans* were almost equally frequent (Table 3.3). However, in case of an *A. nidulans* IFD, osteomyelitis occurred in 74 % ($n=17/23$), whereas *A. fumigatus* had a lower propensity to spread to the bone ($n=18/44$, 41 %). It is very important to recognize that speciation of *Aspergillus* is an evolving field that has been transformed by molecular biological approaches, just as the other fields in microbiology has. The designation of *A. fumigatus* in most clinical laboratories is insentitive to subtle distinctions between *A. fumigatus sensu strictu* and organisms that have roughly similar growth and color characteristics but different clinical features, such as *A. viridinutans* [42].

As previously described by Sponseller et al. [43], fungal osteomyelitis is often the result of

**Table 3.3** Fungal species distribution and localization of invasive infections in CGD patients

| Organ involvement | Cases (%) | Species involved |
|---|---|---|
| Lung | 83(72) | *A. fumigatus* ($n=31$) |
| | | *A. nidulans* ($n=25$) |
| | | *Zygomycetes* ($n=9$) |
| | | *Trichosporon* spp. ($n=7$) |
| | | *Scedosporium* spp., *Aspergillus* other ($n=4$) |
| | | *Paecilomyces* spp. ($n=3$) |
| | | *Penicillium* spp., *Candida* spp., *Acremonium* spp. ($n=2$) |
| | | *Exophilia dermatitis, Chrysosporium zonatum, Hansenula polymorpha* ($n=1$) |
| Bone | 47(41) | *A. nidulans* ($n=17$) |
| | | *A. fumigatus* ($n=18$) |
| | | *Scedosporium* spp. ($n=3$) |
| | | *A. niger, Trichosporon* spp. ($n=2$) |
| | | *Paecilomyces* spp., *Penicillium* spp., *Crysosporium zoantum, Inonotus tropicalis, Exophilia dermatitis* ($n=1$) |
| Brain | 17(15) | *A. fumigatus* ($n=9$) |
| | | *A. nidulans* ($n=2$) |
| | | *Zygomycosis* ($n=3$) |
| | | *Scedosporium prolificans, Exophilia dermatitidis, Alternaria infectoria* ($n=1$) |
| Skin | 11(10) | *A. fumigatus, Paecilomyces* spp. ($n=3$) |
| | | *Fusarium solani, Microascus cinerius, Alternaria alternate, A. nidulans* and *T. pullulans* ($n=1$) |
| Gastro-intestinal | 11(10) | *Rhizopus* spp. ($n=4$) |
| | | *A. fumigatus, C. glabrata* ($n=3$) |
| | | *Paecilomyces* ($n=1$) |
| Lymphadenitis | 11(10) | *A. fumigatus* ($n=4$) |
| | | *C. lusitaniae* ($n=2$) |
| | | *A. flavus, A. nidulans, T. glabarata, C. albicans, H. polymorphae* ($n=1$) |
| Meningitis | 1 | *C. lusitaniae* ($n=1$) |

direct spread from a pulmonary focus towards adjacent chest-wall structures (64 %). Localized bone infection, without other organ involvement, was seen in only four subjects. Of these, *A. fumigatus* was isolated twice: a localized osteomyelitis of the midfoot [44] and an osteomyelitis of the patella and arthritis with positive culture of both tissue and the synovial fluid [45]. *A. nidulans*, later identified as *E. rugulosa*, was responsible for a femoral osteomyelitis [46] and the basiomycetous mold *Inonotus tropicalis* for a destructive sacral bone osteomyelitis with soft tissue abscesses [47]. None of these infections were preceded by trauma or apparent direct inoculation. The mortality rate was 21 %. Surgery was the cornerstone of treatment (72 %) and in combination with antifungal therapy the cure rate was 62 %. However, more prospective research is needed to define optimal treatment strategies.

Few data are available about the clinical, histopathology and radiologic characteristics of fungal bone infections in CGD patients. In a brief report published by Galluzzo et al. [48], 14 CGD patients were compared with control subjects to determine osteomyelitis-related markers suggestive of CGD. Generally, multifocal osteomyelitis and/or simultaneous other organ involvement were the most relevant and statistically significant clinical findings suggesting CGD as an underlying disease. Histopathologic findings of chronic inflammation plus granulomata, multinucleated giant cells, histiocytes, or necrosis were significantly over-represented among CGD patients. In contrast, acute or chronic inflammation plus granulation tissue, remodeled bone, or lymphocytes were significantly under-represented and infrequently found in CGD patients. The presence of extensive destruction in association with minimum sclerosis on the X-ray, almost invariably noted at the time of diagnosis, was also described previously by Wolfson et al. [49]. Whether the particular histopathologic and radiologic patterns described in osteomyelitis occurring in CGD patients are fungus dependent or CGD host dependent is unclear. Overall, the lack of early clinical signs of osteomyelitis in association with advanced destructive lesions—often multi-focal—and the isolation of a likely opportunistic pathogen, is highly suggestive for the presence of CGD as an underlying condition, especially if the osteomyelitis is associated with other organ involvement, or if there are histopathologic signs of chronic inflammation and granulomata-related features.

### 3.4.3  Other Organ Involvement

The skin is less commonly infected by fungal than bacterial pathogens. The pathology is diverse, ranging from erythematous plaques, crusted, papulous, pustulous to purulent ulcers in the absence or presence of generalized symptoms as malaise and fever. The responsible pathogens are listed in Table 3.2. One case of linear IgA dermatosis has been described during IFD caused by *Paecilomyces* spp. [50].

Half of fungal skin lesions were preceded by a superficial abrasion of the skin or a minor trauma. The extremities were the most common locations. Primary cutaneous fungal infections can be the first manifestation of CGD. Mansoory et al. [9] described a chronic cutaneous *Fusarium* infection in a 54-year-old women with undiagnosed CGD. In such patients, milder CGD phenotypes likely due to residual levels of superoxide production are considered to be the reason for the unusually late manifestation [51]. Secondary cutaneous fungal infection, defined as the skin being an end-organ of hematogenous dissemination, is a much more aggressive clinical feature, resulting in massive dissemination and a very high case-fatality rate.

Fungal invasive gastrointestinal (GI) involvement in CGD is very rare, especially when compared to the extensive granulomatous involvement of the GI tract [52]. Etiologic agents are listed in Table 3.3. The clinical picture is that of hematogenous dissemination with abscess formation in liver, spleen and kidney, with exception of one isolated case of gastrointestinal mucormycosis that presented with an abdominal mass in a 10-month-old prematurely born infant [53]. In this boy, the diagnosis of gastrointestinal zygomycosis served as the presenting symptom of the underlying CGD.

Lymphadenitis, one of the classical presenting signs of CGD, has been rarely attributed to infection by fungi. Nevertheless, *Candida* lymphadenitis in the infant is very suspicious for CGD [54–56]. In case of *Aspergillus* infections, the lymph nodes involved are often those draining the site of infection.

Infection of the central nervous system was mainly observed as the consequence of a spread by contiguity from a primary pulmonary focus to the spinal cord. Localized brain abscesses are extremely rare but have been reported caused by *A. fumigatus*, *S. prolificans* and *Alternaria infectoria* [7, 57–59]. *Candida* spp. were responsible as causative agents in meningitis or ventriculitis [60, 61].

### 3.5  Fungal Infections in CGD: by Organism

#### 3.5.1  *Aspergillus fumigatus*

Forty-four invasive *A. fumigatus* infections in CGD patients were reviewed. According to our criteria, 40 cases were proven IA and 4 were categorized as probable. Three out of the four patients with a probable IA presented with signs of symptoms of a mulch pneumonitis [37]. The majority of infections (37/44), were caused by a single fungus. In mixed fungal infections, the associated organisms were, *Rhizopus* spp., *A. niger*, *Penicil-*

*lium* spp., *Absidia corymbifera* and *C. dublinien-sis* [37, 38, 40, 62, 63]. Interestingly, in five out of the seven multiple-fungal infections a clear temporal relationship between massive exposure to molds, especially mulch, and the clinical presentation of pneumonia, was found [37, 38, 40]. The age, sex distribution and genetic profile did not differ from the overall demographic characteristics. Interferon-γ had been administered prophylactically to one third of the patients. However, prophylactic use of itraconazole was documented in only two cases. All of the invasive fungal infections caused by *A. fumigatus* were symptomatic, and accompanied by fever in 61 %. Detailed information on the individual clinical pictures is illustrated in Table 3.4.

Almost three-quarters had a history suggestive for primary lung invasion (30/44). Of these, half remained localized in the lungs [22, 36–38, 63–66] and more than one-third were disseminated, causing metastatic foci in bone, skin and/ or the central nervous system [22, 67–74]. In 14 % of the *A. fumigatus* infections, the fungus had spread by contiguity from the lung to adjacent chest-wall structures [22, 70, 75–77].

The brain and the skin were the most common extra-pulmonary sites affected by *A. fumigatus*. Primary cutaneous aspergillosis due to *A. fumigatus* occurred in three cases, all of them in relation to direct inoculation of disrupted skin [78–80]. In four patients *A. fumigatus* caused isolated brain abscesses [7, 22, 57, 73]. It should be emphasized that the symptoms can be very indolent ranging from mild fever and headache in the absence of any localizing neurologic signs [57], up to isolated seizures mimicking intracranial tumor [58]. Isolated osteomyelitis without evidence for pulmonary involvement was reported twice [44, 45].

Susceptibility data were rarely provided. An elevated MIC for amphotericin B of 1.2 mg/l was found in an *A. fumigatus* isolate causing pneumonia and secondary osteomyelitis of the foot [69] and a MIC of 1.5 mg/l in a dual pulmonary infection with *A. fumigatus* and *C. dubliensis* [63]. A multi-azole resistant *A. fumigatus* isolate (MIC itraconazole > 16 mg/l, voriconazole > 16 mg/l, posaconazole 0.25 mg/l) causing osteomyelitis of the midfoot of a patient who received itracon-

azole prophylaxis, was treated successfully with extensive surgery and long-term oral posaconazole [44]. Patients with CGD may be at risk for azole-resistant fungi as they usually receive life-long azole antifungal prophylaxis [81–84].

Amphotericin B was the most commonly used first line treatment, followed by voriconazole. Half of the cases also underwent surgical interventions. If mentioned, adjuvant immuno-modulatory therapy, in the form of INF-γ and/or granulocyte transfusion was given in 27 %. The overall mortality rate was 30 %, mulch pneumonitis being responsible for more than half of the case-fatalities. However, as *A. fumigatus* has been the most commonly encountered fungal pathogen in CGD, it might be that less severe cases are not published, leading to overrepresentation of severe cases.

### 3.5.2  *Aspergillus nidulans*

Twenty-three cases of invasive fungal infections due to *A. nidulans* in CGD patients were previously reported and fulfilled the inclusion criteria of proven infection ($n = 22$) or probable infection ($n = 1$). In one patient, *A. nidulans*, *A. fumigatus* and *A. restrictus* were detected in sputum. However, no histological confirmation of multiple fungal infections was obtained [85].

The major clinical features of invasive *A. nidulans* infections in CGD patients are summarized in Table 3.5. Twenty-two patients were male, and the median age of the population was 8 years (range 3–21 years). Interestingly, of those patients whose genetic pattern was reported ($n = 19$), 90 % were X-linked gp91[phox], compared to 75 % in all documented cases and 74 % in those caused by *A. fumigatus*. One boy had a defect in p22[phox] and one girl had a defect in the cytoplasmic factor p67[phox]. Four patients had suffered from previous fungal infections by the time the invasive *A. nidulans* infection occurred.

The most common localization (74 %) of the reported *A. nidulans* infections in CGD patient was the lung, with direct spread to adjacent chest-wall structures (Fig. 3.2). In contrast to IFD caused by *A. fumigatus*, the presenting signs

**Table 3.4** The major clinical features of A. *fumigatus* infections in CGD

| Case [Reference] | Sex | Genetic type | Age (years) | Site of disease | Prophylaxis | Treatment | Surgery | Outcome |
|---|---|---|---|---|---|---|---|---|
| 1 [22] | NM | AR | 8 | Lung, chest wall | TMP-SMX, INF-γ | AMB | Yes | Survived |
| 2 [22] | NM | AR | 6 | Lung, liver, brain, ribs | TMP-SMX, INF-γ | AMB | No | Died |
| 3 [22] | M | X-CGD | 2 | Brain | TMP-SMX, INF-γ | AMB, 5-FC | Yes | Survived |
| 4 [22] | M | X-CGD | 5 | Hepatic abscess, chest wall, rib osteomyelitis | TMP-SMX, INF-γ | AMB, ITC | Yes | Died |
| 5 [22] | NM | AR | 4, 5 | Lung | TMP-SMX, INF-γ | AMB, ITC | Yes | Died |
| 6 [62] | M | NA | 5 | Chest-wall, thoracic vertebrae, spinal cord | NM | AMB | NA | Survived |
| 7 [62] | M | NA | 5 | Lung, chest wall, thoracic vertebrae | No | AMB, 5-FC, MIC | No | Survived |
| 8 [36] | M | X-CGD suspected FA | 20 | Mulch pneumonitis | No | AMB, 5-FC | No | Died |
| 9 [67] | M | X-CGD suspected FA | 10 | Lung, lymphadenitis, femoral osteomyelitis | TMP-SMX | AMB, ITC | No | Survived |
| 10 [68] | M | X-CGD | 11 | Lung, brain, bone (mastoid) | TMP, rifampicin | AMB, 5-FC, ITC, gran Tx | Yes | Survived |
| 11 [126] | M | NA | 18 | Anterior chest wall, sternum osteomyelitis, mediastinum, extensive vascular invasement with mycotic pseudoaneurysm | NM | NM | Yes | Survived |
| 12 [69] | M | X-CGD suspected FA | 7 | Lung, osteomyelitis foot, lymphadenitis draining Lnn | Ampicilline | AMB | Yes | Survived |
| 13 [70] | M | X-CGD | 18 | Lung, ribs, humerus, chest wall | TMP-SMX | AMB, 5-FC, ITC | Yes | Survived |
| 14 [71] | M | NA | 15 | Lung, humerus | No | AMB, 5-FC, ITC, gran Tx | Yes | Survived |
| 15 [75] | M | X-CGD | 10 | Lung, rib, vertebrae, epidural soft tissue | TMP-SMX, INF-γ | AMB, ABLC,5-FC, ITC, gran Tx | Yes | Survived |
| 16 [64] | M | X-CGD | 30 | Lung | TMP-SMX, INF-γ | VOR | No | Survived |
| 17 [65] | M | X-CGD | 0 | Lung | No | AMB | No | Survived |
| 18 [57] | M | X-CGD | 11 | Brain abscess | TMP-SMX | AMB, ITC | Yes | Survived |
| 19 [44] | M | X-CGD | 18 | Osteomyelitis midfoot | TMP-SMX, INF-γ | VOR, CAS, POS | Yes | Survived |
| 20 [76] | F | AR | 14 | Lung, pericard, myocard, endocard | No | AMBL, VOR, CAS, gran Tx | Yes | Died |
| 21 [166] | M | X-CGD | 3 | Soft-tissue skull, parietal bone, brain parenchym | No | NM | Yes | NA |
| 22 [80] | M | X-CGD suspected FA | 3 | Skin ''sporotrichoid Aspergillosis'', lung | No | AMB, VOR, ITC | Yes | Still not under control |

**Table 3.4** (continued)

| Case [Reference] | Sex | Genetic type | Age (years) | Site of disease | Prophylaxis | Treatment | Surgery | Outcome |
|---|---|---|---|---|---|---|---|---|
| 23 [79] | M | X-CGD | 13 | Skin | TMP-SMX, INF-γ | AMB, ITC | Yes | Survived |
| 24 [72] | M | NA | 7 | Lung, skin | NM | AMB | NA | Died |
| 25 [78] | M | X-CGD | 19 | Skin | TMP-SMX | ITC | No | Survived |
| 26 [37, 38, 40] | M | p47phox | 14 | Mulch pneumonitis | No | VOR, gran Tx | No | Died |
| 27 [38] | M | X-CGD | 7 | Mulch pneumonitis | No | AML, VOR, CAS | No | Died |
| 28 [37] | F | p47phox | 23 | Mulch pneumonitis | No | VOR, CAS | No | Survived |
| 29 [37] | M | X-CGD | 20 | Mulch pneumonitis | TMP-SMX | AMB | No | Died |
| 30 [37] | M | X-CGD | 23 | Mulch pneumonitis | TMP-SMX, ITC | AMBL | No | Died |
| 31 [37] | M | p47phox | 64 | Mulch pneumonitis | TMP-SMX, INF-γ ITC | AMB | No | Died |
| 32 [37] | M | X-CGD | 10 | Mulch pneumonitis | TMP-SMX, INF-γ | AMB | No | Died |
| 33 [66] | M | NA | 3 | Lung | NA | AMB | No | Survived |
| 34 [66] | M | NA | 8 | Lung | NA | AMB | No | Survived |
| 35 [66] | M | NA | 9 | Lung | NA | AMB | No | Survived |
| 36 [39] | M | X-CGD suspected FA | 32 | Mulch pneumonitis | NA | gran Tx | No | Died |
| 37 [167] | M | AR suspected, sister AR CGD | 17 | Vertebrae, ribs, cervical sinus, spinal cord | No | AMB, AMBL, ITC, CAS, ABLC, VOR | Drainage abscess, vertebral surgery not possible | Survived |
| 38 [58] | M | p47phox | 16 | Brain abscess | No | AMBL, VOR | Yes | Survived |
| 39 [7] | M | X-CGD | 8 | Brain abscesses, multiple | No | AMB, AMBL, VOR | No | Survived |
| 40 [73] | M | X-CGD | 20 | Lung, osteomyelitis skull and subcutaneous abscess | TMP-SMX | AMB, gran Tx | Yes | Survived |
| 41 [74] | M | X-CGD | 4 | Lung, osteomyeltis skull, epidural and intracerebral abscesses | No | AMB, 5-FC | Yes | Survived |
| 42 [45] | M | NA | 17 | Arthritis, osteomyelitis patella | No | AMB, ITC | No | Survived |
| 43 [63] | F | NA | 6 | Lung | No | AMB, VOR | No | Survived |
| 44 [77] | M | X-CGD | 4 | Lung, rib, subphrenic abscess | NA | AMB, ITR | NA | Survived |

*AMB* amphotericin B deoxycholate, *AMBL* amphotericin B liposomal, *ABLC* amphotericin B lipid complex, *ITC* itraconazole, *VOR* voriconazole, *POS* posaconazole, *KTC* ketoconazole, *MIC* miconazole, *CAS* caspofungin, *5-FC* 5-flucytosine, *NA* not available

**Table 3.5** The major clinical features of *A. nidulans* infections in CGD

| Case [Reference] | Sex | Genetic type | Age (years) | Site of disease | Prophylaxis | Treatment | Surgery | Outcome |
|---|---|---|---|---|---|---|---|---|
| 1 [46] | M | NA | 16 | Osteomyelitis, long bone | INF-γ | ABLC, AMBL, ITC | Yes | Survived |
| 2 [168] | M | NA | 6 | Lung, chest wall, vertebrae | TMP-SMX | AMB, ABLC, ITC | Yes | Died |
| 3 [135] | M | X-CGD | 20 | Lung, 3th rib, femur, skull | NA | AMB, ITC, 5-FC | Yes | Survived |
| 4 [43] | M | X-CGD | 4 | Lung, 8th-9th ribs, T6-L1 vertebrae | NA | AMB, AMBL, gran Tx | Yes | Died |
| 5 [43] | M | X-CGD | 9 | Lung, 4th rib | NA | AMB | Yes | Survived |
| 6 [43] | M | X-CGD | 13 | Progression of 4th rib lesion, T3-T4 with paraparesis | NA | AMB | Yes | Survived |
| 7 [23] | M | X-CGD | 6 | Lung, pleura | No INF-γ | AMB, ITC, gran Tx | Yes | Survived |
| 8 [23] | M | X-CGD | 19 | Lung, pleura, chest wall, vertebrae, skin, skull, brain | No | AMB, ABLC, ITC, 5-FC, gran Tx | Yes | Died |
| 9 [23] | M | X-CGD | 16 | Lung, pleura, vertebrae, chest wall | KTC | AMB, ABLC, ITC, gran Tx | Yes | Survived |
| 10 [23] | M | X-CGD | 7 | Lung, pleura, vertebrae, chest wall, sinuses, brain | NA | AMB, ABLC, gran Tx | Yes | Died |
| 11 [23] | M | X-CGD | 4 | Lung | INF-γ until 1 mo before *A. nidulans* infection | AMB, ITC | Yes | Survived |
| 12 [169] | M | X-CGD | 8 | Lung, rib | Clindamycin | AMB, gran Tx | No | Survived |
| 13 [170] | M | NA | 10 | Lung, pleura, axillay abces, 2nd-3rd ribs, vertebrae | No | AMB | Yes | Survived |
| 14 [86] | F | p67phox | 3 | Endocarditis, skin lesions, blood | TMP | AMB | Yes | Died |
| 15 [128] | M | X-CGD | 21 | Lung, chest wall, spinal cord, T5-T7 vertebrae, popliteal abscess | TMP-SMX, ITC, INF-γ | AMB, AMBL, VOR, CAS, gran Tx | Yes | Died |
| 16 [129] | M | X-CGD | 4 | Lung, T2-T5 vertebrae, spinal cord | TMP-SMX, ITC | AMB, AMBL, VOR, POS, CAS, gran Tx | Yes | Survived (*ex-vivo* gene therapy) |
| 17 [85] | M | X-CGD | 13 | Lung | TMP-SMX, ITC but diarrhea and serum levels (−) | AMBL, VOR, CAS | No | Survived |
| 18 [130] | M | p22phox | 5 | Lung, chest wall cutaneous abscess | NA | AMB, VOR, ITC, gran Tx | Yes | Survived |
| 19 [149] | M | X-CGD | 8 | Lung, 6th rib, chest wall fistula over the rib, psoas abcess | TMP-SMX, ITC stopped 4 wk prior to *A. nidulans* | AMB, AMBL, gran Tx | Yes | Survived (BM Tx) |
| 20 [24] | M | X-CGD | 4 | Lung, chest wall, T8-T11 vertebrae, 7th rib | NA | AMB, 5-FC, ITC gran Tx | Yes | Survived |
| 21 [171] | M | NA | 5 | Lung, chest wall, vertebrae, spinal cord syringomyelia | NA | VOR | Yes | Survived |
| 22 [37] | M | X-CGD | 16 | lung, MULCH pneumonitis | TMP-SMX, INF-γ | AMB, ITC | No | Survived |
| 23 [150] | M | X-CGD | 18 | lung, osteomyelitis rib, vertebra | TMP-SMX, ITC | AMBL, VOR, CAS | Yes | Survived (BM Tx) |

*AMB* amphotericin B deoxycholate, *AMBL* amphotericin B liposomal, *ABLC* amphotericin B lipid complex, *ITC* itraconazole, *VOR* voriconazole, *POS* posaconazole, *KTC* ketoconazole, *CAS* caspofungin, *5-FC* 5-flucytosine, *NA* not available, *gran Tx* granulocyte transfusion, *BM Tx* bone marrow transplantation

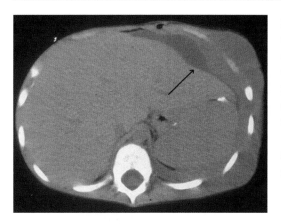

**Fig. 3.2** Computed tomography scan of *A. nidulans* infection in a patient with chronic granulomatous disease. (The extensive chest wall invasion and subcutaneous infiltration (*arrow*))

and symptoms were often mild with low grade fever, local pain or swelling, malaise and cough, but could even be completely asymptomatic, with undetected lung infiltrates. Vertebrae, mostly thoracic, were the most frequent sites of bone involvement (47 % of cases), while the ribs and skull were involved in 39 % and 9 % of the cases, respectively. Invasion of the vertebrae was accompanied in 45 % by clinical signs and symptoms of spinal cord invasion. Two more exclusive cases were noted: Casson et al. [86] reported an *A. nidulans* endocarditis in a 3-year-old girl with CGD who also had an atrial septal defect. Extensive prior investigation by local biopsy and repeated fungal antibody titers were negative and diagnosis was made on a second culture in which the organism grew from a mass obtained in the right atrium. Subsequently, *A. nidulans* was grown from embolic skin lesions and the site of the sternotomy. Despite extensive surgery and antifungal treatment, blood cultures grew *A. nidulans* and she died 24 days post-surgery. In 2003, an unusual mild and localized course of the disease was described in a 16-year-old male with CGD presenting with femoral osteomyelitis, without pulmonary involvement [46]. No blunt or penetrating trauma was noted. Treatment with liposomal amphotericin B and granulocyte colony-stimulating factor as well as extensive surgical debridement, followed by prolonged treatment with itraconazole resulted

in an excellent clinical response. Microscopic characteristics include very short conidiophores and the presence of ascospores with two equatorial crests. However, after molecular typing in 2008, this showed to be caused by *A. rugulosa*, originally identified as *A. nidulans* based on microscopic features [87].

*A. nidulans* has a unique predeliction for CGD. It remains a rare pathogen among all other immunosuppressed patients at high risk for IA [88–90].

### 3.5.3 Other *Aspergillus* Species

*A. fumigatus* is the primary agent of invasive human fungal infections, followed by *A. flavus*, *A. terreus*, and *A. niger* [12, 91–93]. In contrast, among CGD patients, *A. fumigatus* and *A. nidulans* are by far the most commonly isolated *Aspergillus* spp., followed by *A. niger* ($n=5$) and *A. flavus* ($n=1$) [22, 37, 62, 94].

Most of the non-fumigatus non-nidulans *Aspergillus* spp. were isolated in mixed fungal infections, and associated to *A. fumigatus* or *Rhizopus* spp. [37, 62]. The clinical picture related to multiple fungal species was that of mulch pneumonitis in half of the cases.

Kaltenis et al. [94] described a 10-year old child suffering from disseminated aspergillosis associated with systemic amyloidosis. The causative agent was *A. niger*. Despite aggressive treatment with amphotericin B and surgical drainage, intravenous leukocyte transfusions, INF-γ and immunoglobulin infusions, the patient died of multiple organ failure. *A. flavus* was the only causative agent in a 7-year old boy suffering from pulmonary aspergillosis associated with lymphadenitis; infection was controlled by surgical resection, amphotericin B and itraconazole [22]. In vitro susceptibility data concerning these *Aspergillus* spp. were not reported.

Attention has to be drawn to the occurrence of *fumigati*-mimetic molds in unusual, chronic or refractory IA. *Neosartorya udagawae* and *A. viridinutans* have been reported in CGD patients. Phenotypical they resemble *A. fumigatus*, but the course of the disease is distinctive from typical aspergillosis, being chronic and spreading in a contiguous manner across anatomical planes, and relative refractoriness to antimycotic drugs. Although encountered in the CGD patient, *fumigati*-mimetic molds are not confined to CGD; unlike *A. nidulans* that is essentially restricted to this host. Correct genotypic-based identification of *fumigati*-mimetic fungi may have implications for clinical course and management [42, 95].

### 3.5.4   Non-*Aspergillus* Species

#### 3.5.4.1   Zygomycetes

Mucormycosis refers to a group of opportunistic invasive fungal infections caused by fungi of the order Mucorales, predominantly of the genus *Rhizopus*, *Mucor* and *Absidia*.

Eleven cases of mucormycosis in CGD patients have been reported. Among them, nine *Rhizopus* isolates were identified [30, 37, 53], one *A. corymbifera* [40], and in one report, published in 1976, the clinical manifestation was described as "Phycomycosis", indicating the causative species as belonging to the phylum *Zygomycota* [96].

Two major clinical identities were noted: "mulch pneumonitis" ($n=5$) [37, 38, 40] and "invasive pulmonary mucormycosis" ($n=5$) [30, 96]. Dissemination occurred in three patients. *Mucorales* are the second most commonly occurring species found in mulch pneumonitis, mostly in association with *Aspergillus* spp. The association of mucorales and mulch pneumonitis in CGD seems to be related to the fact that both mucorales and aspergilli are found in the same demographic or environmental niche.

Invasive pulmonary mucormycosis can progress to thoracic wall abscess or nodular infiltrates of lung, spleen and brain. Attention has to be drawn to the use of prolonged steroid-based intense immunosuppressive regimes prior to the onset of mucormycosis. Median age was high, 18 years, which might well be explained by the fact that the inflammatory sequelae needing anti-inflammatory therapy develop with age in CGD. The prescription and prolonged use of steroid-based intense immunosuppressive regimes seems

to predispose those CGD patients to "less typical" fungal infections, such as mucormycosis.

A particular case of localized gastrointestinal mucormycosis in an eight-month-old infant was reported by Dekkers et al. [53]. The patient presented with an abdominal tumor associated with fever and diarrhea and *Rhizopus microsporus var. rhizopodiformis* was cultured within 24 h from tissue specimens taken during laparotomy. Analysis of his neutrophil function confirmed CGD. The use of antacids might have contributed to the development of invasive mucormycosis by increasing the pH in the stomach [53]. Voriconazole and itraconazole are not effective treatment options for mucormycosis. The first line treatment was amphotericin B in most of the cases. However, in two patients with mulch pneumonitis, voriconazole was used together with caspofungin, without coverage for zygomycetes. Overall mortality was high (54 %) compared to mortality associated with IA in CGD, likely due to delays in diagnosis and delayed and inadequate use of steroids to down-modulate the acute inflammatory aspects of mulch pneumonitis.

#### 3.5.4.2   *Candida* Species

The clinical spectrum of *Candida* infections in the immunocompromised host ranges from mucocutaneous disease to systemic life-threatening infections. Mucocutaneous involvement is remarkably rare in CGD. One report mentions an achalasia-like esophageal motility disorder as a result of a previously experienced *Candida* esophagitis [97], although this has not been observed in other diseases. Overall, in the CGD patient, *Candida* was the most common cause of meningitis, fungemia and lymphadenitis [54–56, 61, 98], but is not a causative agent of pneumonia, keratitis, hepatic abscesses and disseminated disease [10, 60, 63, 99, 100].

Young children appear more prone to develop *Candida* lymphangitis, blood stream infections and meningitis than older patients. Among the proven cases of reported meningitis, candidemia and lymphadenitis with or without dissemination, the age ranged from 8 weeks to 4 months. Clinical signs were those of a septic infant, febrile, irritable, sometimes associated with local signs of organ involvement like lymphade-nopathy, hepatosplenomegaly or signs of CNS involvement. The isolation of *Candida* spp. in an "otherwise healthy infant" was often the clue to diagnosis of a primary immunodeficiency defect. A possible explanation might be the immature mucosal immunity of the infant in the absence of a well-established tolerance for the *Candida* pathogen superposed to a defect in the phago-cytic defense mechanisms.

The second risk group includes older children and adults in whom the "classic" risk factors, such as prolonged broad spectrum antimicrobial drug use (3/5 for having bacterial co-infection), use of intravascular catheters and prolonged intensive care stay or hospitalization, may be present.

Detailed information on *Candida* spp. distri-bution in CGD patients is not available. Accord-ing to the individual reports, *C. glabrata* and *C. lusitaniae* were each isolated three times [10, 54, 55, 61, 98, 100], *C. albicans* was suspected twice according to microscopic features or positivity of antigen testing and the occurrence of antibodies [56, 99]; in one case no species identification was given [60]. Susceptibility testing was rarely per-formed. Species identification is important with respect to the intrinsic resistance profiles of vari-ous *Candida* spp., most notably the non-*albicans Candida* spp. Amphotericin B resistance has been reported in cases of *C. lusitaniae* infections. Four of the 11 CGD patients with invasive *Can-dida* infections died, three of them in the neonatal period due to a severely disseminated disease.

### 3.5.4.3 *Paecilomyces* Species

In CGD patients, *Paecilomyces* spp. have been reported as the third most common cause of osteo-myelitis after *Serratia* spp. and *Aspergillus* spp. [4]. Cultured-proven infections caused by *Paeci-lomyces* spp. in CGD have been nicely reviewed by Wang et al. [8] in 2005. One additional prob-able pulmonary infection was described in a 12 year old boy receiving cyclosporine treatment for severe colitis and peri-rectal disease [101]. In total six cases have been reported of which four showed primary extrapulmonary disease. One of the extra-pulmonary cases was related to trau-matic inoculation through a minor trauma. All of

them were symptomatic and the clinical picture did not differ from those reported in the non-CGD host.

*P. variotii* has been shown to be the caus-ative agent in four cases: two cases of soft-tissue infection of which one was associated with multifocal osteomyelitis [102, 103], one case developed isolated splenic abscess [8] and one pneumonia [101]. *P. lilacinus* was reported only once in a soft-tissue infection of the abdomi-nal wall [104]. Sillevis Smitt [50] implicated a *Paecilomyces* spp. in a pulmonary infection of a CGD patient also suffering from bullous Ig-A dermatosis.

Differentiation between the two species is extremely important, since *P. lilacinus* and *P. var-iotii* seem to present marked differences in their in vitro susceptibilities. Amphotericin B, itra-conazole, and echinocandins showed poor activ-ity against *P. lilacinus*, while the new triazoles, voriconazole and posaconazole show in vitro activity. In contrast, *P. variotii* was susceptible to most antifungal agents apart from voriconazole and ravuconazole [105]. All of the patients recov-ered; however, two were still under treatment by the time their report was published and one patient died after 4 months due to an *Aspergillus* pneumonia.

De Ravin et al. [106] raise the possibility that some of the *Paecilomyces* infection in CGD patients in previously published reports may have been caused by the emerging pathogen *Geosmi-thia argillacea*. When *P. variotii* is identified phenotypically, molecular approaches should be used to exclude *G. argillacea*, as infections observed in CGD are often aggressive, refractory to medical treatment and associated with a high fatality rate.

### 3.5.4.4 *Scedosporium* Species

*S. apiospermum* (previously known as *Pseu-doallescheria boydii*) and *S. prolificans* repre-sent important multidrug-resistant opportunistic pathogens. Both pathogens have been described in CGD patients as breakthrough infections dur-ing treatment with amphotericin B and voricon-azole, or after long-term treatment for *Aspergillus* infections [107, 108].

In these patients, *S. apiospermum* caused pulmonary infections with both contiguous spread to adjacent bone and central nervous system, and/or hematogenous dissemination. Outcome was favorable after extensive surgery of the infected tissues and antifungal therapy including azole derivatives, itraconazole and voriconazole, and INF-γ. *S. apiospermum* is generally considered resistant to amphotericin B, while extended-spectrum triazoles are active *in vitro* and voriconazole as well as posaconazole have been used successfully for the treatment of CNS abscesses [109–111].

Bhat et al. [107] described the development of a *S. prolificans* brain abscess in a 23-year-old CGD patient with a history of recurrent pulmonary aspergillosis who was on maintenance voriconazole and interferon-γ. MRI revealed a right frontal lobe lesion and CT-guided biopsy yielded non-specific yeast forms without hyphal elements. Cultures identified *S. prolificans*. This pathogen is considered resistant to virtually all antifungal agents, including the extended-spectrum triazoles and the echinocandins [109]. The isolate showed high MICs for amphotericin B (>16 mg/l) and 5-FC (>16 mg/l). Although terbinafine does not appear to be active alone [112], the combination of voriconazole and terbinafine appeared synergistic and was used to treat the case, resulting in complete resolution of the brain abscess [107].

### 3.5.4.5 *Trichosporon* Species

The eight cases of *Trichosporon* infection described in patients suffering from CGD include three due to *T. pullulans*, although one isolate was subsequently identified as *Cryptococcus albinus*, a likely contaminant [113, 114]. *T. pullullans* has been reported as being a human pathogen, but controversy exists regarding whether *T. pullulans* causes invasive infection in patients with CGD [114]. It is unlikely that *T. pullulans* causes invasive disease in patients with CGD, nor is it a significant invasive infectious agent in other immune compromised humans.

The case described by Lestini et al. [115], highlighted how difficult it may be to treat CGD colitis with a high dose of immunosuppressive drugs, including prednisone and infliximab:

opportunistic organisms with high mortality and morbidity soon occur in such patients.

The four remaining *Trichosporon* spp. were identified as *T. inkin*, based on microscopic characteristics and inoculation on commercially available yeast identification systems. Clinically, *T. inkin* infection presented with pulmonary infiltrates. In three cases, the individuals were asymptomatic; the pulmonary infiltrates were found on routine chest radiographs. In one case, the lung lesions spread directly to adjacent ribs and penetrated the chest wall. Variable susceptibility to amphotericin B has described, and in general this agent lacks fungicidal activity against *Trichosporon*. The outcomes with voriconazole have been curative (S. M. Holland, unpublished).

### 3.5.4.6 Miscellaneous

A few other species of fungi have also been reported to cause invasive infections in CGD patients. Among these species, *Penicillium* spp. have been reported most commonly (*n*=4) [37, 116, 117]. However, it is now clear that many if not most of the isolates identified as invasive *Penicillium* in CGD are in fact due to the morphologically similar but treatment resistant *Geosmithia argillacea* [106]. *Acremonium* spp. (*n*=2) [118, 119] and *Alternaria* spp. (*n*=2) [59, 120] cause occasional infections as well. *Inonotus, Chrysosporium, Exophiala, Fusarium, Microascus* and *Hansenula* spp. have been reported once [9, 47, 121–125]. All of these cases fulfilled the criteria of a proven IFD, with the exception of one mulch pneumonitis, in which a bronchoscopic specimen grew *A. fumigatus*, *A. niger* and *Penicillium* spp. [37].

The majority of patients presented with localized infections, two of them with spread to adjacent structures. The lung was the most commonly involved organ, skin and soft-tissue infections were observed in three cases, osteomyelitis and cerebral abscess were each reported once. With exception of a 4-month-old infant who suffered from a localized pneumonia caused by *Acremonium kiliense* [119], the age at onset was quite high. Half of the patients were adults (median 18, range 0–69). Seven patients had X-linked CGD, four were diagnosed with autosomal recessive CGD and in three patients the genetic mutation was not reported. Five patients each had received itraconazole prophylaxis and interferon-gamma prophylaxis. Amphotericin B was used in 12/14 cases as first line treatment, 2 of them also received miconazole or voriconazole combination therapy. One patient received itraconazole. The mortality rate was 21 %. Two patients died of multiple organ failure related to mulch pneumonitis and one patient succumbed to cerebral alternariosis [59].

## 3.6   Diagnosis

### 3.6.1   Culture and Biopsy of Clinical Specimens

One hundred and eight of the 116 patients (93 %) reviewed in this study fulfilled our criteria for proven IFD and 8 patients for probable IFD. The proportion of proven cases was much higher than observed in other risk groups, such as patients with hematological malignancy. The tendency to prefer proven cases for publication might cause a bias leading to a high proportion of proven cases. Invasive procedures may be higher in CGD patients compared to other patient groups because they are relatively in a better clinical condition and rarely have thrombocytopenia.

Recovery of fungi by culture of sterile material in association with consistent radiologic or clinical infectious disease process made the diagnosis in 101/108 cases. Blood cultures were positive in one mycotic pseudoaneurysm due to *A. fumigatus* [126], one *A. nidulans* endocarditis [86], fungemia due to *C. lusitaniae* [54] and *T. pullulans* [115]. Histopathologic and direct microscopic examination of a biopsy taken during open surgery was conclusive in the 7 additional proven cases [39, 59, 60, 85, 96, 108, 113]. Percutaneous fine needle biopsy was performed in 16 % (*n* = 19), including 11 fine needle lung biopsies. Of these lung biopsies, all were conclusive, either on microscopic examination or culture. Molecular studies were helpful in diagnosing *Inonotus tropicalis* osteomyelitis of the sacrum using ultrasound guided bone biopsy [47].

In an era of multiple classes of antifungal agents, it is crucial to identify causative fungi to tailor therapy appropriately. Since 1970 numerous changes in the taxonomy of fungi have been implemented, especially through the use of molecular sequence information. Multilocus sequence-based phylogenetic analyses have emerged as the primary tool for inferring phylogenetic species boundaries and relationships within subgenera and sections. Recently, an international group of experts gathered for a workshop entitled "Aspergillus Systematics in the Genomic Era" and reviewed research data presented from research groups worldwide on recent genomic investigations, secondary metabolite analyses, multi locus phylogenetic analyses of the genus *Aspergillus*, and sequence based identification schemes for previously recognized human pathogens within the genus [127].

Sequence analyses of the ITS region appear to be appropriate for identification of *Aspergillus* isolates to the subgenus/section level [127]. Use of the ITS sequence will not provide sufficient sensitivity to discriminate among individual species within the section. At present partial β-tubulin or calmodulin are the most promising loci for *Aspergillus* identification to the species level. For the section *Nidulanti* the identification of clinical isolates by use of β-tubulin sequencing revealed that clinical isolates were commonly misidentified when solely relied on morphologic characteristics [87]. Differences in drug activity against the different species of the section *Nidulanti* was shown, underscoring the need for identification to the species level [87]. In addition to *Aspergillus* spp. the taxonomy of many other genera involved in fungal diseases in CGD has changed including *Candida, Paecilomyces, Scedosporium,* and *Trichosporon*. These changes will have impact on the epidemiology of fungal diseases in CGD patients, and the use of molecular techniques for fungal strain identification will increase our insight into specific pathogens and their susceptibilities in the near future.

### 3.6.2   Nucleic Acid Amplification, Antigens and Serology

In seven of the reported CGD cases, information about circulating *Aspergillus* antigen galactomannan could be extracted [46, 65, 77, 99, 128–130]. In five of seven patients (71 %) circulating galactomannan could not be detected despite extensive invasive infection [46, 65, 77, 128, 129]. Interestingly, Verweij et al. [77] reported a CGD patient who developed invasive pulmonary aspergillosis and a subphrenic abscess. Fine needle aspiration of the subphrenic abscess recovered *A. fumigatus* as the etiologic agent. During treatment, high levels of *Aspergillus* antigen were detected in the abscess, but circulating antigen and *Aspergillus* DNA were undetectable in the serum.

van't Hek et al. [130] described circulating galactomannan in a 5-year-old boy with X-linked CGD suffering from invasive *A. nidulans* infection and chest-wall invasion in whom treatment with amphotericin B failed. The galactomannan ratio in serum remained high during amphotericin B treatment but started to decline 6 weeks after therapy with voriconazole was initiated and finally became negative after 4 months of treatment.

There has been a progressive increase in the understanding of the diagnostic utility of galactomannan which has enabled its incorporation into the EORTC/MSG consensus definitions and the ECIL guidelines [35]. Excellent performance characteristics in patients with hematological malignancies have been reported, with sensitivity up to 92.6 % and specificity of 95.4 % when serial monitoring of galactomannan was performed [131, 132], but the kinetics and predictive values of the antigen as a diagnostic marker in non-neutropenic hosts have not yet been defined [133]. The performance of galactomannan detection is the highest in the neutropenic host, while in non-neutropenic patients circulating galactomannan is detected in only a quarter of patients [134]. Possibly the rapid angio-invasive growth of *Aspergillus* in neutropenic patients facilitates the release of galactomannan into the blood. The poor performance of galactomannan detection in patients with CGD may be due to the fact that they are not neutropenic and the well-formed abscesses may preclude the release of the antigen into the blood [46, 77]. Furthermore, since galactomannan is an immunogenic molecule, it could be rapidly removed from the circulation in the non-neutropenic host.

In our review of IFD in CGD, information on diagnostic PCR was retrieved in six proven cases [46, 47, 59, 77, 85, 129]. Nucleic acid amplification of plasma and cerebrospinal fluid samples were all negative. In contrast, PCR performed on material obtained by biopsy of a brain lesion [59] and in the case of an osteomyelitis [47] revealed the etiological fungus.

Six cases mentioned strongly positive anti-*Aspergillus* antibodies [24, 64, 69, 77, 128, 135], even while circulating galactomannan was undetectable in repeated serum samples [128]. Except for the latter, no information was given about the assays used or about the epitopes recognized by these antibodies or corresponding antigen testing. In addition, antibody detection against *Scedosporium* spp. was helpful in two diagnoses [108], and once in a *C. albicans* sepsis [99]. Antibody responses to *Aspergillus* spp. may result from environmental exposure in the absence of disease, occur in the course of either aspergilloma or invasive disease, or be detected as part of a hypersensitivity syndrome. Secreted, cell wall and cytosolic fractions of *A. fumigatus* hyphae have been reported as potent B cell antigens [136]. Furthermore, work using the recombinant antigen mitogillin (*Aspergillus* protein toxin) suggests that specific antibody may be more prevalent in severely immunocompromised patients than has been described previously [137]. However at the moment, antibody detection is not considered useful for the diagnosis of IFD. Combined detection of *Candida* mannanaemia and anti-mannan antibodies may substantially contribute to the early diagnosis of candidosis in high-risk patients [138]. Nevertheless, little is known about the sensitivity of diagnostic modalities for IFD within specific immunocompromised patient groups as CGD. More work is required to define which diagnostic methods—or combination of methods—confer optimal predictive values in diagnosing IFD at an early stage in the CGD patient.

### 3.6.3  Radiologic Findings

Although pulmonary involvement was present in 72 % of the reported cases, beyond the description of "pulmonary consolidation" and "lung infiltrates", no specific signs as nodules, air crescent formation, halo signs, necrosis or cavitations were described or could be detected in the published images. Computed tomography and magnetic resonance scanning established extra-pulmonary extension to soft tissues, spinal cord injury and the extension of the osteolytic lesions. Local extension of disease from lung parenchyma to adjacent structures and osteomyelitis of the thoracic cage have been found particularly associated with underlying CGD. Unfortunately, not a single feature was pathogen-specific [139]. The absence of the characteristic diagnostic radiologic findings of IA in pediatric patients has been previously described by Burgos [93] in a multicentre retrospective analysis of 139 children with malignancy with or without hematopoietic stem cell transplantation. Thomas [139] performed a retrospective review of imaging performed in 27 consecutive patients with documented invasive aspergillosis, 12 of whom had primary immunodeficiency syndromes (including 6 CGD patients). Different pathophysiology in children compared to adults or, the exuberant granuloma formation in CGD in response to infection might explain the observed radiographic differences [64].

Pulmonary involvement is a common clinical feature of invasive fungal infections. This highlights the importance of identifying early, host-specific radiologic features on CT scan early in the course of disease. In the case of new or rapidly changing lesions, it is usually necessary and highly desirable to move to invasive diagnostic methods to identify the causative organisms.

### 3.7  Treatment

Overall, the combination of antifungal treatment and extensive and early surgical debridement was used in most patients. Almost all of the patients received presumptive initial treatment with amphotericin B (86 %). However, it is important to keep in mind that this series is accumulated from the era well before the advent of azole antifungals, and therefore is heavily weighted to amphotericin as the sole antifungal agent available. In general, the treatment guidelines for patients with aspergillosis and candidiosis can be followed in patients with CGD [140, 141] keeping in mind that non-fumigatus *Aspergillus* species, in particular *A. nidulans*, may require a different antifungal treatment. Furthermore, the use of adjunctive immunomodulatory treatment options as INF-γ and granulocyte transfusions, while ceonceptually robust in CGD is unproven and remains quite anecdotal.

Amphotericin B is active against the majority of clinically relevant yeasts and moulds isolated in the CGD patient, including *Candida* spp., *Aspergillus* spp. and Zygomycetes. However, the susceptibility of *Trichosporon* spp. to amphotericin B is variable and *Scedosporium* spp. are generally considered resistant [109].

Acquired and intrinsic resistance of *Aspergillus* spp. has been documented. Acquired azole-resistance is an emerging problem in *A. fumigatus* and may develop during azole therapy. This observation is of particular interest for CGD patients, as they should receive life-long itraconazole prophylaxis [82–84]. Development of resistance in *A. fumigatus* during azole therapy has recently been described in a patient suffering from CGD [81]. Non-*fumigatus Aspergillus* species, especially *A. terreus* and *A. nidulans*, may be intrinsically resistant to specific classes of antifungal agents [142]. Of particular interest is the resistance of *A. nidulans* to amphotericin B which might contribute to the high rate of treatment failure of these IFD in the era before azole antifungals [143]. In a clinical CGD study, the total dose of amphotericin B used in patients with *A. nidulans* infections was not only higher than for those with *A. fumigatus* infections (231 mg versus 56 mg, respectively), but also the duration of amphotericin B therapy was substantially longer, 220 days for *A. nidulans* infections compared to 65 days for *A. fumigatus* infections [23].

Voriconazole has recently been introduced as drug of choice of invasive aspergillosis in children and has been shown as successful salvage therapy in patients with CGD [144]. Among the described cases, use of voriconazole was first reported in 1998 and prescribed in 23 % of the cases. In comparison, 86 % of the cases received amphotericin B.

In an open-label trial of oral posaconazole, Segal et al. evaluated the effect in eight CGD patients suffering from refractory invasive mold infection including *P. variotii, S. apiospermum* and *A. fumigatus* infections. Posaconazole led to complete response in seven of the eight infections. In addition, Notheis et al. [129] described the beneficial effect in a deteriorating *A. nidulans* infection refractory to voriconazole and caspofungin treatment. Posaconazole shows promising results as safe and effective salvage therapy in CGD patients [145, 146].

Granulocyte transfusions were prescribed in 22 % of our cases, and/or 28 % received INF-γ. The use of granulocyte transfusions in CGD is supported by the observation that in vitro a small number of normal phagocytes may complement the oxidative defect and restore the killing ability towards *A. fumigatus* hyphae [147]. However, the efficacy of granulocyte transfusions is poorly documented and limited to case reports. The use of INF-γ as adjunctive therapy of IFD in CGD patients has not been investigated by controlled studies and remains controversial [148]. Three uncontrolled *A. nidulans* invasive fungal infections were cured either by ex-vivo gene therapy [129] or by BMT [149, 150].

The utility of in vitro susceptibility testing for filamentous fungi is poorly defined, as acquired resistance was uncommon and species identification was sufficient to guide antifungal therapy. However, both CLSI and EUCAST are developing and validating reference methods for in vitro susceptibility testing of *Aspergillus* and other conidium-forming fungi [151]. For *A. fumigatus* breakpoints have been proposed, while awaiting the recommendations by CLSI and EUCAST [152]. With the emergence of acquired resistance in *A. fumigatus* the need to determine the susceptibility in order to guide antifungal therapy has

increased. In general, voriconazole is appropriate first line therapy for most IFD in CGD, with the important exception of mucormycosis, for which lipid formulations of amphotericin are first line and posaconazole a possible oral substitute.

In 104 of the 116 cases information on prophylactic treatment was retrieved. Itraconazole or ketoconazole was prophylactically prescribed in 19 cases (18 %). In addition, 3 patients were still under antifungal treatment at the onset of a new IFD: 2 developed a *Scedosporium* breakthrough infection (*S. prolificans* during voriconazole and *S. apiospermum* during amphotericin B treatment) [107, 108] and one *Trichosporon* sp. was cultured during amphotericin B, caspofungin, and voriconazole treatment for active *A. nidulans* infection [113]. Twenty-eight patients received INF-γ (27 %), in almost half in combination with antifungal prophylaxis.

Interferon-gamma (INF-γ) is a macrophage-activating cytokine produced by T cells and natural killer cells and has a key role in the innate and adaptive host response against fungi [153]. A subgroup of variant X-linked CGD patients (i.e., with very low, but detectable, baseline superoxide-generating activity), who have splice site mutations, has been shown to be responsive to INF-γ [154, 155]. Treatment resulted in improved splicing efficiency and an increase in cytochrome b expression, allowing near normal levels of superoxide production and bactericidal activity of neutrophils and monocytes [156–158]. However, subsequent reports have not confirmed the enhancement of superoxide production as the principal mechanism of INF-γ activity in CGD individuals [159–161]. To date placebo-controlled studies are still limited and controversy remains about the routine prophylactic administration in CGD patients. INF-γ prophylaxis is offered only in selected CGD cases by most European physicians, while it is more universally prescribed in the USA and Japan [162, 163]. In contrast to the multiple studies confirming value for the use of INF-γ prophylaxis to reduce infections in CGD, there are no prospective data to show any benefit of INF-γ treatment during acute infection.

One of the cornerstones of clinical care for CGD patients is the lifelong prophylactic treatment with intracellular active microbicidal agents. After the introduction of antibacterial prophylaxis, fungal infections persisted with an incidence of 0.15 episodes per patient year [4]. In a prospective, open-label study with 30 patients, administration of itraconazole reduced the rate of *Aspergillus* infections to one-third in comparison with historical controls [164]. Gallin et al. [165] performed a randomized, double-blind, placebo-controlled crossover study in 39 CGD patients, finding only one serious fungal infection in the itraconazole group compared to seven cases in the placebo arm. Serious fungal infections developing during long-term itraconazole prophylaxis may well be due to molds that are resistant to itraconazole. Further studies are warranted to evaluate potential alternatives to the currently used antifungal prophylactic strategies.

## 3.8  Clinical Outcome

Unambiguous data on clinical outcome in accordance to the definition of complete response (defined as resolution of all clinical signs and symptoms attributable to IFD and complete or nearly complete radiological resolution) or a partial response (improvement or resolution of all clinical signs and symptoms attributable to IFD and ≥50 % improvement in radiological findings) were lacking and follow-up ranged from "still under treatment" to 9 years after treatment. At the time cases were published, the mortality rate related to the fungal infection was 26 % (Table 3.1). Mixed fungal infections (50 %) and mucormycosis (55 %) seem to be responsible for the highest mortality rates, followed by *Candida* spp. (40 %), *A. fumigatus* (30 %) and *A. nidulans* (27 %). However, the reported small numbers of the individual groups, the occurrence of particular species like *A. nidulans*, or severe clinical presentations of the more commonly encountered *A. fumigatus* may substantially influence these mortality rates. This underscores again the need of more prospective surveillance cohort studies of invasive fungal infections in this unique host.

## 3.9  Conclusions

To summarize our current understanding of fungal epidemiology and clinical management of invasive fungal disease in CGD patients we have to conclude that there is a paucity of data. Consequently, current management strategies may not be optimal to protect CGD patients from these life-threatening infectious complications.

Host defense mechanisms influence the manifestations and severity of infections; the clinical presentation of fungal disease depends on the patient's immune response. As a result, there are important clinical and diagnostic differences between IFD in CGD patients compared to patients with other immunocompromised conditions.

First of all, the CGD patient is a non-neutropenic host suffering from impaired phagocytic killing and a dysregulated inflammatory response. As a result, IFD can present with a huge variety of often non-specific symptoms, ranging from indolent and asymptomatic up to acute onset of hypoxia and respiratory failure, or isolated symptoms of specific organ involvement. Timely diagnosis of IFD in CGD requires a high degree of suspicion. This is in contrast to the neutropenic host, in whom the risk depends mainly on the duration of the neutropenia and in whom persistent febrile neutropenia is one of the early signs of a possible invasive fungal infection.

Secondly, invasive mould infections in neutropenic patients are characterized by hyphal angioinvasion, necrosis and paucity of inflammatory cells. In contrast, angioinvasion is not a feature of invasive mould infections in CGD. As a consequence, the characteristic radiologic halo-signs, air-crescents and cavities within areas of consolidation (one of the EORTC/MSG diagnostic criteria of invasive fungal infections) are typically lacking in CGD infections.

Furthermore, indirect laboratory markers of IFD, like the detection of *Aspergillus* antigen (galactomannan) in serum, have a low sensitivity in this patient group. Molecular diagnostic modalities, like the *Aspergillus* PCR, either have not yet been standardized or their sensitivity and specificity in CGD patients have not been studied. As a result, the recognition and diagnosis of IFD in CGD patients is distinct from that in cancer and transplant patients.

Due to the rather long and obscure list of unusual invasive fungal infections in CGD patients, it is of the utmost importance to identify the causative fungus in a particular patient. In general, the clinical condition of CGD patients often can tolerate invasive diagnostic procedures and they should be performed to isolate the causative species, subsequently followed by targeted antifungal therapy. On the other hand, it is critical to keep in mind that the identification of an invasive infection in a previously healthy host is highly suggestive of CGD as the underlying disease.

IA caused by *A. nidulans* is virtually unique to CGD patients and needs more aggressive antifungal therapy than the more commonly encountered *A. fumigatus*. Invasive infections by *A. nidulans* are very seldomly reported in other immunocompromised patients and indicate a unique interaction between this fungus and the CGD host. The clinical entity of 'mulch pneumonitis' is a characteristic presentation of fungal pneumonia in the CGD patient in which multiple causative fungi may be recovered. The occurrence of infectious complications by Zygomycetes is mainly noted in the setting of inflammatory complications being treated with immunosuppressive drugs. *Candida* infections are less frequently seen but show an age-dependent clinical presentation; meningitis, fungemia and lymphadenitis with or without dissemination, were reported in neonates and young infants, and were not observed in older patients.

Details regarding strain identification and in vitro susceptibility testing are scarce in CGD patients. Accurate identification of the causative fungus to species level, due to a species dependent susceptibility to the currently used antifungal drugs for treatment and prophylaxis will definitely affect patient management. In vitro susceptibility testing may also have a role, but is as yet undefined.

Extensive and early surgery has been the cornerstone of treatment and amphotericin B was mostly used as the first-line drug. Since the introduction of the new triazoles, the occurrence of invasive fungal infections as a cause of death seems to decrease dramatically. It remains to be seen if the combination of an early diagnosis with an accurate specification of the causative fungus and subsequently directed antifungal therapy will improve outcome without the need for aggressive surgery. Adjunctive immunomodulatory therapy was used in about 25 % of the patients. The role for immunomodulatory agents in addition to antifungal therapy is not well studied. Due to lack of a profound understanding of the pathogenesis of invasive fungal infections in the CGD host, it is currently hard to define to what extent specific immunomodulatory therapies will be of benefit for this specific patient population.

Antifungal prophylaxis is not universally prescribed and has been associated with invasive infections caused by intrinsically or acquired resistant fungal organisms. However, since the introduction of azole prophylaxis the incidence of invasive aspergillosis seems to have decreased.

An infection-related mortality of 26 % was noted, but underestimation cannot be ruled out because of the huge variability in follow-up. Furthermore, large variability in the mortality was noted based on the clinical presentation and the causative fungus.

Management and outcome data should be gathered to guide optimal treatment decisions for future patients. At present, international and national databases are used to collect clinical data prospectively in children suffering from invasive fungal infections. Although not specifically designed for patients with CGD, it will be a first step to improve our knowledge about these devastating infections.

## References

1. Jones LB, McGrogan P, Flood TJ, Gennery AR, Morton L, Thrasher A et al (2008) Special article: chronic granulomatous disease in the United Kingdom and Ireland: a comprehensive national patient-based registry. Clin Exp Immunol 152(2):211–218
2. Hasui M (1999) Chronic granulomatous disease in Japan: incidence and natural history. The Study Group of Phagocyte Disorders of Japan. Pediatr Int 41(5):589–593

3. Ahlin A, De Boer M, Roos D, Leusen J, Smith CI, Sundin U et al (1995) Prevalence, genetics and clinical presentation of chronic granulomatous disease in Sweden. Acta Paediatr 84(12):1386–1394

4. Winkelstein JA, Marino MC, Johnston RB Jr, Boyle J, Curnutte J, Gallin JI et al (2000) Chronic granulomatous disease. Report on a national registry of 368 patients. Medicine 79(3):155–169

5. Mouy R, Fischer A, Vilmer E, Seger R, Griscelli C (1989) Incidence, severity, and prevention of infections in chronic granulomatous disease. J Pediatr 114(4 Pt 1):555–560

6. Kosut JS, Kamani NR, Jantausch BA (2006) One-month-old infant with multilobar round pneumonias. Pediatr Infect Dis J 25(1):95–97

7. Alsultan A, Williams MS, Lubner S, Goldman FD (2006) Chronic granulomatous disease presenting with disseminated intracranial aspergillosis. Pediatr Blood Cancer 47(1):107–110

8. Wang SM, Shieh CC, Liu CC (2005) Successful treatment of *Paecilomyces variotii* splenic abscesses: a rare complication in a previously unrecognized chronic granulomatous disease child. Diagn Microbiol Infect Dis 53(2):149–152

9. Mansoory D, Roozbahany NA, Mazinany H, Samimagam A (2003) Chronic *Fusarium* infection in an adult patient with undiagnosed chronic granulomatous disease. Clin Infect Dis 37(7):e107–e108

10. Cohen MS, Isturiz RE, Malech HL, Root RK, Wilfert CM, Gutman L et al (1981) Fungal infection in chronic granulomatous disease. The importance of the phagocyte in defense against fungi. Am J Med 71(1):59–66

11. Segal BH, Leto TL, Gallin JI, Malech HL, Holland SM (2000) Genetic, biochemical, and clinical features of chronic granulomatous disease. Medicine 79(3):170–200

12. Denning DW (1998) Invasive aspergillosis. Clin Infect Dis 26(4):781–803

13. Zaoutis TE, Heydon K, Chu JH, Walsh TJ, Steinbach WJ (2006) Epidemiology, outcomes, and costs of invasive aspergillosis in immunocompromised children in the United States, 2000. Pediatrics 117(4):e711–e716

14. Steinbach WJ (2010) Epidemiology of Invasive Fungal Infections in Neonates and Children. Clin Microbiol Infect 16(9):1321–1327

15. Groll AH, Kurz M, Schneider W, Witt V, Schmidt H, Schneider M et al (1999) Five-year-survey of invasive aspergillosis in a paediatric cancer centre. Epidemiology, management and long-term survival. Mycoses 42(7–8):431–442

16. Crassard N, Hadden H, Piens MA, Pondarre C, Hadden R, Galambrun C et al (2008) Invasive aspergillosis in a paediatric haematology department: a 15-year review. Mycoses 51(2):109–116

17. Hovi L, Saarinen-Pihkala UM, Vettenranta K, Saxen H (2000) Invasive fungal infections in pediatric bone marrow transplant recipients: single center experience of 10 years. Bone Marrow Transplant 26(9):999–1004

18. van den Berg JM, van Koppen E, Ahlin A, Belohradsky BH, Bernatowska E, Corbeel L et al (2009) Chronic granulomatous disease: the European experience. PLoS ONE 4(4):e5234

19. Kobayashi S, Murayama S, Takanashi S, Takahashi K, Miyatsuka S, Fujita T et al (2008) Clinical features and prognoses of 23 patients with chronic granulomatous disease followed for 21 years by a single hospital in Japan. Eur J Pediatr 167(12):1389–1394

20. Martire B, Rondelli R, Soresina A, Pignata C, Broccoletti T, Finocchi A et al (2008) Clinical features, long-term follow-up and outcome of a large cohort of patients with Chronic Granulomatous Disease: an Italian multicenter study. Clin Immunol 126(2):155–164

21. Almyroudis NG, Holland SM, Segal BH (2005) Invasive aspergillosis in primary immunodeficiencies. Med Mycol 43(Suppl 1):S247–S259

22. Mamishi S, Parvaneh N, Salavati A, Abdollahzadeh S, Yeganeh M (2007) Invasive aspergillosis in chronic granulomatous disease: report of 7 cases. Eur J Pediatr 166(1):83–84

23. Segal BH, DeCarlo ES, Kwon-Chung KJ, Malech HL, Gallin JI, Holland SM (1998) *Aspergillus nidulans* infection in chronic granulomatous disease. Medicine 77(5):345–354

24. White CJ, Kwon-Chung KJ, Gallin JI (1988) Chronic granulomatous disease of childhood. An unusual case of infection with *Aspergillus nidulans var. echinulatus*. Am J Clin Pathol 90(3):312–316

25. Dotis J, Roilides E (2004) Osteomyelitis due to *Aspergillus* spp. in patients with chronic granulomatous disease: comparison of *Aspergillus nidulans* and *Aspergillus fumigatus*. Int J Infect Dis 8(2):103–110

26. Antachopoulos C, Walsh TJ, Roilides E (2007) Fungal infections in primary immunodeficiencies. Eur J Pediatr 166(11):1099–1117

27. Roilides E, Zaoutis TE, Walsh TJ (2009) Invasive zygomycosis in neonates and children. Clin Microbiol Infect 15(Suppl 5):50–54

28. Roden MM, Zaoutis TE, Buchanan WL, Knudsen TA, Sarkisova TA, Schaufele RL et al (2005) Epidemiology and outcome of zygomycosis: a review of 929 reported cases. Clin Infect Dis 41(5):634–653

29. Zaoutis TE, Roilides E, Chiou CC, Buchanan WL, Knudsen TA, Sarkisova TA et al (2007) Zygomycosis in children: a systematic review and analysis of reported cases. Pediatr Infect Dis J 26(8):723–727

30. Vinh DC, Freeman AF, Shea YR, Malech HL, Abinun M, Weinberg GA et al (2009) Mucormycosis in chronic granulomatous disease: association with iatrogenic immunosuppression. J Allergy Clin Immunol 123(6):1411–1413

31. Baehner RL, Nathan DG (1967) Leukocyte oxidase: defective activity in chronic granulomatous disease. Science 155(764):835–836

32. Vowells SJ, Fleisher TA, Sekhsaria S, Alling DW, Maguire TE, Malech HL (1996) Genotype-dependent variability in flow cytometric evaluation of reduced nicotinamide adenine dinucleotide phosphate oxidase function in patients with chronic granulomatous disease. J Pediatr 128(1):104–107

33. Vowells SJ, Sekhsaria S, Malech HL, Shalit M, Fleisher TA (1995) Flow cytometric analysis of the granulocyte respiratory burst: a comparison study of fluorescent probes. J Immunol Methods 178(1):89–97

34. Berendes H, Bridges RA, Good RA (1957) A fatal granulomatosus of childhood: the clinical study of a new syndrome. Minn Med 40(5):309–312

35. de Pauw B, Walsh TJ, Donnelly JP, Stevens DA, Edwards JE, Calandra T et al (2008) Revised definitions of invasive fungal disease from the European Organization for Research and Treatment of Cancer/Invasive Fungal Infections Cooperative Group and the National Institute of Allergy and Infectious Diseases Mycoses Study Group (EORTC/MSG) Consensus Group. Clin Infect Dis 46(12):1813–1821

36. Kelly JK, Pinto AR, Whitelaw WA, Rorstad OP, Bowen TJ, Matheson DS (1986) Fatal *Aspergillus* pneumonia in chronic granulomatous disease. Am J Clin Pathol 86(2):235–240

37. Siddiqui S, Anderson VL, Hilligoss DM, Abinun M, Kuijpers TW, Masur H et al (2007) Fulminant mulch pneumonitis: an emergency presentation of chronic granulomatous disease. Clin Infect Dis 45(6):673–681

38. Gupta A, McKean M, Haynes S, Wright C, Barge D, Flood TJ et al (2009) Chronic granulomatous disease presenting as fulminant *Aspergillus* pneumonitis: a lethal combination? Pediatr Crit Care Med 10(4):e43–e45

39. Conrad DJ, Warnock M, Blanc P, Cowan M, Golden JA (1992) Microgranulomatous aspergillosis after shoveling wood chips: report of a fatal outcome in a patient with chronic granulomatous disease. Am J Ind Med 22(3):411–418

40. Abinun M, Wright C, Gould K, Flood TJ, Cassidy J (2007) *Absidia corymbifera* in a patient with chronic granulomatous disease. Pediatr Infect Dis J 26(12):1167–1168

41. Eppinger TM, Greenberger PA, White DA, Brown AE, Cunningham-Rundles C (1999) Sensitization to *Aspergillus* species in the congenital neutrophil disorders chronic granulomatous disease and hyper-IgE syndrome. J Allergy Clin Immunol 104(6):1265–1272

42. Vinh DC, Shea YR, Jones PA, Freeman AF, Zelazny A, Holland SM (2009) Chronic invasive aspergillosis caused by *Aspergillus viridinutans*. Emerg Infect Dis 15(8):1292–1294

43. Sponseller PD, Malech HL, McCarthy EF Jr, Horowitz SF, Jaffe G, Gallin JI (1991) Skeletal involvement in children who have chronic granulomatous disease. J Bone Joint Surg Am 73(1):37–51

44. Hodiamont CJ, Dolman KM, Ten Berge IJ, Melchers WJ, Verweij PE, Pajkrt D (2009) Multiple-azole-resistant *Aspergillus fumigatus* osteomyelitis in a patient with chronic granulomatous disease successfully treated with long-term oral posaconazole and surgery. Med Mycol 47(2):217–220

45. Bodur H, Ozoran K, Colpan A, Balaban N, Tabak Y, Kulacoglu S (2003) Arthritis and osteomyelitis due to *Aspergillus fumigatus*: a 17 years old boy with chronic granulomatous disease. Ann Clin Microbiol Antimicrob 2:2

46. Dotis J, Panagopoulou P, Filioti J, Winn R, Toptsis C, Panteliadis C et al (2003) Femoral osteomyelitis due to *Aspergillus nidulans* in a patient with chronic granulomatous disease. Infection 31(2):121–124

47. Davis CM, Noroski LM, Dishop MK, Sutton DA, Braverman RM, Paul ME et al (2007) Basidiomycetous fungal *Inonotus tropicalis* sacral osteomyelitis in X-linked chronic granulomatous disease. Pediatr Infect Dis J 26(7):655–656

48. Galluzzo ML, Hernandez C, Davila MT, Perez L, Oleastro M, Zelazko M et al (2008) Clinical and histopathological features and a unique spectrum of organisms significantly associated with chronic granulomatous disease osteomyelitis during childhood. Clin Infect Dis 46(5):745–749

49. Wolfson JJ, Kane WJ, Laxdal SD, Good RA, Quie PG (1969) Bone findings in chronic granulomatous disease of childhood. A genetic abnormality of leukocyte function. J Bone Joint Surg Am 51(8):1573–1583

50. Sillevis Smitt JH, Leusen JH, Stas HG, Teeuw AH, Weening RS (1997) Chronic bullous disease of childhood and a paecilomyces lung infection in chronic granulomatous disease. Arch Dis Child 77(2):150–152

51. Kuhns DB, Alvord WG, Heller T, Feld JJ, Pike KM, Marciano BE et al (2010) Residual NADPH oxidase and survival in chronic granulomatous disease. N Engl J Med 363(27):2600–2610

52. Marciano BE, Rosenzweig SD, Kleiner DE, Anderson VL, Darnell DN, naya-O'Brien S et al (2004) Gastrointestinal involvement in chronic granulomatous disease. Pediatrics 114(2):462–468

53. Dekkers R, Verweij PE, Weemaes CM, Severijnen RS, Van Krieken JH, Warris A (2008) Gastrointestinal zygomycosis due to *Rhizopus microsporus* var. *rhizopodiformis* as a manifestation of chronic granulomatous disease. Med Mycol 46(5):491–494

54. Levy O, Bourquin JP, McQueen A, Cantor AB, Lachenauer C, Malley R (2002) Fatal disseminated *Candida lusitaniae* infection in an infant

with chronic granulomatous disease. Pediatr Infect Dis J 21(3):262–264

55. Estrada B, Mancao MY, Polski JM, Figarola MS (2006) *Candida lusitaniae* and chronic granulomatous disease. Pediatr Infect Dis J 25(8):758–759

56. Khoo BH, Cho CT, Smith SD, Dudding BA (1975) Letter: cervical lymphadenitis due to *Candida albicans* infection.920c. J Pediatr 86(5):812–813

57. Saulsbury FT (2001) Successful treatment of aspergillus brain abscess with itraconazole and interferon-gamma in a patient with chronic granulomatous disease. Clin Infect Dis 32(10):E137–E139

58. Patiroglu T, Unal E, Yikilmaz A, Koker MY, Ozturk MK (2010) Atypical presentation of chronic granulomatous disease in an adolescent boy with frontal lobe located *Aspergillus* abscess mimicking intracranial tumor. Childs Nerv Syst 26(2):149–154

59. Hipolito E, Faria E, Alves AF, De Hoog GS, Anjos J, Goncalves T et al (2009) *Alternaria infectoria* brain abscess in a child with chronic granulomatous disease. Eur J Clin Microbiol Infect Dis 28(4):377–380

60. Agus S, Spektor S, Israel Z (2000) CNS granulomatosis in a child with chronic granulomatous disease. Br J Neurosurg 14(1):59–61

61. Fleischmann J, Church JA, Lehrer RI (1986) Primary *Candida* meningitis and chronic granulomatous disease. Am J Med Sci 291(5):334–341

62. Ferris B, Jones C (1985) Paraplegia due to aspergillosis. Successful conservative treatment of two cases. J Bone Joint Surg Br 67(5):800–803

63. Miron D, Horowitz Y, Lumelsky D, Hanania S, Colodner R (2005) Dual pulmonary infection with *Candida dubliniensis* and *Aspergillus fumigatus* in a child with chronic granulomatous disease. J Infect 50(1):72–75

64. de Sevaux RG, Kullberg BJ, Verweij PE, van de Nes JA, Meis JF, van der Meer JW (1998) Microgranulomatous aspergillosis in a patient with chronic granulomatous disease: cure with voriconazole. Clin Infect Dis 26(4):996–997

65. Mouy R, Ropert JC, Donadieu J, Hubert P, de Blic J, Revillon Y et al (1995) Chronic septic granulomatosis revealed by neonatal pulmonary aspergillosis. Arch Pediatr 2(9):861–864

66. Chusid MJ, Sty JR, Wells RG (1988) Pulmonary aspergillosis appearing as chronic nodular disease in chronic granulomatous disease. Pediatr Radiol 18(3):232–234

67. Pasic S, Abinun M, Pistignjat B, Vlajic B, Rakic J, Sarjanovic L et al (1996) *Aspergillus* osteomyelitis in chronic granulomatous disease: treatment with recombinant gamma-interferon and itraconazole. Pediatr Infect Dis J 15(9):833–834

68. Kloss S, Schuster A, Schroten H, Lamprecht J, Wahn V (1991) Control of proven pulmonary and suspected CNS aspergillus infection with itracon-

azole in a patient with chronic granulomatous disease. Eur J Pediatr 150(7):483–485

69. Elgefors B, Haugstvedt S, Brorsson JE, Esbjorner E (1980) Disseminated aspergillosis treated with amphotericin B and surgery in a boy with chronic granulomatous disease. Infection 8(4):174–176

70. van 't Wout JW, Raven EJ, van der Meer JW (1990) Treatment of invasive aspergillosis with itraconazole in a patient with chronic granulomatous disease. J Infect 20(2):147–150

71. Heinrich SD, Finney T, Craver R, Yin L, Zembo MM (1991) *Aspergillus* osteomyelitis in patients who have chronic granulomatous disease. Case report. J Bone Joint Surg Am 73(3):456–460

72. Galimberti R, Kowalczuk A, Hidalgo P I, Gonzalez RM, Flores V (1998) Cutaneous aspergillosis: a report of six cases. Br J Dermatol 139(3):522–526

73. Watanabe C, Yajima S, Taguchi T, Toya K, Fujii Y, Hongo T et al (2001) Successful unrelated bone marrow transplantation for a patient with chronic granulomatous disease and associated resistant pneumonitis and *Aspergillus* osteomyelitis. Bone Marrow Transplant 28(1):83–87

74. Pollack IF, Pang D, Schuit KE (1987) Chronic granulomatous disease with cranial fungal osteomyelitis and epidural abscess. Case report. J Neurosurg 67(1):132–136

75. Kline MW, Bocobo FC, Paul ME, Rosenblatt HM, Shearer WT (1994) Successful medical therapy of *Aspergillus* osteomyelitis of the spine in an 11-year-old boy with chronic granulomatous disease. Pediatrics 93(5):830–835

76. Yeh J, Culbertson C, Petru AM (2009) Purulent pericardial effusion in a 14-year-old girl. Pediatr Infect Dis J 28(12):1140, 1147–1148

77. Verweij PE, Weemaes CM, Curfs JH, Bretagne S, Meis JF (2000) Failure to detect circulating *Aspergillus* markers in a patient with chronic granulomatous disease and invasive aspergillosis. J Clin Microbiol 38(10):3900–3901

78. Chatzipanagiotou S, Takou K, Perogamvros A (1998) Cutaneous purulent aspergillosis in a young man with chronic granulomatous disease. Mycoses 41(9–10):379–382

79. Dohil M, Prendiville JS, Crawford RI, Speert DP (1997) Cutaneous manifestations of chronic granulomatous disease. A report of four cases and review of the literature. J Am Acad Dermatol 36(6 Pt 1):899–907

80. Palmero ML, Pope E, Brophy J (2009) Sporotrichoid aspergillosis in an immunocompromised child: a case report and review of the literature. Pediatr Dermatol 26(5):592–596

81. Arendrup MC, Mavridou E, Mortensen KL, Snelders E, Frimodt-Moller N, Khan H et al (2010) Development of azole resistance in *Aspergillus fumigatus* during azole therapy associated with change in virulence. PLoS ONE 5(4):e10080

82. Snelders E, van der Lee HA, Kuijpers J, Rijs AJ, Varga J, Samson RA et al (2008) Emergence of

azole resistance in *Aspergillus fumigatus* and spread of a single resistance mechanism. PLoS Med 5(11):e219

83. Verweij PE, Mellado E, Melchers WJ (2007) Multiple-triazole-resistant aspergillosis. N Engl J Med 356(14):1481–1483

84. Warris A, Weemaes CM, Verweij PE (2002) Multidrug resistance in *Aspergillus fumigatus*. N Engl J Med 347(26):2173–2174

85. Sallmann S, Heilmann A, Heinke F, Kerkmann ML, Schuppler M, Hahn G et al (2003) Capofungin therapy for *Aspergillus* lung infection in a boy with chronic granulomatous disease. Pediatr Infect Dis J 22(2):199–200

86. Casson DH, Riordan FA, Ladusens EJ (1996) *Aspergillus* endocarditis in chronic granulomatous disease. Acta Paediatr 85(6):758–759

87. Verweij PE, Varga J, Houbraken J, Rijs AJ, Verduynlunel FM, Blijlevens NM et al (2008) *Emericella quadrilineata* as cause of invasive aspergillosis. Emerg Infect Dis 14(4):566–572

88. Mohammedi I, Eckert A, Thiebaut A, Piens MA, Malhiere S, Robert D (2005) Fatal *Aspergillus nidulans* pneumonia. Rev Med Interne 26(3):249–250

89. Peyrade F, Boscagli A, Roa M, Taillan B, Gary-Toussain M, Dujardin P (1997) *A. nidulans* bronchial aspergillosis after treatment of low-grade lymphoma with fludarabine. Rev Med Interne 18(3):235–236

90. Lucas GM, Tucker P, Merz WG (1999) Primary cutaneous *Aspergillus nidulans* infection associated with a Hickman catheter in a patient with neutropenia. Clin Infect Dis 29(6):1594–1596

91. Pappas PG, Alexander BD, Andes DR, Hadley S, Kauffman CA, Freifeld A et al (2010) Invasive fungal infections among organ transplant recipients: results of the Transplant-Associated Infection Surveillance Network (TRANSNET). Clin Infect Dis 50(8):1101–1111

92. Kontoyiannis DP, Marr KA, Park BJ, Alexander BD, Anaissie EJ, Walsh TJ et al (2010) Prospective surveillance for invasive fungal infections in hematopoietic stem cell transplant recipients, 2001–2006: overview of the Transplant-Associated Infection Surveillance Network (TRANSNET) Database. Clin Infect Dis 50(8):1091–1100

93. Burgos A, Zaoutis TE, Dvorak CC, Hoffman JA, Knapp KM, Nania JJ et al (2008) Pediatric invasive aspergillosis: a multicenter retrospective analysis of 139 contemporary cases. Pediatrics 121(5):e1286–e1294

94. Kaltenis P, Mudeniene V, Maknavicius S, Seinin D (2008) Renal amyloidosis in a child with chronic granulomatous disease and invasive aspergillosis. Pediatr Nephrol 23(5):831–834

95. Vinh DC, Shea YR, Sugui JA, Parrilla-Castellar ER, Freeman AF, Campbell JW et al (2009) Invasive aspergillosis due to *Neosartorya udagawae*. Clin Infect Dis 49(1):102–111

96. Bruun JN, Solberg CO, Hamre E, Janssen CJ Jr, Thunold S, Eide J (1976) Acute disseminated phycomycosis in a patient with impaired neutrophil granulocyte function. Acta Pathol Microbiol Scand C 84(2):93–99

97. Bode CP, Schroten H, Koletzko S, Lubke H, Wahn V (1996) Transient achalasia-like esophageal motility disorder after candida esophagitis in a boy with chronic granulomatous disease. J Pediatr Gastroenterol Nutr 23(3):320–323

98. Orson JM, Greco RG (1985) Chronic granulomatous disease and *Torulopsis* infections. Am J Dis Child 139(6):545

99. Malmvall BE, Follin P (1993) Successful interferon-gamma therapy in a chronic granulomatous disease (CGD) patient suffering from *Staphylococcus aureus* hepatic abscess and invasive *Candida albicans* infection. Scand J Infect Dis 25(1):61–66

100. Djalilian AR, Smith JA, Walsh TJ, Malech HL, Robinson MR (2001) Keratitis caused by *Candida glabrata* in a patient with chronic granulomatous disease. Am J Ophthalmol 132(5):782–783

101. Rosh JR, Tang HB, Mayer L, Groisman G, Abraham SK, Prince A (1995) Treatment of intractable gastrointestinal manifestations of chronic granulomatous disease with cyclosporine. J Pediatr 126(1):143–145

102. Williamson PR, Kwon-Chung KJ, Gallin JI (1992) Successful treatment of *Paecilomyces varioti* infection in a patient with chronic granulomatous disease and a review of *Paecilomyces* species infections. Clin Infect Dis 14(5):1023–1026

103. Cohen-Abbo A, Edwards KM (1995) Multifocal osteomyelitis caused by *Paecilomyces varioti* in a patient with chronic granulomatous disease. Infection 23(1):55–57

104. Silliman CC, Lawellin DW, Lohr JA, Rodgers BM, Donowitz LG (1992) *Paecilomyces lilacinus* infection in a child with chronic granulomatous disease. J Infect 24(2):191–195

105. Castelli MV, astruey-Izquierdo A, Cuesta I, Monzon A, Mellado E, Rodriguez-Tudela JL et al (2008) Susceptibility testing and molecular classification of *Paecilomyces* spp. Antimicrob Agents Chemother 52(8):2926–2928

106. De Ravin SS, Challipalli M, Anderson V, Shea YR, Marciano B, Hilligoss D et al (2011) *Geosmithia argillacea*: an emerging cause of invasive mycosis in human chronic granulomatous disease. Clin Infect Dis 52(6):e136–e143

107. Bhat SV, Paterson DL, Rinaldi MG, Veldkamp PJ (2007) *Scedosporium prolificans* brain abscess in a patient with chronic granulomatous disease: successful combination therapy with voriconazole and terbinafine. Scand J Infect Dis 39(1):87–90

108. Jabado N, Casanova JL, Haddad E, Dulieu F, Fournet JC, Dupont B et al (1998) Invasive pulmonary infection due to *Scedosporium apiosper-*

*mum* in two children with chronic granulomatous disease. Clin Infect Dis 27(6):1437–1441

109. Pfaller MA, Diekema DJ (2004) Rare and emerging opportunistic fungal pathogens: concern for resistance beyond *Candida albicans* and *Aspergillus fumigatus*. J Clin Microbiol 42(10):4419–4431

110. Nesky MA, McDougal EC, Peacock JE Jr (2000) *Pseudallescheria boydii* brain abscess successfully treated with voriconazole and surgical drainage: case report and literature review of central nervous system pseudallescheriasis. Clin Infect Dis 31(3):673–677

111. Mellinghoff IK, Winston DJ, Mukwaya G, Schiller GJ (2002) Treatment of *Scedosporium apiospermum* brain abscesses with posaconazole. Clin Infect Dis 34(12):1648–1650

112. Meletiadis J, Meis JF, Mouton JW, Rodriquez-Tudela JL, Donnelly JP, Verweij PE (2002) In vitro activities of new and conventional antifungal agents against clinical *Scedosporium* isolates. Antimicrob Agents Chemother 46(1):62–68

113. Moylett EH, Chinen J, Shearer WT (2003) *Trichosporon pullulans* infection in 2 patients with chronic granulomatous disease: an emerging pathogen and review of the literature. J Allergy Clin Immunol 111(6):1370–1374

114. Holland SM, Shea YR, Kwon-Chung J (2004) Regarding "*Trichosporon pullulans* infection in 2 patients with chronic granulomatous disease". J Allergy Clin Immunol 114(1):205–206

115. Lestini BJ, Church JA (2006) *Trichosporon pullulans* as a complication of chronic granulomatous disease in a patient undergoing immunosuppressive therapy for inflammatory bowel disease. Pediatr Infect Dis J 25(1):87–89

116. Santos PE, Piontelli E, Shea YR, Galluzzo ML, Holland SM, Zelazko ME et al (2006) *Penicillium piceum* infection: diagnosis and successful treatment in chronic granulomatous disease. Med Mycol 44(8):749–753

117. Lyratzopoulos G, Ellis M, Nerringer R, Denning DW (2002) Invasive infection due to penicillium species other than *P. marneffei*. J Infect 45(3):184–195

118. Boltansky H, Kwon-Chung KJ, Macher AM, Gallin JI (1984) *Acremonium strictum*-related pulmonary infection in a patient with chronic granulomatous disease. J Infect Dis 149(4):653

119. Pastorino AC, Menezes UP, Marques HH, Vallada MG, Cappellozi VL, Carnide EM et al (2005) *Acremonium kiliense* infection in a child with chronic granulomatous disease. Braz J Infect Dis 9(6):529–534

120. Uenotsuchi T, Moroi Y, Urabe K, Fukagawa S, Tsuji G, Matsuda T et al (2005) Cutaneous alternariosis with chronic granulomatous disease. Eur J Dermatol 15(5):406–408

121. Sutton DA, Thompson EH, Rinaldi MG, Iwen PC, Nakasone KK, Jung HS et al (2005) Identification and first report of *Inonotus (Phellinus)* *tropicalis* as an etiologic agent in a patient with chronic granulomatous disease. J Clin Microbiol 43(2):982–987

122. Roilides E, Sigler L, Bibashi E, Katsifa H, Flaris N, Panteliadis C (1999) Disseminated infection due to *Chrysosporium zonatum* in a patient with chronic granulomatous disease and review of non-*Aspergillus* fungal infections in patients with this disease. J Clin Microbiol 37(1):18–25

123. Kenney RT, Kwon-Chung KJ, Waytes AT, Melnick DA, Pass HI, Merino MJ et al (1992) Successful treatment of systemic *Exophiala dermatitidis* infection in a patient with chronic granulomatous disease. Clin Infect Dis 14(1):235–242

124. Marques AR, Kwon-Chung KJ, Holland SM, Turner ML, Gallin JI (1995) Suppurative cutaneous granulomata caused by *Microascus cinereus* in a patient with chronic granulomatous disease. Clin Infect Dis 20(1):110–114

125. McGinnis MR, Walker DH, Folds JD (1980) *Hansenula polymorpha* infection in a child with chronic granulomatous disease. Arch Pathol Lab Med 104(6):290–292

126. Sanchez FW, Freeland PN, Bailey GT, Vujic I (1985) Embolotherapy of a mycotic pseudoaneurysm of the internal mammary artery in chronic granulomatous disease. Cardiovasc Intervent Radiol 8(1):43–45

127. Balajee SA, Houbraken J, Verweij PE, Hong SB, Yaghuchi T, Varga J et al (2007) *Aspergillus* species identification in the clinical setting. Stud Mycol 59:39–46

128. Dellepiane RM, Tortorano AM, Liotto N, Laicini E, Di LG, Carnelli V et al (2008) Invasive *Aspergillus nidulans* infection in a patient with chronic granulomatous disease. Mycoses 51(5):458–460

129. Notheis G, Tarani L, Costantino F, Jansson A, Rosenecker J, Friederici D et al (2006) Posaconazole for treatment of refractory invasive fungal disease. Mycoses 49(Suppl 1):37–41

130. van 't Hek LG, Verweij PE, Weemaes CM, van Dalen R, Yntema JB, Meis JF (1998) Successful treatment with voriconazole of invasive aspergillosis in chronic granulomatous disease. Am J Respir Crit Care Med 157(5 Pt 1):1694–1696

131. Maertens J, Verhaegen J, Demuynck H, Brock P, Verhoef G, Vandenberghe P et al (1999) Autopsy-controlled prospective evaluation of serial screening for circulating galactomannan by a sandwich enzyme-linked immunosorbent assay for hematological patients at risk for invasive Aspergillosis. J Clin Microbiol 37(10):3223–3228

132. Maertens J, Verhaegen J, Lagrou K, Van Eldere J, Boogaerts M (2001) Screening for circulating galactomannan as a noninvasive diagnostic tool for invasive aspergillosis in prolonged neutropenic patients and stem cell transplantation recipients: a prospective validation. Blood 97(6):1604–1610

133. Leeflang MM, Debets-Ossenkopp YJ, Visser CE, Scholten RJ, Hooft L, Bijlmer HA et al (2008)

Galactomannan detection for invasive aspergillosis in immunocompromized patients. Cochrane Database Syst Rev 4:CD007394

134. Meersseman W, Vandecasteele SJ, Wilmer A, Verbeken E, Peetermans WE, Van WE (2004) Invasive aspergillosis in critically ill patients without malignancy. Am J Respir Crit Care Med 170(6):621–625

135. Neijens HJ, Frenkel J, de Muinck Keizer-Schrama SM, Dzoljic-Danilovic G, Meradji M, van Dongen JJ (1989) Invasive *Aspergillus* infection in chronic granulomatous disease: treatment with itraconazole. J Pediatr 115(6):1016–1019

136. Denikus N, Orfaniotou F, Wulf G, Lehmann PF, Monod M, Reichard U (2005) Fungal antigens expressed during invasive aspergillosis. Infect Immun 73(8):4704–4713

137. Weig M, Frosch M, Tintelnot K, Haas A, Gross U, Linsmeier B et al (2001) Use of recombinant mitogillin for improved serodiagnosis of *Aspergillus fumigatus*-associated diseases. J Clin Microbiol 39(5):1721–1730

138. Sendid B, Poirot JL, Tabouret M, Bonnin A, Caillot D, Camus D et al (2002) Combined detection of mannanaemia and antimannan antibodies as a strategy for the diagnosis of systemic infection caused by pathogenic *Candida* species. J Med Microbiol 51(5):433–442

139. Thomas KE, Owens CM, Veys PA, Novelli V, Costoli V (2003) The radiological spectrum of invasive aspergillosis in children: a 10-year review. Pediatr Radiol 33(7):453–460

140. Walsh TJ, Anaissie EJ, Denning DW, Herbrecht R, Kontoyiannis DP, Marr KA et al (2008) Treatment of aspergillosis: clinical practice guidelines of the Infectious Diseases Society of America. Clin Infect Dis 46(3):327–360

141. Pappas PG, Kauffman CA, Andes D, Benjamin DK Jr, Calandra TF, Edwards JE Jr et al (2009) Clinical practice guidelines for the management of candidiasis: 2009 update by the Infectious Diseases Society of America. Clin Infect Dis 48(5):503–535

142. Van Der Linden JW, Warris A, Verweij PE (2010) *Aspergillus* species intrinsically resistant to antifungal agents. Med Mycol

143. Kontoyiannis DP, Lewis RE, May GS, Osherov N, Rinaldi MG (2002) *Aspergillus nidulans* is frequently resistant to amphotericin B. Mycoses 45(9–10):406–407

144. Walsh TJ, Lutsar I, Driscoll T, Dupont B, Roden M, Ghahramani P et al (2002) Voriconazole in the treatment of aspergillosis, scedosporiosis and other invasive fungal infections in children. Pediatr Infect Dis J 21(3):240–248

145. Torres HA, Hachem RY, Chemaly RF, Kontoyiannis DP, Raad II (2005) Posaconazole: a broad-spectrum triazole antifungal. Lancet Infect Dis 5(12):775–785

146. Segal BH, Barnhart LA, Anderson VL, Walsh TJ, Malech HL, Holland SM (2005) Posaconazole as salvage therapy in patients with chronic granulomatous disease and invasive filamentous fungal infection. Clin Infect Dis 40(11):1684–1688

147. Rex JH, Bennett JE, Gallin JI, Malech HL, Melnick DA (1990) Normal and deficient neutrophils can cooperate to damage *Aspergillus fumigatus* hyphae. J Infect Dis 162(2):523–528

148. Seger RA (2008) Modern management of chronic granulomatous disease. Br J Haematol 140(3):255–266

149. Ozsahin H, von Planta M, Muller I, Steinert HC, Nadal D, Lauener R et al (1998) Successful treatment of invasive aspergillosis in chronic granulomatous disease by bone marrow transplantation, granulocyte colony-stimulating factor-mobilized granulocytes, and liposomal amphotericin-B. Blood 92(8):2719–2724

150. Sastry J, Kakakios A, Tugwell H, Shaw PJ (2006) Allogeneic bone marrow transplantation with reduced intensity conditioning for chronic granulomatous disease complicated by invasive *Aspergillus* infection. Pediatr Blood Cancer 47(3):327–329

151. Clinical and Laboratory Standards Institute (2008) Reference method for broth dilution antifungal susceptibility testing of filamentous fungi; Approved standard—Second edition. CLSI document M38-A2, 2nd edn. Clinical and Laboratory Standards Institute, Wayne

152. Verweij PE, Howard SJ, Melchers WJ, Denning DW (2009) Azole-resistance in Aspergillus: proposed nomenclature and breakpoints. Drug Resist Updat 12(6):141–147

153. Segal BH (2007) Role of macrophages in host defense against aspergillosis and strategies for immune augmentation. Oncologist 12(Suppl 2):7–13

154. Condino-Neto A, Newburger PE (2000) Interferon-gamma improves splicing efficiency of CYBB gene transcripts in an interferon-responsive variant of chronic granulomatous disease due to a splice site consensus region mutation. Blood 95(11):3548–3554

155. Ishibashi F, Mizukami T, Kanegasaki S, Motoda L, Kakinuma R, Endo F et al (2001) Improved superoxide-generating ability by interferon gamma due to splicing pattern change of transcripts in neutrophils from patients with a splice site mutation in CYBB gene. Blood 98(2):436–441

156. Ezekowitz RA, Dinauer MC, Jaffe HS, Orkin SH, Newburger PE (1988) Partial correction of the phagocyte defect in patients with X-linked chronic granulomatous disease by subcutaneous interferon gamma. N Engl J Med 319(3):146–151

157. Ezekowitz RA, Orkin SH, Newburger PE (1987) Recombinant interferon gamma augments phagocyte superoxide production and X-chronic granulomatous disease gene expression in X-linked

variant chronic granulomatous disease. J Clin Invest 80(4):1009–1016

158. Sechler JM, Malech HL, White CJ, Gallin JI (1988) Recombinant human interferon-gamma reconstitutes defective phagocyte function in patients with chronic granulomatous disease of childhood. Proc Natl Acad Sci USA 85(13):4874–4878

159. Muhlebach TJ, Gabay J, Nathan CF, Erny C, Dopfer G, Schroten H et al (1992) Treatment of patients with chronic granulomatous disease with recombinant human interferon-gamma does not improve neutrophil oxidative metabolism, cytochrome b558 content or levels of four anti-microbial proteins. Clin Exp Immunol 88(2):203–206

160. A controlled trial of interferon gamma to prevent infection in chronic granulomatous disease (1991) The International Chronic Granulomatous Disease Cooperative Study Group. N Engl J Med 324(8):509–516

161. Woodman RC, Erickson RW, Rae J, Jaffe HS, Curnutte JT (1992) Prolonged recombinant interferon-gamma therapy in chronic granulomatous disease: evidence against enhanced neutrophil oxidase activity. Blood 79(6):1558–1562

162. Gallin JI (1991) Interferon-gamma in the treatment of the chronic granulomatous diseases of childhood. Clin Immunol Immunopathol 61(2 Pt 2):S100–S105

163. Weening RS, Leitz GJ, Seger RA (1995) Recombinant human interferon-gamma in patients with chronic granulomatous disease–European follow up study. Eur J Pediatr 154(4):295–298

164. Mouy R, Veber F, Blanche S, Donadieu J, Brauner R, Levron JC et al (1994) Long-term itraconazole prophylaxis against *Aspergillus infections* in thirty-two patients with chronic granulomatous disease. J Pediatr 125(6 Pt 1):998–1003

165. Gallin JI, Alling DW, Malech HL, Wesley R, Koziol D, Marciano B et al (2003) Itraconazole to prevent fungal infections in chronic granulomatous disease. N Engl J Med 348(24):2416–2422

166. Zambrano E, Esper F, Rosenberg R, Kim J, Reyes-Mugica M (2003) Chronic granulomatous disease. Pediatr Dev Pathol 6(6):577–581

167. Al-Tawfiq JA, Al-Abdely HM (2010) Vertebral osteomyelitis due to *Aspergillus fumigatus* in a patient with chronic granulomatous disease successfully treated with antifungal agents and interferon-gamma. Med Mycol 48(3):537–541

168. Kim M, Shin JH, Suh SP, Ryang DW, Park CS, Kim C et al (1997) *Aspergillus nidulans* infection in a patient with chronic granulomatous disease. J Korean Med Sci 12(3):244–248

169. Bujak JS, Kwon-Chung KJ, Chusid MJ (1974) Osteomyelitis and pneumonia in a boy with chronic granulomatous disease of childhood caused by a mutant strain of *Aspergillus nidulans*. Am J Clin Pathol 61(3):361–367

170. Altman AR (1977) Thoracic wall invasion secondary to pulmonary aspergillosis: a complication of chronic granulomatous disease of childhood. Am J Roentgenol 129(1):140–142

171. Bukhari E, Alrabiaah A (2009) First case of extensive spinal cord infection with *Aspergillus nidulans* in a child with chronic granulomatous disease. J Infect Dev Ctries 3(4):321–323

172. Soler-Palacin P, Margareto C, Llobet P, Asensio O, Hernandez M, Caragol I, Espanol T (2007) Chronic granulomatous disease in pediatric patients: 25 years of experience. Allergol Immunopathol (Madr) 35:83–89

# Clostridium difficile in Children: A Review of Existing and Recently Uncovered Evidence

Oliver Morris, Marc Tebruegge, Ann Pallett, Steve M. Green, Andrew D. Pearson, Andrew Tuck, Stuart C. Clarke, Paul Roderick and Saul N. Faust

**Abstract**

The clinical significance of the presence of *Clostridium difficile* in children's faeces remains uncertain using current diagnostic procedures. *Clostridium difficile* is a relatively common finding in infants with no symptoms of gastrointestinal disease, suggesting it may be an incidental finding and form part of the normal gut micro-flora in this age group. On the other hand, particularly in older children or those with significant co-morbidity, there are examples where *C. difficile* causes disease and exerts considerable morbidity and even mortality (*C. difficile* infection, CDI). Between these extremes lie a substantial group of children who have both diarrhoea and *C. difficile* in their stools but where the nature of the association is not clear: *Clostridium difficile* associated disease (CDAD). We review the significance of *C. difficile* in children presenting recently uncovered paediatric data from a large UK epidemiological study that informs some key unanswered questions.

S. N. Faust (✉) · O. Morris
Academic Unit of Clinical & Experimental Sciences,
Faculty of Medicine, University of Southampton,
University Hospital Southampton NHS Foundation
Trust, NIHR Wellcome Trust Clinical Research Facility,
Southampton, UK
e-mail: s.faust@soton.ac.uk

O. Morris
e-mail: omorris@doctors.org.uk

M. Tebruegge
Academic Unit of Clinical & Experimental Sciences,
Faculty of Medicine, University of Southampton,
University Hospital Southampton NHS Foundation Trust,
Southampton, UK
e-mail: M.Tebruegge@soton.ac.uk

A. Pallett
University Hospital Southampton NHS Foundation Trust,
Southampton, UK
e-mail: Ann.Pallett@uhs.nhs.uk

S. M. Green
Health Protection Agency, Southampton, UK
e-mail: S.Green@soton.ac.uk

A. D. Pearson
Department of Advanced Computational Biology,
University of Maryland, USA
e-mail: drandrewpearson@googlemail.com

A. Tuck · S. C. Clarke
Academic Unit of Clinical & Experimental Sciences,
Faculty of Medicine, University of Southampton and
Health Protection Agency, Southampton, UK
e-mail: A.C.Tuck@soton.ac.uk

S. C. Clarke
e-mail: S.C.Clarke@soton.ac.uk

P. Roderick
Academic Unit of Primary Care and Population Sciences,
Faculty of Medicine, University of Southampton, UK
e-mail: pjr@soton.ac.uk

N. Curtis et al. (eds.), *Hot Topics in Infection and Immunity in Children IX*,
Advances in Experimental Medicine and Biology 764, DOI 10.1007/978-1-4614-4726-9_4,
© Springer Science+Business Media New York 2013

## 4.1 Introduction

The clinical significance of *Clostridium difficile* colonisation in children is crucial in determining testing and treatment policies. *C. difficile* is the leading cause of nosocomial diarrhoea in adults and one of the Healthcare Acquired Infections (HAI) specifically targeted by the National Health Service (NHS) in England [1]. Mandatory monitoring in the UK currently requires reporting of all diarrhoeal samples positive for *C. difficile* in those over the age of 2 years [2, 3]. *C. difficile* infection (CDI) in adults has been extensively studied and there are clear guidelines for its diagnosis and management [3–8]. The situation is much less clear in children because [9, 10] of the following issues:

a. ***Clostridium difficile* carriage rates are high in young children, especially infants:** Asymptomatic carriage of *C. difficile* in children is much more frequent than in adults. Carriage can be found in up to 64 % of neonates (under 4 weeks old), who appear to acquire *C. difficile* environmentally in the first few weeks of life [10]. Asymptomatic carriage is so high in infants under 1 year of age that testing for *C. difficile* is of questionable benefit [11]. Carriage declines with age, [12, 13] whilst the proportion with symptomatic disease increases to approach levels in adults (5–8 %, [10]).

b. ***Clostridium difficile* may be an incidental finding:** Acute diarrhoeal illness is common in young children, and due to a wide range of potential pathogens [12, 14–17], although in a substantial proportion no pathogen can be identified even after extensive testing for a range of bacteria, viruses and parasites. Given that so many children suffer diarrhoea without an identifiable cause, it is possible that even in those cases where *C. difficile is* identified, it is in fact not the causative agent but an incidental finding. It is even possible that colonisation with *C. difficile* may be protective [18, 19] against CDI.

c. **The relationship between prior antibiotic usage and *C. difficile* is considerably weaker in younger children:** Not only do many

children with *C. difficile* in their stool sample not have a recent history of antibiotic exposure [10, 20, 21], but a significant proportion of them have concurrent infection with other pathogenic organisms [14, 20, 21]. Thus, two of the original diagnostic criteria for CDI in adults [6] would exclude a large proportion of symptomatic children with *C. difficile* positive stools. Although there are recent guidelines for diagnosing CDI in adults [7, 8], no equivalent guidelines exist for children, where the evidence base is much more limited.

Despite this, there is a consensus that *C. difficile* can cause serious disease in children [10, 22]. It is helpful to maintain a distinction between examples of CDI, where *C. difficile* is the accepted cause of children's symptoms, and "*C. difficile associated* disease" (CDAD) where the exact nature of the association between *C. difficile* and gastrointestinal symptoms remains unclear. Although variably defined, for example "gastrointestinal symptoms in a child with *C. difficile* toxin positive stools" [23], or "children with a clinical diagnosis of *C. difficile* infection on discharge who had *C. difficile* testing and antibiotic therapy for CDAD" [24], the incidence of CDAD and/or CDI in children appears to be on the increase, particularly in cases arising from the community [25]. Recent reviews have highlighted the importance of considering CDAD/CDI in children and outside the hospital setting [26–28]. Difficult questions are raised for clinicians faced with children suffering from diarrhoea [9, 10]:

1. When should *C. difficile* be considered as a cause for the child's symptoms and what is the appropriate testing strategy?
2. Should the presence of *C. difficile* in diarrhoea in children be treated, a*nd if so how*?

With current technologies, testing for *C. difficile* in children is controversial [11]. This is especially true for children under the age of 1, who are the most likely to suffer from diarrhoea, have the highest prevalence of asymptomatic carriage of *C. difficile*, and also form a group for whom the incidence of CDAD is increasing [23]. Devising the most appropriate policy has significant implications for all acute NHS Hospital Trusts in England, who face financial penalties if their

**Table 4.1**  Key features of different testing methods for *C. difficile*

| | Tests requiring organism culture | | Tests used directly on stool specimens | | | |
|---|---|---|---|---|---|---|
| | Selective culture | TGC | GDH EIA | PCRCDT | Toxin A/B EIA | CCA |
| Test target | Whole organism | Products/genes of isolate | Enzyme product | Toxin gene (s) | Toxin (s) | Toxin (s) |
| **Detects** | | | | | | |
| Non-toxigenic *C. difficile* | + | – | + | – | – | – |
| Toxigenic *C. difficile* | + | + | + | + | ±[a] | ±[a] |
| Turnaround time | 2–5 days | 2–5 days | Minutes | 1–3 h | Minutes | 1–3 days |
| Performance in adults[c] | Sensitivity | cf. CCA | 0.80–0.97 | 0.87–1.00 | 0.31–0.99 | |
| | | cf. TGC | 0.60–0.74 | 0.86–0.94 | 0.32–0.79 | 0.55–0.67 |
| | Specificity | cf. CCA | 0.75–0.97 | 0.94–1.00 | 0.65–1.00 | |
| | | cf. TGC | 0.76–0.95 | 0.94–0.97 | 0.84–1.00 | 0.98 |
| Performance in children[d] | Sensitivity[b] | 0.75 (by EIA) 0.90 (by PCR) | | 0.95 | 0.35 | |
| | Specificity[b] | 1.00 | | 1.00 | 1.00 | |

[a] Tests are only positive if sufficient toxin present in faeces

[b] The "reference standard" was any stool specimen where at least four of the six tests used were positive (stool EIA, postculture EIA, stool PCRCDT for *tcdA*, postculture PCRCDT for *tcdA*, stool PCRCDT for *tcdB*, postculture PCRCDT for *tcdA*)

[c] Crobach et al. [7]; Tenover et al. [39]; Carroll [40]; Stamper [100]

[d] Luna [42]

rates of *C. difficile* infection fail to meet national targets under a system that has been criticised as unfair due to arbitrary thresholds [29]. This is a potential disincentive to test in populations expected to have higher prevalence of colonisation with *C. difficile*, or to improve the accuracy of testing strategies in ways that increase reportable rates of CDI [30]. Such conflicts can only be resolved by a better understanding of the significance of *C. difficile* colonisation and infection in children that can inform new guidelines.

## 4.2  The Complexity of Testing for *C. difficile* and Diagnosing *C. difficile* Infection

Conventional detection of *C. difficile* relies on selective culture of the organism from faecal samples or detection of toxin by cell culture cytotoxicity assay (CCA) or enzyme immunoassay (EIA) [4]. *Clostridium difficile*'s pathogenicity is associated primarily with expression of toxin B (cytotoxin), with the role of toxin A

(enterotoxin) being less certain [31, 32]. Not all strains are capable of producing toxin A [33, 34] so tests directed against this alone may miss CDI. Molecular assays based on amplification of *C. difficile* toxin gene sequences (PCRCDT) provide rapid and sensitive results but may not be as specific for CDI as phenotypic testing methods (e.g. CCA) which detect toxin production [35]. *Clostridium difficile* exhibits considerable genomic variation [36], allowing development of genetic typing schemes using techniques such as PCR ribotyping [37]. Genetic variation within the toxin genes may affect the performance of both phenotypic and genetic detection methods [38, 39]. Detection of *C. difficile* will therefore depend on the strain present and the test used (Table 4.1) [4, 7, 40]. A two-stage testing strategy has recently been implemented in the UK, [3] based on an observational diagnostic study in four UK laboratories [41] that confirmed that *C. difficile* toxin EIAs are not suitable as stand alone tests for the diagnosis of CDI or detection of *C. difficile*. The new UK recommendation contains a two test screening protocol comprising a

GDH EIA (or toxin gene PCR) followed by a sensitive toxin EIA. If the first test (GDH or toxin gene test) is negative, the second test (sensitive toxin EIA) does not need to be performed.

Unfortunately, nearly all of the studies comparing the performance of different testing strategies have been conducted in adults and the clinical significance of the results will be subject to different interpretations in paediatric and neonatal populations. One study that investigated both children and young adults is a recent prospective study from Texas that compared direct stool EIA toxin A/B and direct stool PCRCDT with toxigenic culture (TGC, testing for toxin with either EIA or PCRCDT after the organism has been cultured selectively from faeces) [42]. Stools from 96 patients (age 15 days–25 years, median 4 years) suspected of having CDAD were tested using EIA toxin A/B and PCRCDT, both before (direct) and after (TGC) isolation and selective culture. Although lacking an independent gold standard, the "reference standard" for calculating sensitivities and specificities was any stool specimen where at least four of the six tests used (stool EIA, postculture EIA, stool PCRCDT (*tcdA*), stool PCRCDT (*tcdB*), postculture PCRCDT (*tcdA*), postculture PCRCDT (*tcdB*)) were positive. Direct stool PCRCDT had the greatest sensitivity of all methods used (95 %, compared to only 35 % for EIA; specificity of both PCRCDT and EIA: 100 %). Interestingly, positivity rates for *C. difficile* have doubled (from ~8 to 16 %) since the introduction of PCRCDT as the standard testing strategy at this US institution (while the number of samples sent for testing has stayed the same).

## 4.3 Epidemiology of *C. difficile* in Children

Few published studies are well placed to establish when *C. difficile* should be considered the cause of a child's diarrhoea, or when it should be treated. We therefore reanalysed paediatric data from the English community-based Infectious Intestinal Disease Study (IIDS) [12, 43] which have not previously been reported within the peer reviewed literature. The IIDS data is informative as the study attempted to: (a) test each sample for a range of viral, bacterial and parasitic pathogens to identify co-infection and the possibility that *C. difficile* is an "innocent bystander"; (b) use an appropriately sensitive testing strategy for *C. difficile* (including selective culture); (c) test both symptomatic and asymptomatic children to assess the association between *C. difficile* and symptoms; and (d) present data within clinically relevant age ranges. A second study (IID2) is currently in progress [44].

### 4.3.1 *Clostridium difficile* in Children in the Community

Several community-based studies have demonstrated that detection of *C. difficile* and/or *C. difficile* toxin is as common in asymptomatic children as in those suffering with diarrhoea (Table 4.2). Conducted between 1993 and 1996, the IIDS collected data and stools from over 6,000 cases of infectious intestinal diseases in children and adults, as well as from controls [12, 43]. The IIDS included two components: a community cohort recruited randomly from 70 General Practitioner (GP) practices across England and a GP-based case-control study which included subjects presenting spontaneously with symptoms of diarrhoea to one of 34 practices. All faecal samples were subject to extensive testing for a range of potential pathogens including *C. difficile*. Asymptomatic carriage of *C. difficile* was highest in infants under 1 year of age: 21 % in those recruited as part of the community cohort and 16.6 % in the age and sex-matched controls for those presenting to their GP with diarrhoea. Carriage rates in infants *with* diarrhoea were not much greater (28.6 % in the community cohort) or even less (7.2 % in infants presenting to their GP). 1 % of children aged 1–4 years of age had *C. difficile* in their faeces, and *C. difficile* was rarely found in children 5 years or older. There was no significant association of *C. difficile* with diarrhoeal symptoms for children under the age of 5 in either the GP study (derived odds ratio for diarrhoea if *C. difficile* positive 0.59; 95 % confi-

**Table 4.2** Prevalence of *C. difficile* in children in the community: asymptomatic children compared to those with diarrhoea for different groups*

| Group | Country | Dates | Age, years (median) | Prevalence (n) | Test Methods | Setting |
|---|---|---|---|---|---|---|
| Community Non-diarrhoea | U.K. | 1993-6 | <1<br>1-4<br>5-14 | 21.0% (19)<br>1.0% (98)<br>0% (72) | CCA (also selective culture (CCFA), TGC via Toxin A EIA, and PCR ribotyping) | Age- and sex-matched controls to community diarrhoea cohort (12) |
| | U.K. | 1993-6 | <1<br>1-4<br>5-14 | 16.6% (199)<br>1.0% (510)<br>0.6% (177) | | Age- and sex-matched controls to GP presentations of diarrhoea, families not part of cohort (12) |
| | U.S.A. | 1981 | 0.04 - 16 | 14.8% (135) | Selective culture (CCFA) | Outpatient attendees (45) |
| | | 2001-2 | (1.27) | 3.5% (484) | Toxin A/B EIA | Community (17) |
| | Japan | 2004 | <1<br>1<br>2<br>3<br>4<br>5 | 100% (12)<br>75% (20)<br>45.5% (11)<br>24% (25)<br>38.5% (13)<br>23.5% (17) | Selective culture, toxin gene typing (PCR), PCR ribotyping and pulsed-field gel electrophoresis | 2 day nurseries and a kindergarten (13) |
| Community Diarrhoea | U.K. | 1993-6 | <1<br>1-4<br>5-14 | 28.6% (21)<br>0.9% (116)<br>0% (94) | CCA (also selective culture (CCFA), TGC via Toxin A EIA, and PCR ribotyping) | Diarrhoeal episode within community (asked to provide samples when have diarrhoea), recruited to cohort via GP practice lists (12) |
| | U.K. | 1993-6 | <1<br>1-4<br>5-14 | 7.2% (180)<br>1.7% (468)<br>0.6% (167) | | Presentations to GP with diarrhoea, families not part of cohort (12) |
| | U.S.A. | 1981 | 0.04 - 16 | 7% (171) | Selective culture (CCFA) | Outpatient attendees (45) |
| | | 2001-2 | (1.27) | 1.9% (431) | Toxin A/B EIA | Community (17) |
| | | 1998-2001 | 0.01-28.6 (1.3) | 6.7% (688) | CCA | Paediatric emergency admissions (14) |
| | | 1998-01<br>2000-01 | 0.05-18.75 (1.61) | 9.0% (89)<br>0% (15) | CCA | Private paediatric outpatient clinic Clinic serving mainly immigrants (15) |
| | Austria | 2007 | <5<br>5-19 | 5% (20)<br>6.5% (46) | Toxin A/B EIA and TGC (Toxin A/B EIA) | Presentations to GP with acute gastroenteritis (16) |

\* Where available, matching shades show data from the same study to facilitate comparison

dence interval (CI) 0.34–1.02) or the community cohort (odds ratio 1.21; 95 % CI 0.37–3.9).

Two other studies comparing *C. difficile* in symptomatic children with asymptomatic controls showed similar results. Boenning et al. [45] showed that the prevalence of *C. difficile* amongst paediatric outpatients was actually higher in those with non-diarrhoeal illnesses (14.8 %) than those with diarrhoea (7.0 %; Odds ratio 0.43, 95 % CI 0.20–0.92), and found no association with antibiotic exposure in the preceding month. More recently, Vernacchio et al. [17] carried out a prospective cohort study of healthy children in the community in which baseline normal stools were cultured and compared to those from the same children who subsequently developed diarrhoea during the study period. Of healthy baseline stools, 3.5 % contained *C. difficile* compared to just 1.9 % in diarrhoeal specimens (matched relative risk 0.54, 95 % CI 0.20–1.50). Contrast-

ing results were obtained from an earlier study of diarrhoeal outbreaks in community day centres [46], where rates of *C. difficile* isolation were greater amongst children with diarrhoea (57 %) than in their healthy classmates (9 %; OR 13.3, 95 % CI 3.5–51).

Other community-based studies have either tested only asymptomatic children [13] or children with diarrhoea [14–16], and therefore cannot provide comparative data regarding the association of *C. difficile* with symptoms. They nevertheless demonstrate that *C. difficile* is a relatively common finding (5–9 %, Table 4.2) in children presenting from the community.

Although the prevalence in different age ranges have varied between studies, a high prevalence of *C. difficile* has largely been found in young children and infants, with an average age in these studies typically 18 months or less. For example, in the recent US study [24] the mean

**Table 4.3** Prevalence (%) of identified microorganisms in diarrhoeal samples from children in the community (IIDS GP data shown, cohort data not shown)

| Putative pathogen | | Setting | | | | | | | |
|---|---|---|---|---|---|---|---|---|---|
| Class | Organism | GP presentations [12] 1993–1996 (UK) | | | Paediatric outpatients[a] [15] 1998–2001 (USA) | Emergency department [14] 1998–2001 (USA) | Community diarrhoea [17] 2001–2002 (USA) | GP presentations [16] 2007 (Austria) | |
| | | <1 year | 1–4 years | 5–14 years | 0.05–18.75 years (1.6 years) | 1.3 years | 0.5–3 years (1.25 years) | <5 years | 5–19 years |
| Bacteria | C. difficile (toxin) | 7.2 | 1.7 | 0.6 | 9.0 | 6.7 | 1.9 | 5 | 6.5 |
| | C. perfringens | 4.0 | 5.6 | 4.5 | | | 2.3 | | |
| | Aeromonas spp. | 10.3 | 4.3 | 7.2 | | | 1.0 | | |
| | Campylobacter spp. | 2 | 5.4 | 11.3 | 4.1 | 1.5 | 0.7 | 0 | 4.3 |
| | Pathogenic E. Coli | | | | | | | | |
| | DAEC | 3.6 | 2.9 | 2.3 | | | | | |
| | AEEC | 6.1 | 9.8 | 3.3 | | | | | |
| | EAggEC | 4.0 | 5.5 | 4.7 | | | 4.1 | | |
| | EPEC | 0 | 0.2 | 0.5 | | | 12.9[b] | | |
| | ETEC | 0 | 0.5 | 0.5 | | | 0.2 | | |
| | EHEC/STEC | 0 | 0.3 | 0.5 | 0 | 2.4 | 0.2 | 0 | 0 |
| | Salmonella spp. | 2.3 | 2.5 | 5.9 | 0 | 2.4 | 0 | 0 | 0 |
| | Staphylococcus aureus | 1.3 | 0 | 0 | | | | | |
| | Yersinia | 0.3 | 2.0 | 1.8 | 0.4 | 0.1 | 0.2 | 0 | 0 |
| | Shigella spp. | 0 | 0.3 | 0.4 | 0 | 0.9 | 0 | | |
| Viruses | Rotavirus | 21.3 | 17.6 | 8.1 | 14.2 | 20.4 | 5.2 | 5 | 6.5 |
| | Norovirus/SRSV | 8.9 | 11.0 | 5.5 | | | 1.9 | 20 | 8.7 |
| | Adenovirus | 6.9 | 10.3 | 1.0 | 6.3 | 4.3 | 5.7 | 10 | 0 |
| | Astrovirus | 1.9 | 6.7 | 3.0 | 3.1 | 6.5 | 3.5 | 0 | 0 |
| | Sapovirus/Calicivirus | 5.8 | 3.8 | 0.5 | | | 3.0 | | |
| Parasites | Giardia | 1.0 | 1.2 | 1.8 | | 0.6 | 0.2 | 0 | 0 |
| | Cryptosporidium | 0.7 | 2.8 | 5.0 | | 0.3 | 0.5 | 0 | 0 |
| No pathogen identified | | 42.0 | 30.9 | 45.7 | 52 | 53 | 57.5 | 60 | 74 |

[a] Data from Site A only (private clinic)
[b] 'Atypical' EPEC

age was 15.2 (±7.5) months, with most children under the age of 3 years. In studies that report prevalence for infants separately this is universally higher than for older children.

A small study in Japan was notable for the particularly high prevalence of C. difficile in asymptomatic children at two day nurseries and a kindergarten [13]. All the infants tested had C. difficile in their faeces but carriage rates declined steadily with increasing age to less than 25 % by 5 years of age. Nearly all (21/22) of the C. difficile strains (by PCR ribotype and PFGE type)

isolated from environmental swabs at one nursery were identical to those found in the faeces of infants in that nursery, suggesting acquisition from the day-care environment.

Community onset diarrhoea is relatively common in young children and infants, with a range of potential pathogens other than C. difficile. Table 4.3 summarises data from five studies which carried out extensive testing for pathogens in children with community-acquired diarrhoea. A number of trends are revealed across the different settings, methodologies and even countries where the studies were based. Despite the investigators' best efforts to test for a large range of pathogens, none could be identified in 30–75 % of the cases of diarrhoea. Viruses were the commonest identified pathogen in all studies, especially in infants and pre-school age children. Rotavirus was most frequently encountered (~14–20 %, except in an Austrian study, where there is routine vaccination against rotavirus [16]), followed by norovirus (where tested, ~9–11 %), adenovirus (~4–10 %) and astrovirus (~2–7 %). *Clostridium difficile* was the commonest identified bacterium in most studies. Other common isolates were *C. perfringens*, pathogenic *E. coli*, *Aeromonas* spp., *Salmonella* spp. and *Campylobacter* spp. (notably the latter two were more commonly isolated in older children).

A proportion of *C. difficile* positive children were co-infected with other known pathogens. In Klein et al.'s [14] study, 25 (6.7 %) of the 372 specimens that underwent extensive testing were positive for *C. difficile* toxin B, but nine of these had concomitant viral or bacterial pathogens. Only 4.3 % of children with diarrhoea had *C. difficile* as the only identifiable pathogen [14] and some common causes (e.g. norovirus) were not tested for, so it is possible that there may have been underestimation of co-infection with *C. difficile*. Denno et al. tested for the same range of organisms, and found one case of co-infection out of 75 children undergoing 'complete' analysis (one of the eight stools positive for *C. difficile* was also positive for adenovirus) [15]. Co-infection is also common in hospitalised children, with other pathogens reported in 36 % [21], 44 % [20] 27 % [47] and 23–38 % of *C. difficile* culture positive stools [48]. In three recently reported cases of severe CDAD in children, it was noted that two had co-infection with viruses (rotavirus and calicivirus), although these two also had underlying chronic medical conditions (Hirschsprung's disease and Down's syndrome) [49].

### 4.3.1.1   Explanations for High Levels of Asymptomatic Colonisation

The high rates of carriage of *C. difficile* may be explained by a lack of 'colonisation resistance' in infants as the intestinal micro-flora is in the process of becoming established [50], but this cannot explain the absence of symptomatic disease (CDI). Asymptomatic carriage does not simply occur because the strains of *C. difficile* present are non-toxigenic, for most of the studies discussed previously demonstrated the presence of toxin in faeces from asymptomatic children. One hypothesis is that the strains typically found in children have reduced virulence compared to those in adults. Factors other than toxin expression, such as those influencing adherence and intestinal colonisation, are also recognised contributors to virulence and may play a part in childhood disease [51].

The hypervirulent strain of *C. difficile* associated with outbreaks of severe CDI in adults (PCR ribotype 027) has greatly increased *in vitro* levels of expression of *tcdA* (16-fold) and *tcdB* (23-fold) [51]. This strain was responsible for a high proportion (19.4 %) of *C. difficile* toxin positive stools in symptomatic children in a recent study in the USA [52]. A study of hospitalised but asymptomatic children in Thailand demonstrated relatively high carriage rates for *C. difficile* (11.0 % of infants, and 21.1 % of 1–11 year olds) but low rates of toxin A gene detection (0.9 % of infants and none of the older children; toxin B not tested) [53]. 87.2 % of the strains isolated in asymptomatic children in the Japanese day-care study were also non-toxigenic (*tcdA–/tcdB–*) [13]. Unfortunately neither study examined rates of carriage or toxin expression in symptomatic children for comparison. A small study in Brazil attempted to compare strains of *C. difficile* in hospitalised children with acute diarrhoea with asymptomatic children recruited from day-care centres, but none of the controls were culture positive for *C. difficile*. Nine out of the ten strains isolated from symptomatic patients were toxigenic: six were *tcdA+/tcdB+* and three were *tcdA–/tcdB+* [54].

Only the IID study provides direct data comparing *C. difficile* colonisation and toxin B detection for cases of diarrhoea and asymptomatic controls [12]. Below 2 years of age, children presenting to their GP with diarrhoea had slightly higher rates of *C. difficile* colonisation (24 % of 374 cf. 19.5 % of 385 cultured) but lower rates of toxin B detection than asymptomatic age- and sex-matched controls (4.3 % of 391 cf. 9.0 % of 423 tested). For children 2 years or older, con-

**Table 4.4** Prevalence of *C. difficile* in hospitalised children for different groups*

| Group | Country | Year | Age Years (median) | Prevalence (n) | Test Method | Setting |
|---|---|---|---|---|---|---|
| Hospital Diarrhoea | Germany | 1989 | 0-1<br>>1 | ~30%<br>~5%<br>(418) | Culture and Y-1 cell culture (Toxin B) and rabbit ileal loop test (Toxin A) | Hospitalised children (55) |
| | Canada | 1991-99 | 0.03-17 (1.3) | 18.0% | Toxin A EIA | Paediatric and Women's Hospital (56) |
| | USA | 1998-99<br><br>2001-03 | (1.3)<br><br>(2.4) | 26% (676)<br>(Toxigenic: 15%)<br>49% (301)<br>(Toxigenic: 14%) | Selective culture (CCFA) and TGC (PCRCDT), PCR fingerprinting | Stand alone paediatric hospital<br><br>Paediatric department within general hospital (48) |
| | Turkey | 2001 | 0.08-13 (Mean 3.2) | 22% (100) | Toxin A/B EIA | Inpatients treated with antibiotics with diarrhoea (47) |
| Hospital Non-diarrhoea | Germany | 1989 | 0-1<br>>1 | ~30%<br>~5%<br>(348) | Culture and Y-1 cell culture (Toxin B) and rabbit ileal loop test (Toxin A) | Hospitalised children (55) |
| | Thailand | 1998-99 | 0-1<br>1-11yrs | 11.9% (235)<br>21.1% (76) | CCFA culture, PCR *tcdA* and PFGE analysis for type | Hospitalised children (53) |
| | Turkey | 2001 | 0.08-13 (Mean 4.1) | 10% (50) | Selective culture (CCFA) and Toxin A/B EIA | Inpatients treated with antibiotics but asymptomatic for diarrhoea (47) |

*Matching shading shows data from the same study to facilitate comparison

trols had lower rates of both toxin B detection (0.2 % of 1,616 cf. 1.1 % of 1,866 tested) and *C. difficile* colonisation (0.4 % of 1,613 cf. 1.0 % of 1,845 cultured). These data suggest that above the age of 2 years, rates of colonisation drop dramatically, and toxin production is more likely to be associated with disease.

### 4.3.2 *Clostridium difficile* in Hospitalised Children

The prevalence of *C. difficile* in hospitalised children is higher than in the community but appears similar in children with and without diarrhoea (Table 4.4). Karsch et al. [55] found high rates of carriage, particularly in infants (30 %), with most isolates producing toxin (82 % toxin A and 43 % toxin B), but no significant difference between symptomatic children and controls, and no clear association with previous antibiotic therapy. A prospective study in Denmark also showed significantly higher isolation of *C. difficile* in infants and no relationship to antibiotic exposure ($p < 0.001$), but it was the only identified pathogen in 12 % of children of all ages with acute gastroenteritis, and this was significantly ($p < 0.01$) higher than for asymptomatic controls [21]. As 78 % of the positive

cultures were obtained within 2 days of admission this suggests acquisition in the community rather than nosocomial infection. A retrospective cohort study in the US reported that 26 % of cases of CDAD occured in infants [24], but it remains unclear whether *C. difficile* was truly the underlying cause. Of the 56 % of specimens from paediatric inpatients in Canada with nosocomial diarrhoea where a pathogen was identified, most were viruses (38 % of episodes, viral diagnosis typically in younger children with mean age 0.8 years) [56]. *Clostridium difficile* was identified in 18 % of all episodes (mainly in older children, mean age 3.9 years). A retrospective case control study in Canada was unable to demonstrate any difference in clinical characteristics between infants with *C. difficile* toxin in their stool and those without toxin present, nor could it identify a significant treatment effect of metronidazole [11]. Colonisation with *C. difficile* was relatively high at two US institutions (49 % and 26 %), but toxigenic colonisation was less common (14 % and 15 %), and many isolates were unique (92 % within the general hospital), indicating that they did not arise from a common source [48]. Underlying medical conditions and exposure to two or more antibiotics were associated with increased toxigenic strain colonisation. In Turkey, although *C. difficile* toxin was found more frequently in hospitalised children with nosocomial diarrhoea (22 %) than asymptomatic controls (10 %), this was not significant (odds ratio 2.54, 95 % CI 0.90–7.17) and co-infection was found in six cases (27 %) of CDAD (rotavirus was the only viral pathogen tested) [47]. Children with CDAD were older (mean 5.4 years) than those asymptomatic controls with *C. difficile* toxin in their stools (all under 2 years).

### 4.3.3 *Clostridium difficile* in Neonates

Neonates acquire *C. difficile* from the environment, resulting in high rates of carriage within the first few weeks of birth [57–60]. In one study, no neonatal faeces cultured *C. difficile* on day 1 after birth but 17 % did by day 4, and most strains were toxigenic (58–65 %) [60]. None of the maternal rectal swabs and only one vaginal swab cultured *C. difficile,* whereas 13 % of environmental samples did and these were all of the same strain (matching 11 of the 31 neonatal strains typed). On one neonatal intensive care unit (NICU), *C. difficile* acquisition reached 33 % after 2 weeks with all cultures toxigenic [61]. In common with previous studies [58, 59], there was no association between *C. difficile* acquisition and gastrointestinal symptoms. In another NICU, 90 % of samples cultured *C. difficile* after only 6 days, and although toxin was detected directly in only 36 % of these, 94 % of the isolates were found to be toxigenic *in vitro* [57]. One small study suggested that neonates with toxin A positive stools are likely to experience increased numbers of days with frequent and abnormal stools [62], but most demonstrate asymptomatic carriage of *C. difficile* in neonates, even when a high proportion of stools test positive for toxin.

### 4.4 Burden of CDAD in Children

The high asymptomatic prevalence of *C. difficile* in young children initially led to the conclusion that CDI was not a problem in this age group [45, 63]. However, several studies have suggested that CDAD (and by implication CDI) is an increas-

ing problem in children. A retrospective cohort study in the USA showed that from 2001 to 2006 the incidence rate of CDAD (defined as "clinical symptoms, such as diarrhoea or bloody stools, in a patient whose stool specimen tested positive for *C. difficile* toxin") amongst outpatients increased by 11 % (from 1.24 to 1.38 cases per 1,000 visits) [23]. The incidence of CDAD in patients attending the emergency department increased 2.5 fold, largely due to an increase in community-associated CDAD (from 0.84 to 2.04 cases per 1,000 visits), while inpatient incidence of CDAD decreased over the same period [23]. In another study the annual incidence of CDAD (identified by the combination of discharge diagnosis of CDI, positive test assay for CDI and treated with antibiotics against CDI) increased from 2.6 to 4.0 cases/1,000 admissions over the same time period [24]. Hospitalisation rates for children with CDAD almost doubled between 1997 and 2006 [64]. Complications of CDAD can be severe, including pseudomembranous colitis [22, 49, 65, 66], rectal prolapse [67], osteomyelitis [68] and reactive arthritis [69, 70].

### 4.5 When to Consider CDAD and How to Test for it

The new UK Department of Health testing algorithm [41] contains a two test screening protocol comprising a GDH EIA (or toxin gene PCR) followed by a sensitive toxin EIA. If the first test (GDH or toxin gene test) is negative, the second test (sensitive toxin EIA) does not need to be performed.

### 4.5.1 Children Without Co-morbidities

Because the prevalence of *C. difficile* in children is so dependent on age it might be more appropriate to tailor the testing and management strategy accordingly (Table 4.5). Given the very limited data on the performance of different testing strategies in children (Table 4.1), adult guidelines may eventually be shown to perform poorly in children, where PCRCDT might be

**Table 4.5** Suggested investigation and management of children with suspected *C. difficile* infection based on current evidence

| Age | Rationale | Management |
|---|---|---|
| Neonates (0–4 weeks) | High incidence of carriage, true CDI very rare/indeterminate | Do not test for *C. difficile*. If very unwell, treatment for necrotising enterocolitis will include cover for *C. difficile* |
| Infants (4 weeks–1 year) | Relatively high prevalence of asymptomatic carriage of *C. difficile* and diarrhoeal illness common (likely CDAD rather than CDI) | UK 2012: GDH EIA (or toxin gene) followed by a sensitive toxin EIA[a]. If strong clinical suspicion of CDI, test faeces for *C. difficile* and for common viral and bacterial pathogens. Where community acquired and treatment *not considered urgent*, treat for CDI only if GDH EIA (or toxin gene) *and* sensitive toxin EIA positive *and* viral and other bacterial testing negative. If hospital acquired or urgent treatment considered necessary, treat for CDI if GDH EIA (or toxin gene) *and* sensitive toxin EIA positive |
| Pre-school (1–4 years) | Incidence of community acquired CDAD declines with age; incidence of CDI increases above the age of 2 years. Significant risk of viral pathogens causing disease | |
| School age (5–18 years) | Community acquired diarrhoea; risk of CDI increases with increasing age, also risk of other bacterial and viral pathogens. Hospital acquired diarrhoea: CDI incidence similar to adults: investigate and manage as for adults | |

For a child presenting with a mild or recent onset diarrhoeal illness then this is likely to be viral and self-limiting: *no testing is indicated* and the child should be managed supportively according to recent NICE guidance [71]

For a child with a significant chronic disease (IBD, cystic fibrosis, cancer, HIV) and significant diarrhoea there is a higher risk of CDI requiring treatment. PCRCDT is the most appropriate first line or only test for paediatric samples where available. *Treat for CDI if PCRCDT positive.* (see footnote[a] )

Secondary care management *where diarrhoea is severe or protracted (>7 days)*

[a] Current paediatric evidence suggests a more prominent role of PCRCDT may be justified in future testing strategies

considered a more appropriate formal second line test after GDH EIA due to its rapid turnaround time and good correlation with toxigenic culture (Table 4.1).

Infants have high asymptomatic prevalence, so testing stools for *C. difficile* is not recommended unless illness is severe and there is a high level of clinical suspicion [71]. Serious cases of CDI (pseudomembranous colitis) have been reported in infants, but have generally been associated with other conditions such as prematurity, Hirschprung's disease, obstruction or necrotising enterocolitis [65, 72, 73], and the role of *C. difficile* in the pathogenesis has not been confirmed. These reports suggest symptomatic infants should only be tested for *C. difficile* in carefully selected circumstances. Other organisms may also be responsible for disease and viral testing in particular should be undertaken alongside testing for *C. difficile*.

It would be helpful to identify reliable predictors for development of CDI in children. Despite the conflicting evidence for an association between antibiotic usage and CDI in children (against association: [11, 20, 21, 45, 55, 74, 75];

for association: [14, 47, 48, 63, 76–79]), severe diarrhoea in the context of recent antibiotic therapy is likely to remain one such predictor that clinicians will use.

## 4.5.2 Children with Co-morbidities

Certain co-morbitities are associated with higher rates of *C. difficile* colonisation, and some of these children may be at particular risk of CDI: inflammatory bowel disease (IBD), Hirschprung's disease, cystic fibrosis, cancer patients and organ transplant recipients.

The prevalence of *C. difficile* is significantly greater in children with IBD compared to controls (indicating an increased risk of colonisation) and also greater in children with IBD experiencing active disease compared to those with inactive disease (indicating a potential role of the organism in the symptom exacerbation) [80]. It is important to use an appropriate testing strategy to prevent missing CDI in the context of inflammatory bowel disease, as the diagnosis may be missed in up to 41 % of IBD patients if a single-toxin assay is used [81]. Such errors could lead to misattribution of symptoms to an exacerbation of underlying IBD and result in inappropriate treatment, potentially even colectomy. Steroids given to treat IBD without appropriate antibiotics to treat CDI are likely to exacerbate symptoms rather than resolve them [82].

Case reports of severe CDI associated with Hirschsprung's disease suggest that this may form another group of children at increased risk [66, 83]. Increased *C. difficile* was found in Hirschsprung's patients with enterocolitis compared with asymptomatic Hirschsprung's patients or healthy controls [84].

Carriage rates of *C. difficile* in patients with cystic fibrosis have been reported to be 22–46 %, double that of control patients receiving antibiotics [85–87]. These high carriage rates were described in older patients (median age 18.5, youngest 15 years) in the absence of symptoms of diarrhoea or abdominal pain, despite the presence of toxigenic strains [87]. Nevertheless, severe cases of CDI do occur in children with cystic fibrosis, emphasizing the importance of considering the diagnosis [88]. Patients may present atypically, without watery diarrhoea but rather abdominal distension and reduced bowel motions, risking confusion with faecal impaction or meconium ileus equivalent [89, 90]. There also appears to be a greater risk of CDI following lung transplantation for cystic fibrosis, with patients experiencing a fulminant course resulting in high mortality [91, 92].

Cancer patients are thought to have increased carriage of *C. difficile* due to their chemotherapy treatment and increased exposure to antimicrobials [74, 77]. Most reported cases have been associated with haematological malignancies, but a recent Italian study demonstrated that 6 % of children with solid tumours had *C. difficile* toxin A in their stools and gastrointestinal symptoms, with three out of nine of these being under 1 year old [93]. An earlier prospective study of oncology patients demonstrated higher rates of toxin detection in asymptomatic (19 %) than in symptomatic (8.7 %) children, questioning the

significance of *C. difficile* as a pathogen in this patient group [74].

## 4.6  Treatment Strategies

If *C. difficile* is the only pathogen identified and seems the most likely cause of disease in a child over the age of 1 then treatment should be considered [10]. In adults, where exposure to broad spectrum antibiotics is a recognised trigger for CDI, cessation of the antibiotic, if clinically possible, is an important step. As previously discussed, the association between antibiotic use and CDAD in children is much weaker, but it is sensible to discontinue broad spectrum antibiotics where possible. Oral metronidazole and, if this fails, vancomycin are the standard treatments for adults with severe CDI requiring treatment and are also used in children. Up to 25 % of patients experience a further episode of CDI within 2 months, and 50–65 % of these will suffer from further recurrences [94].

More recently, interest has moved to alternative treatment strategies for CDI, in particular to prevent recurrence of disease [10, 95]. Alternative strategies may be divided into those that attempt to re-establish colonisation resistance (use of probiotics, faecotherapy), direct chemical neutralisation of the toxins (toxin binding by ion exchange resins and polymers) and immunotherapy (passive immunoglobulin therapy) [94, 95]. Unfortunately, few randomised controlled trials have been conducted in adult populations, and none have been done in children. A recent Cochrane review [96] found little evidence that probiotics were useful in adults, identifying only one study that showed a beneficial effect when added to antibiotic therapy [97]. Extrapolation of such data to children is problematic given that in a child's early years the natural intestinal flora is different from adults and evolves over time. None of the alternative strategies explored so far has demonstrated sufficient success to recommend incorporation into general clinical practice [95].

**Table 4.6** Outstanding questions for research

| |
|---|
| Are children suffering with CDAD colonised with different strains of *C. difficile* from those found in asymptomatic, age-matched controls? |
| What are the virulence factors other than toxin expression that influence pathogenicity in *C. difficile* and how do these explain the varying prevalence of symptoms with the age of the host? |
| Are there markers that correlate more closely with disease than toxin expression? Such markers might be specific to the strain of *C. difficile* present (e.g. other virulence genes) or the host specific response to infection (e.g. cytokine profile) |
| How do the new 'gold standard' tests for diagnosing *C. difficile* infection in adults perform in children at different ages? |
| What relationships exist between *C. difficile* and other members of the gut micro-flora, in both health and disease? |
|     How common is co-infection in CDAD in children? |
|     Are other bacterial or viral pathogens a risk factor for the development of CDI or *vice versa*? |
|     Is colonisation with specific strains of *C. difficile* protective against other pathogens in children? |
| What factors underlie geographical variations in the prevalence of *C. difficile* colonisation and infection in children? |

## 4.7 Conclusions

*Clostridium difficile* is a relatively common finding in the faeces of infants under 1 year of age, and is very unlikely to signify disease, even when toxins are produced. Children with diarrhoea may have *C. difficile* in their stools, but viruses are a more likely cause of symptoms. It is not possible to identify the aetiological agent in a large proportion of childhood diarrhoea, but the disease is typically self-limiting and requires only supportive care [71].

CDAD appears to be increasing in children and should be considered as adult-type CDI whenever symptoms are particularly severe, protracted or the child belongs to a known at risk group such as inflammatory bowel disease or cystic fibrosis following lung transplantation. The use of predictors for CDAD such as recent antibiotic usage or prior hospitalisation is generally unhelpful in children, where these associations are much weaker and colonisation appears to occur in the community.

Few diagnostic tests in routine clinical use for *C. difficile* infection have been evaluated in children, where the true sensitivities and specificities are likely to be different from adult populations. The evidence on which to base decisions of when and how to treat CDAD in children remains limited and many important, interrelated research questions are yet unanswered (Table 4.6).

## 4.8 Search Strategy and Selection Criteria

This review was prompted by re-appraisal of data contained in a large, UK government-funded investigation report (The Infectious Intestinal Diseases (IID) Study in England [12]) in the light of a growing appreciation that *C. difficile* in children differs substantially from adults [9, 10, 98]. The IID study produced substantial data on the incidence of *C. difficile* in children in the community not contained in the original journal article [43], which combined data for children and adults. Other data and references were obtained by searching PubMed using "infant", "child(ren)", "p(a)ediatric" and "*C. difficile*" as search terms. Retrieved titles and abstracts were screened and full text versions obtained of suitable articles in English. Where necessary, odds ratios and confidence intervals were derived from original data presented in the manuscript by a standard method [99].

**Acknowledgements** This review was carried out as part of OM's UK National Institute of Health Research (NIHR) Academic Foundation placement at the University of Southampton and was supported by the University of Southampton NIHR Wellcome Trust Clinical Research Facility (WTCRF). SNF and MT were funded by the UK NIHR via the NIHR WTCRF and and NIHR Clinical Lectureship respectively.

# References

1. (2008) National target to reduce Clostridium difficile infections: SHA envelopes. Department of Health

2. (2008) Changes to the mandatory healthcare associated infection surveillance system for Clostridium difficile infection (CDI) from 1st January 2008. Department of Health

3. Infection ACoARaHA. UK Department of Health Updated Guidance on the Diagnosis and Reporting of *Clostridium difficile* 2012. http://www.dh.gov.uk/prod_consum_dh/groups/dh_digitalassets/@dh/@en/documents/di gitalasset/dh_133016.pdf

4. Brazier JS (1998) The diagnosis of Clostridium difficile-associated disease. J Antimicrob Chemother 41(Suppl C):29–40

5. Delmee M (2001) Laboratory diagnosis of Clostridium difficile disease. Clin Microbiol Infect 7(8):411–416

6. Fekety R (1997) Guidelines for the diagnosis and management of Clostridium difficile-associated diarrhea and colitis. American College of Gastroenterology, Practice Parameters Committee. Am J Gastroenterol 92(5):739–750

7. Crobach MJ, Dekkers OM, Wilcox MH, Kuijper EJ (2009) European society of clinical microbiology and infectious diseases (ESCMID): data review and recommendations for diagnosing Clostridium difficile-infection (CDI). Clin Microbiol Infect 15(12):1053–1066

8. Cohen SH, Gerding DN, Johnson S, Kelly CP, Loo VG, McDonald LC, Pepin J, Wilcox MH (2010) Clinical practice guidelines for Clostridium difficile infection in adults: 2010 update by the society for healthcare epidemiology of America (SHEA) and the infectious diseases society of America (IDSA). Infect Control Hosp Epidemiol 31(5):431–455

9. Wilson ME (2006) Clostridium difficile and childhood diarrhea: cause, consequence, or confounder. Clin Infect Dis 43(7):814–816

10. McFarland LV, Brandmarker SA, Guandalini S (2000) Pediatric Clostridium difficile: a phantom menace or clinical reality? J Pediatr Gastroenterol Nutr 31(3):220–231

11. Tang P, Roscoe M, Richardson SE (2005) Limited clinical utility of Clostridium difficile toxin testing in infants in a pediatric hospital. Diagn Microbiol Infect Dis 52(2):91–94

12. (2000) A Report of the Study of Infectious Intestinal Disease in England. The Stationery Office, London

13. Matsuki S, Ozaki E, Shozu M, Inoue M, Shimizu S, Yamaguchi N, Karasawa T, Yamagishi T, Nakamura S (2005) Colonization by Clostridium difficile of neonates in a hospital, and infants and children in three day-care facilities of Kanazawa, Japan. Int Microbiol 8(1):43–48

14. Klein EJ, Boster DR, Stapp JR, Wells JG, Qin X, Clausen CR, Swerdlow DL, Braden CR, Tarr PI (2006) Diarrhea etiology in a Children's Hospital Emergency Department: a prospective cohort study. Clin Infect Dis 43(7):807–813

15. Denno DM, Stapp JR, Boster DR, Qin X, Clausen CR, Del Beccaro KH, Swerdlow DL, Braden CR, Tarr PI (2005) Etiology of diarrhea in pediatric outpatient settings. Pediatr Infect Dis J 24(2):142–148

16. Huhulescu S, Kiss R, Brettlecker M, Cerny RJ, Hess C, Wewalka G, Allerberger F (2009) Etiology of acute gastroenteritis in three sentinel general practices, Austria 2007. Infection 37(2):103–108

17. Vernacchio L, Vezina RM, Mitchell AA, Lesko SM, Plaut AG, Acheson DW (2006) Diarrhea in American infants and young children in the community setting: incidence, clinical presentation and microbiology. Pediatr Infect Dis J 25(1):2–7

18. Kyne L, Warny M, Qamar A, Kelly CP (2000) Asymptomatic carriage of Clostridium difficile and serum levels of IgG antibody against toxin A. N Engl J Med 342(6):390–397

19. Jangi S, Lamont JT (2010) Asymptomatic colonization by Clostridium difficile in infants: implications for disease in later life. J Pediatr Gastroenterol Nutr 51(1):2–7

20. Niyogi SK, Dutta P, Dutta D, Mitra U, Sikdar S (1991) Clostridium difficile and its cytotoxin in hospitalized children with acute diarrhea. Indian Pediatr 28(10):1129–1132

21. Tvede M, Schiotz PO, Krasilnikoff PA (1990) Incidence of Clostridium difficile in hospitalized children. A prospective study. Acta Paediatr Scand 79(3):292–299

22. Brook I (2005) Pseudomembranous colitis in children. J Gastroenterol Hepatol 20(2):182–186

23. Benson L, Song X, Campos J, Singh N (2007) Changing epidemiology of Clostridium difficile-associated disease in children. Infect Control Hosp Epidemiol 28(11):1233–1235

24. Kim J, Smathers SA, Prasad P, Leckerman KH, Coffin S, Zaoutis T (2008) Epidemiological features of Clostridium difficile-associated disease among inpatients at children's hospitals in the United States, 2001–2006. Pediatrics 122(6):1266–1270

25. Pituch H (2009) Clostridium difficile is no longer just a nosocomial infection or an infection of adults. Int J Antimicrob Agents 33(Suppl 1):S42–45

26. DuPont HL, Garey K, Caeiro JP, Jiang ZD (2008) New advances in Clostridium difficile infection: changing epidemiology, diagnosis, treatment and control. Curr Opin Infect Dis 21(5):500–507

27. Cohen MB (2009) Clostridium difficile infections: emerging epidemiology and new treatments. J Pediatr Gastroenterol Nutr 48(Suppl 2):S63–65

28. McFarland LV, Beneda HW, Clarridge JE, Raugi GJ (2007) Implications of the changing face of

Clostridium difficile disease for health care practitioners. Am J Infect Control 35(4):237–253

29. Walker AS, Spiegelhalter D, Crook DW, Wyllie D, Morris J, Peto TE (2008) Fairness of financial penalties to improve control of Clostridium difficile. BMJ 337:a2097

30. Goldenberg SD, Price NM, Tucker D, Wade P, French GL (2011) Mandatory reporting and improvements in diagnosing Clostridium difficile infection: An incompatible dichotomy? J Infect 62(5):363–370

31. Lyras D, O'Connor JR, Howarth PM, Sambol SP, Carter GP, Phumoonna T, Poon R, Adams V, Vedantam G, Johnson S, Gerding DN, Rood JI (2009) Toxin B is essential for virulence of Clostridium difficile. Nature 458(7242):1176–1179

32. Kuehne SA, Cartman ST, Heap JT, Kelly ML, Cockayne A, Minton NP (2010) The role of toxin A and toxin B in Clostridium difficile infection. Nature 467(7316):711–713

33. Drudy D, Fanning S, Kyne L (2007) Toxin A-negative, toxin B-positive Clostridium difficile. Int J Infect Dis 11(1):5–10

34. Kyne L, Farrell RJ, Kelly CP (2001) Clostridium difficile. Gastroenterol Clin North Am 30(3):753–77

35. Wilcox MH, Planche T, Fang FC, Gilligan P (2010) What is the current role of algorithmic approaches for diagnosis of Clostridium difficile infection? J Clin Microbiol. 48(12):4347–4353

36. Sebaihia M, Wren BW, Mullany P, Fairweather NF, Minton N, Stabler R, Thomson NR, Roberts AP, Cerdeno-Tarraga AM, Wang H, Holden MT, Wright A, Churcher C, Quail MA, Baker S, Bason N, Brooks K, Chillingworth T, Cronin A, Davis P, Dowd L, Fraser A, Feltwell T, Hance Z, Holroyd S, Jagels K, Moule S, Mungall K, Price C, Rabinowitsch E, Sharp S, Simmonds M, Stevens K, Unwin L, Whithead S, Dupuy B, Dougan G, Barrell B, Parkhill J (2006) The multidrug-resistant human pathogen Clostridium difficile has a highly mobile, mosaic genome. Nat Genet 38(7):779–786

37. Stubbs SL, Brazier JS, O'Neill GL, Duerden BI (1999) PCR targeted to the 16S-23S rRNA gene intergenic spacer region of Clostridium difficile and construction of a library consisting of 116 different PCR ribotypes. J Clin Microbiol 37(2):461–463

38. Rupnik M (2001) How to detect Clostridium difficile variant strains in a routine laboratory. Clin Microbiol Infect 7(8):417–420

39. Tenover FC, Novak-Weekley S, Woods CW, Peterson LR, Davis T, Schreckenberger P, Fang FC, Dascal A, Gerding DN, Nomura JH, Goering RV, Akerlund T, Weissfeld AS, Baron EJ, Wong E, Marlowe EM, Whitmore J, Persing DH (2010) Impact of strain type on detection of toxigenic Clostridium difficile: comparison of molecular diagnostic and enzyme immunoassay approaches. J Clin Microbiol 48(10):3719–3724

40. Carroll KC (2011) Tests for the diagnosis of Clostridium difficile infection: The next generation. Anaerobe 17(4):170–174

41. Wilcox MH, Planche T Defining a testing algorithm to improve the laboratory diagnosis of CDI. http://www.hpa.org.uk/webc/HPAwebFile/HPAweb_C/1317132979562

42. Luna RA, Boyanton BL, Jr, Mehta S, Courtney EM, Webb CR, Revell PA, Versalovic J (2011) Rapid Stool-Based Diagnosis of Clostridium difficile Infection by Real-Time PCR in a Children's Hospital. J Clin Microbiol 49(3):851–857

43. Wheeler JG, Sethi D, Cowden JM, Wall PG, Rodrigues LC, Tompkins DS, Hudson MJ, Roderick PJ (1999) Study of infectious intestinal disease in England: rates in the community, presenting to general practice, and reported to national surveillance. The Infectious Intestinal Disease Study Executive. BMJ 318(7190):1046–1050

44. Sandora TJ, Fung M, Flaherty K, Helsing L, Scanlon P, Potter-Bynoe G, Gidengil CA, Lee GM (2011) Epidemiology and risk factors for Clostridium difficile infection in children. Pediatr Infect Dis J 30(7):580–584

45. Boenning DA, Fleisher GR, Campos JM, Hulkower CW, Quinlan RW (1982) Clostridium difficile in a pediatric outpatient population. Pediatr Infect Dis 1(5):336–338

46. Kim K, DuPont HL, Pickering LK (1983) Outbreaks of diarrhea associated with Clostridium difficile and its toxin in day-care centers: evidence of person-to-person spread. J Pediatr 102(3):376–382

47. Oguz F, Uysal G, Dasdemir S, Oskovi H, Vidinlisan S (2001) The role of Clostridium difficile in childhood nosocomial diarrhea. Scand J Infect Dis 33(10):731–733

48. Rexach CE, Tang-Feldman YJ, Cantrell MC, Cohen SH (2006) Epidemiologic surveillance of Clostridium difficile diarrhea in a freestanding pediatric hospital and a pediatric hospital at a university medical center. Diagn Microbiol Infect Dis 56(2):109–114

49. Pokorn M, Radsel A, Cizman M, Jereb M, Karner P, Kalan G, Grosek S, Andlovic A, Rupnik M (2008) Severe Clostridium difficile-associated disease in children. Pediatr Infect Dis J 27(10):944–946

50. Wilson KH (1993) The microecology of Clostridium difficile. Clin Infect Dis 16(Suppl 4):S214–218

51. Deneve C, Janoir C, Poilane I, Fantinato C, Collignon A (2009) New trends in Clostridium difficile virulence and pathogenesis. Int J Antimicrob Agents 33(Suppl 1):S24–28

52. Toltzis P, Kim J, Dul M, Zoltanski J, Smathers S, Zaoutis T (2009) Presence of the epidemic North American Pulsed Field type 1 Clostridium difficile strain in hospitalized children. J Pediatr 154(4):607–608

53. Wongwanich S, Pongpech P, Dhiraputra C, Huttayananont S, Sawanpanyalert P (2001) Charac-

teristics of Clostridium difficile strains isolated from asymptomatic individuals and from diarrheal patients. Clin Microbiol Infect 7(8):438–441

54. Ferreira CE, Nakano V, Durigon EL, Avila-Campos MJ (2003) Prevalence of Clostridium spp. and Clostridium difficile in children with acute diarrhea in Sao Paulo city, Brazil. Mem Inst Oswaldo Cruz 98(4):451–454

55. Karsch W, Strelau E, Grahlow WD, Fischer E, Schulz R (1989) Occurrence and significance of Clostridium difficile in faecal specimens of hospitalized children. Zentralbl Bakteriol Mikrobiol Hyg A 270(3):441–448

56. Langley JM, LeBlanc JC, Hanakowski M, Goloubeva O (2002) The role of Clostridium difficile and viruses as causes of nosocomial diarrhea in children. Infect Control Hosp Epidemiol 23(11):660–664

57. Al-Jumaili IJ, Shibley M, Lishman AH, Record CO (1984) Incidence and origin of Clostridium difficile in neonates. J Clin Microbiol 19(1):77–78

58. Zedd AJ, Sell TL, Schaberg DR, Fekety FR, Cooperstock MS (1984) Nosocomial Clostridium difficile reservoir in a neonatal intensive care unit. Pediatr Infect Dis 3(5):429–432

59. Delmee M, Verellen G, Avesani V, Francois G (1988) Clostridium difficile in neonates: serogrouping and epidemiology. Eur J Pediatr 147(1):36–40

60. Martirosian G, Kuipers S, Verbrugh H, van Belkum A, Meisel-Mikolajczyk F (1995) PCR ribotyping and arbitrarily primed PCR for typing strains of Clostridium difficile from a Polish maternity hospital. J Clin Microbiol 33(8):2016-2021

61. el-Mohandes AE, Keiser JF, Refat M, Jackson BJ (1993) Prevalence and toxigenicity of Clostridium difficile isolates in fecal microflora of preterm infants in the intensive care nursery. Biol Neonate 63(4):225–229

62. Enad D, Meislich D, Brodsky NL, Hurt H (1997) Is Clostridium difficile a pathogen in the newborn intensive care unit? A prospective evaluation. J Perinatol 17(5):355–359

63. Vesikari T, Isolauri E, Maki M, Gronroos P (1984). Clostridium difficile in young children. Association with antibiotic usage. Acta Paediatr Scand 73(1):86–91

64. Zilberberg MD, Shorr AF, Kollef MH (2008) Increase in Clostridium difficile-related hospitalizations among infants in the United States, 2000-2005. Pediatr Infect Dis J 27(12):1111–1113

65. Zwiener RJ, Belknap WM, Quan R (1989) Severe pseudomembranous enterocolitis in a child: case report and literature review. Pediatr Infect Dis J 8(12):876–882

66. Qualman SJ, Petric M, Karmali MA, Smith CR, Hamilton SR (1990) Clostridium difficile invasion and toxin circulation in fatal pediatric pseudomembranous colitis. Am J Clin Pathol 94(4):410–416

67. Harris PR, Figueroa-Colon R (1995) Rectal prolapse in children associated with Clostridium difficile infection. Pediatr Infect Dis J 14(1):78–80

68. Gaglani MJ, Murray JC, Morad AB, Edwards MS (1996) Chronic osteomyelitis caused by Clostridium difficile in an adolescent with sickle cell disease. Pediatr Infect Dis J 15(11):1054–1056

69. Durand CL, Miller PF (2009) Severe Clostridium difficile colitis and reactive arthritis in a 10-year-old child. Pediatr Infect Dis J 28(8):750–751

70. Loffler HA, Pron B, Mouy R, Wulffraat NM, Prieur AM (2004) Clostridium difficile-associated reactive arthritis in two children. Joint Bone Spine 71(1):60–62

71. Clinical guideline 84 (CG84) (2009) Diarrhoea and vomiting in children under 5: full guideline. National Institute for Health and Clinical Excellence, London

72. Singer DB, Cashore WJ, Widness JA, Campognone P, Hillemeier C (1986) Pseudomembranous colitis in a preterm neonate. J Pediatr Gastroenterol Nutr 5(2):318–320

73. Adler SP, Chandrika T, Berman WF (1981) Clostridium difficile associated with pseudomembranous colitis. Occurrence in a 12-week-old infant without prior antibiotic therapy. Am J Dis Child 135(9):820–822

74. Burgner D, Siarakas S, Eagles G, McCarthy A, Bradbury R, Stevens M (1997) A prospective study of Clostridium difficile infection and colonization in pediatric oncology patients. Pediatr Infect Dis J 16(12):1131–1134

75. Tullus K, Aronsson B, Marcus S, Mollby R (1989) Intestinal colonization with Clostridium difficile in infants up to 18 months of age. Eur J Clin Microbiol Infect Dis 8(5):390–393

76. Brady MT, Pacini DL, Budde CT, Connell MJ (1989) Diagnostic studies of nosocomial diarrhea in children: assessing their use and value. Am J Infect Control 17(2):77–82

77. Schuller I, Saha V, Lin L, Kingston J, Eden T, Tabaqchali S (1995) Investigation and management of Clostridium difficile colonisation in a paediatric oncology unit. Arch Dis Child 72(3):219–222

78. Ferroni A, Merckx J, Ancelle T, Pron B, Abachin E, Barbut F, Larzul J, Rigault P, Berche P, Gaillard JL (1997) Nosocomial outbreak of Clostridium difficile diarrhea in a pediatric service. Eur J Clin Microbiol Infect Dis 16(12):928–933

79. Gogate A, De A, Nanivadekar R, Mathur M, Saraswathi K, Jog A, Kulkarni MV (2005) Diagnostic role of stool culture & toxin detection in antibiotic associated diarrhoea due to Clostridium difficile in children. Indian J Med Res 122(6):518–524

80. Pascarella F, Martinelli M, Miele E, Del Pezzo M, Roscetto E, Staiano A (2009) Impact of Clostridium difficile infection on pediatric inflammatory bowel disease. J Pediatr 154(6):854–858

81. Markowitz JE, Brown KA, Mamula P, Drott HR, Piccoli DA, Baldassano RN (2001) Failure of single-toxin assays to detect clostridium difficile infection in pediatric inflammatory bowel disease. Am J Gastroenterol 96(9):2688–2690

82. Issa M, Ananthakrishnan AN, Binion DG (2008) Clostridium difficile and inflammatory bowel disease. Inflamm Bowel Dis 14(10):1432–1442

83. Parsons SJ, Fenton E, Dargaville P (2005) Clostridium difficile associated severe enterocolitis: a feature of Hirschsprung's disease in a neonate presenting late. J Paediatr Child Health 41(12):689–690

84. Thomas DF, Fernie DS, Bayston R, Spitz L, Nixon HH (1986) Enterocolitis in Hirschsprung's disease: a controlled study of the etiologic role of Clostridium difficile. J Pediatr Surg 21(1):22–25

85. Welkon CJ, Long SS, Thompson CM Jr Gilligan PH (1985). Clostridium difficile in patients with cystic fibrosis. Am J Dis Child 139(8):805–808

86. Yahav J, Samra Z, Blau H, Dinari G, Chodick G, Shmuely H (2006) Helicobacter pylori and Clostridium difficile in cystic fibrosis patients. Dig Dis Sci 51(12):2274–2279

87. Peach SL, Borriello SP, Gaya H, Barclay FE, Welch AR (1986) Asymptomatic carriage of Clostridium difficile in patients with cystic fibrosis. J Clin Pathol 39(9):1013–1018

88. Rivlin J, Lerner A, Augarten A, Wilschanski M, Kerem E, Ephros MA (1998) Severe Clostridium difficile-associated colitis in young patients with cystic fibrosis. J Pediatr 132(1):177–179

89. Binkovitz LA, Allen E, Bloom D, Long F, Hammond S, Buonomo C, Donnelly LF 1999 Atypical presentation of Clostridium difficile colitis in patients with cystic fibrosis. AJR Am J Roentgenol 172(2):517–521

90. Barker HC, Haworth CS, Williams D, Roberts P, Bilton D (2008) Clostridium difficile pancolitis in adults with cystic fibrosis. J Cyst Fibros 7(5):444–447

91. Yates B, Murphy DM, Fisher AJ, Gould FK, Lordan JL, Dark JH, Corris PA (2007) Pseudo-membranous colitis in four patients with cystic fibrosis following lung transplantation. Thorax 62(6):554–556

92. Theunissen C, Knoop C, Nonhoff C, Byl B, Claus M, Liesnard C, Estenne MJ, Struelens MJ, Jacobs F (2008) Clostridium difficile colitis in cystic fibrosis patients with and without lung transplantation. Transpl Infect Dis 10(4):240–244

93. Castagnola E, Battaglia T, Bandettini R, Caviglia I, Baldelli I, Nantron M, Moroni C, Garaventa A (2009) Clostridium difficile-associated disease in children with solid tumors. Support Care Cancer 17(3):321–324

94. McFarland LV (2005) Alternative treatments for Clostridium difficile disease: what really works? J Med Microbiol 54(Pt 2):101–111

95. Bauer MP, van Dissel JT (2009) Alternative strategies for Clostridium difficile infection. Int J Antimicrob Agents 33(Suppl 1):S51–56

96. Pillai A, Nelson R (2008) Probiotics for treatment of Clostridium difficile-associated colitis in adults. Cochrane Database Syst Rev 23(1):CD004611

97. McFarland LV, Surawicz CM, Greenberg RN, Fekety R, Elmer GW, Moyer KA, Melcher SA, Bowen KE, Cox JL, Noorani Z et al (1994) A randomized placebo-controlled trial of Saccharomyces boulardii in combination with standard antibiotics for Clostridium difficile disease. JAMA 271(24):1913–1918

98. Bryant K, McDonald LC (2009) Clostridium difficile infections in children. Pediatr Infect Dis J 28(2):145–146

99. Bland JM, Altman DG (2000) Statistics notes. The odds ratio. BMJ 320(7247):1468

100. Stamper PD, Alcabasa R, Aird D, Babiker W, Wehrlin J, Ikpeama I, Carroll KC (2009) Comparison of a commercial real-time PCR assay for tcdB detection to a cell culture cytotoxicity assay and toxigenic culture for direct detection of toxin-producing Clostridium difficile in clinical samples. J Clin Microbiol 47(2):373–378

# Diarrhea Among Children in Developing Countries

**5**

James P. Nataro

## Abstract

Diarrhea continues to stand among the most important causes of global morbidity and mortality in children under 5 years of age. Although the introduction of oral rehydration and other case-management strategies have reduced acute diarrhea fatalities, many of the survivors develop persistent diarrhea and/or deficiencies of growth and cognition. Thus understanding the true global burden of diarrhea requires attention to acute diarrhea as well is its sequelae. To understand the etiology of moderate to severe diarrhea among children in high mortality areas of sub-Saharan Africa and south Asia we performed a comprehensive case-control study of children under 5 years of age at seven sites. Each site employed an identical case-control study design and each utilized a uniform comprehensive set of microbiological assays to identify the likely bacterial, viral and protozoal etiologies. Results of the studies will inform diarrhea prevention and management efforts worldwide.

## 5.1 Introduction

Diarrhea as a clinical complaint has been with hominids long before our species emerged, but in man's most natural state, infectious diarrhea may not have been a major health problem. Among small bands and tribes, infectious agents could have spread through the local population, engendered widespread immunity, then submerged to a subclinical existence, or perhaps failed to be sustained at all. With the rise of human cities, however, enteric pathogens may have found an opportunity for persistence, given continual replenishment of susceptible individuals, along with new opportunities provided by sewage systems and close proximity to domesticated livestock.

But historical records suggest that diarrhea emerged as a global scourge with widespread confluence of human populations brought about by colonization and frequent inter-continental transportation. Cholera spread dramatically across the globe in the ninteenth century in a series of pandemics [1]. No civilized area

J. P. Nataro (✉)
Department of Pediatrics, University of Virginia School of Medicine, University of Virginia Children's Hospital, Charlottesville, USA
e-mail: JPN2R@hscmail.mcc.virginia.edu

N. Curtis et al. (eds.), *Hot Topics in Infection and Immunity in Children IX,*
Advances in Experimental Medicine and Biology 764, DOI 10.1007/978-1-4614-4726-9_5,
© Springer Science+Business Media New York 2013

remained untouched. At the same time, cholera in the headlines drove breakthroughs in science and medicine from which we daily benefit. John Snow's investigations into the source of London cholera epidemics founded the science of epidemiology [2]. Robert Koch's seminal investigations, coupled with the infamous opposition by von Pettenkofer [3] and others, drove the final nails in the coffin of the miasma theory and established modern microbiology.

Realization that cholera derived from a microorganism spread through sewerage eventually drove development of modern urban sanitation systems. Sanitation finally stemmed the tide: deaths due to enteric diseases in industrialized cities declined dramatically around the turn of the twentieth century, long before we knew of antibiotics or oral rehydration therapy [4]. Unfortunately, however, while cities in the industrialized world reaped the benefits of cleaner living, the sprawling metropolae of the under-developed world continued to provide ripe opportunities for propagation of enteric diseases. In many parts of the world, this heritage persists.

## 5.2 Toward Effective Control of the Global Diarrhea Burden

The legacy of cholera also provided the first breakthrough in diarrhea therapy. In the 1950s and 1960s, physiologists described the phenomenon of coupled sodium-glucose co-transport across the mammalian intestine [5]. In the presence of glucose, animal perfusion models showed dramatically improved sodium uptake. Capt. Robert Phillips and others quickly exploited this fundamental breakthrough [6]. Working in Bangladesh (while others in Calcutta were making similar strides), Phillips' team demonstrated that stool output of cholera patients was substantially reduced when oro-gastric sodium chloride infusions included glucose [7]. With astonishing rapidity, this observation led to the miracle of oral rehydration therapy (ORT) [8].

ORT was propagated across the face of the globe throughout the decades of the 1970s, 1980s and 1990s, with remarkable success. Everywhere

ORT reached, mortality rates from dehydrating diarrhea declined. Global data reflected the local effects. From an estimated 5 million deaths per year attributable to diarrhea among children under 5 years globally in 1980, this rate fell steadily to below 2 million, and data from the Child Health Research Group (CHERG) place the number of diarrhea-related deaths at 1.32 million in 2008 [9]. This still represents ca. 18 % of all deaths among children under 5 years (Fig. 5.1).

After rapid decline soon after the introduction of ORT, the rate of decline of diarrhea-related deaths has plateaued. Several causes can be postulated. First, the vigor accompanying early ORT implementation programs could not be sustained indefinitely, in view of new public health priorities. In addition, many enteric infections may be intrinsically refractory to ORT, or perhaps their pathophysiologic assault may involve derangements other than dehydration. Studies focused on the etiologic agents of childhood diarrhea now suggest that the story is complex.

## 5.3 Pathogenesis of Infectious Diarrhea

The fundamental derangement in human diarrhea is the reprogramming of the gastrointestinal tract from a net absorptive organ to one that eliminates feces with abnormally high water content. Molecular mechanisms may involve up-regulation of secretion channels and pumps, reduction of absorptive functions, and/or erosion of intestinal barrier function. Remarkably, pathogenesis research suggests that most enteric pathogens effect more than one (quite often all three) of these assaults. Enteropathogenic E. coli (EPEC), a common pathogen of infants and toddlers in developing countries, provides a prominent example (reviewed in references [10–13]) (Fig. 5.2). EPEC attaches tightly to the brush border of small intestinal enterocytes, and injects a set of protein toxins directly into the cellular cytoplasm. The net effects of these toxins are astonishing, and only partly understood [14]. A subset of the proteins trigger signal transduction pathways in the afflicted cell, resulting in

**Fig. 5.1** Major causes of death among children less than 5 years of age worldwide, 2008. (From [9])

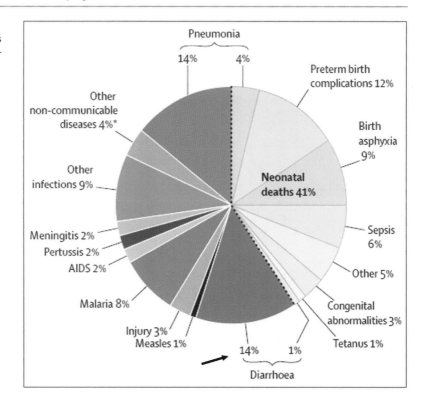

**Fig. 5.2** Pathogenetic effects of enteropathogenic *E. coli* infection on small bowel epithelial cells. Attachment of the pathogen to the apical membrane of enterocytes results in decreased absorptive and enhanced secretory mechanisms, along with impairment of normal barrier function. (Adapted from [13])

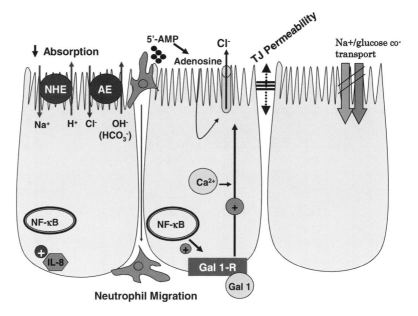

release of pro-inflammatory cytokines, including the chemokine IL-8, though pro-inflammatory responses are finely modulated [12]. Release of these cytokines results in recruitment of leukocytes to the tissue, leading to tight junction opening and tissue damage. Emerging evidence also suggests that some of the EPEC effector proteins may have direct pathologic effects on cellular absorptive and secretion channels in the apical membrane. Perhaps more remarkably, several proteins act in synergistic fashion to open tight junctions and stimulate leakage of fluid via the

**Fig. 5.3** The cycle of malnutrition and enteric infection. (Redrawn from [18])

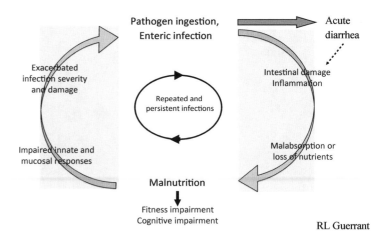

intercellular junctions [13]. One of the more intriguing effects of EPEC intoxication is the inhibition of the sodium-glucose co-transport apparatus [15]. It can be envisioned that damage to the epithelial layer, loss of absorptive driving force, and reduced function of the sodium-glucose contransporter could render EPEC infection relatively resistant to the effects of ORT. This hypothesis requires clinical validation, but there can be little doubt that the impact of some enteric pathogens extends well beyond loss of water and salts.

## 5.4 Diarrhea, Growth and Cognition

The notion that diarrhea does more than simply dry out the host has gained increasing support from epidemiologic and clinical studies. Studies in Fortaleza Brazil and elsewhere show clearly that patients with diarrhea persisting longer than 7–14 days demonstrate shortfalls in growth and cognitive development, as well as increased mortality that may be due to factors other than diarrhea per se. Guerrant and others have accordingly suggested that enteric infection may trigger a vicious cycle in which damage to the intestinal mucosa leads to malabsorption of nutrients that are required for growth, cognitive development, maintenance of gut integrity and immune function [16, 17]. This relationship has complex and potentially devastating implications. Fig. 5.3 illustrates the diarrhea vicious

cycle: infection leads to malabsorption, which predisposes to further infection. Each individual infection exacts its toll in terms of growth and development; eventually, the ability of the child to "catch up" is lost. The effect has been documented since the 1970s, starting with the famous studies of Leonardo Mata in Guatemala [18]. In growth curves of children in Mata's study villages, he observed growth flattening accompanying each infection, followed by catch up growth that is increasingly meager as the child ages. The result is a child who is stunted, and likely abnormal in many more subtle ways. And as we rescue more and more of the word's children from acute diarrhea-related death, this silent toll of diarrhea and malnutrition may emerge as the principle threat of enteric infection on human health.

## 5.5 The Global Enteric Multisite Study (GEMS)

The urgency to understand the persistence of diarrhea-related death among the world's poorest children has motivated the Bill & Melinda Gates Foundation to embark on the largest diarrhea case-control study every attempted. The study, coordinated by Dr. Myron M. Levine at the University of Maryland School of Medicine, has implemented an identical case-control design at seven sites in sub-saharan Africa and south Asia (Table 5.1). For a period of 3 years, the study enrolls children under 5 years of age who pres-

**Table 5.1** Sites represented in the global enteric multi-site research study

| Country | Site | Pincipal investigator | Characteristics |
|---|---|---|---|
| The Gambia | Medical Research Center, Basse | Richard Adegbola | Rural, low HIV prevalence |
| Mali | Center for Vaccine Development, Bamako | Samba Sow | Urban, low HIV prevalence |
| Kenya | Centers for Disease Control, Kenya Medical Research Institute | Robert Breiman | Rural, high HIV prevalence |
| Mozambique | Centro de Investigaçao em Saude da Manhiça | Pedro Alonso | Suburban, high HIV prevalence |
| Pakistan | Aga Khan University Hospital, Karachi | Anita Zaidi | Urban coastal |
| India | National Institute of Cholera and Enteric Diseases, Kolkata | Sujit Bhattacharya | Urban slum |
| Bangladesh | International Center for Diarrheal Disease Research, Dhaka. (Mirzapur field site) | ASM Faruque | Rural and town |

ent to sentinel health centers with a complaint of acute moderate-to-severe diarrhea, defined as diarrhea within 7 days of onset accompanied by clinical dehydration, fecal blood, need for intravenous hydration, and/or admission to a medical facility. For each of the subjects enrolled, a matched control is recruited from the subject's village.

Each GEMS case and control is studied in comprehensive fashion. A detailed questionnaire is administered to elucidate risk factors for enteric infection. Precise anthropometric data are acquired. A stool sample is collected and analyzed comprehensively for the major enteric pathogens, using validated microbiological methods. A household visit is conducted at 60 days post-enrollment for both cases and controls to ascertain survival, length of diarrhea, and growth during the follow-up period. Results of the study are expected in 2012.

## 5.6  The Future of Diarrhea Control

Future diarrhea prevention and therapy strategies must remain focused on the last ca. 1 million diarrhea-related deaths among children under 5 years of age, but also on diarrhea-related morbidities. The most fruitful approach to avoiding diarrhea-related deaths will require a multi-pronged approach. Rotavirus remains the most common cause of dehydrating diarrhea, and implementation of the new rotavirus vaccines has become an urgent global health priority. Successful deploy-

ment of the rotavirus vaccines to the world's poorest will save many thousands of lives, though the precise impact will depend on complex epidemiologic and political factors [19, 20]. Beyond the rotavirus vaccine, it is essential to elucidate other specific etiologic agents that may be over-represented among fatal cases. These pathogens may become amenable to vaccines or control strategies targeted at their modes of transmission. The GEMS study should provide vital illumination of these important microorganisms. GEMS may also reveal information to guide sanitation and hygiene programs, particularly important in areas where pathogen-specific interventions are impractical.

Along with prevention of diarrhea-related deaths, we must become increasingly attuned to the silent morbidity attending diarrhea. Provision of zinc supplementation to children with diarrhea has already been shown to diminish the duration of diarrhea episodes. Providing anticipatory zinc to at risk populations may provide significant benefits in terms of diarrhea related morbidity and mortality [16, 21–31]. Several studies have examined the provision of zinc along with other nutrients (e.g. vitamin A), and inclusion of zinc in oral rehydration salts may be a practical way to target populations in greatest need [21, 23, 25, 30].

Understanding the effects of diarrhea on human cognition will become increasingly important as we prevent death and growth shortfalls. The studies in Fortaleza reveal not only that children with diarrhea are more likely to mani-

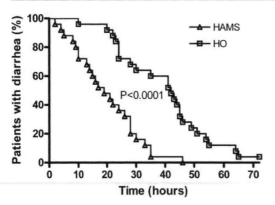

**Fig. 5.4** Compared with standard ORS, high amylase maize starch (HAMS-ORS) reduced diarrhea duration by 55 % and significantly reduced fecal weight after the first 12 h of ORS therapy in adults with cholera-like diarrhea. (From [36])

**Table 5.2** Seven point plan for diarrhea control

| Package | Treatment and prevention |
|---|---|
| Treatment package | Fluid replacement |
| | Zinc treatment |
| Prevention package | Rotavirus and measles vaccination |
| | Promotion of early and exclusive breast-feeding and vitamin A supplementation |
| | Promotion of handwashing with soap |
| | Improved water supply quantity and quality, including treatment and safe storage of household water |
| | Community-wide sanitation promotion |

United Nations Children's Fund/World Health Organization, 2009

fest cognitive disability, but that there may be a genetic predisposition to this adverse outcome [32]. These studies need to be replicated more widely, and the relative contribution of predisposition vs. insult needs to be elucidated. Beyond genetic predisposition, it will be important to understand the roles of intercurrent infections and the intestinal microbiome in conferring susceptibility to enteric-related morbidity and mortality.

A great deal of research remains devoted to improving oral rehydration. Oral rehydration salts (ORS) with low osmolarity [33], complex starch [34], amino acids [35], and other additives may ultimately provide a material improvement in our ability to avoid adverse outcomes in children who can receive these interventions. Ramakrishna et al. [36] conducted a prospective randomized clinical trial at a tertiary referral hospital in southern India to compare the effects of low osmolarity ORS incorporating amylase-resistant starch instead of glucose (Fig. 5.4). These investigators reported significantly shorter duration of diarrhea in adult males given the maize-based ORT.

Integration of multiple interventions is likely to produce the greatest benefit in terms of lives saved [37]. However, each additional intervention adds complexity to the problems of deployment. Despite decades of international effort, only an estimated 38 % of the world's poorest children receive ORS when they develop diarrhea [38]. The causes for this are many, and

include lack of delivery infrastructure, lack of resources for purchase and dissemination, and focus on other public health priorities. It is likely that developing countries will be able to leverage ORS dissemination programs to introduce other interventions against enteric diseases, providing programmatic synergy.

Fischer-Walker et al. [37] have recently applied a Lives Saved Tool (LiST) analysis to estimate the impact of various interventions in the management of diarrhea. The authors analysed the impact of implementing the following interventions in 68 high child mortality countries: ORS, zinc, antibiotics for dysentery, rotavirus vaccine, vitamin A supplementation and breastfeeding promotion, in addition to basic water, sanitation and hygiene. They found that diarrhea mortality would be reduced by 78 % and 92 %, respectively under an "ambitious" (feasible improvement in coverage of all interventions) and "universal" (assumes near 100 % coverage of all interventions) scale-up impact scenarios. With universal coverage, nearly 5 million diarrheal deaths could be averted during the 5-year scale-up period for an additional cost of US$ 12.5 billion invested across 68 priority countries for individual-level prevention and treatment interventions, and an additional US$ 84.8 billion would be required for the addition of all water and sanitation interventions.

Following similar analyses, UNICEF has developed a comprehensive seven-point plan for diarrhea control (Table 5.2). Given the recognition that the burden of enteric disease extends far beyond death from dehydration, success in the UNICEF plan may realize dramatic improvements in child health.

Perhaps the final frontier in enteric pathogenesis research will be understanding the true cost of exposure to enteric pathogens. Studies in Guinea-Bissau [39] have documented frequent exposure of infants and children to enteric pathogens, with only a minority of the exposures resulting in frank diarrhea. But this observation compels us to ask the key question: Is asymptomatic exposure truly benign? It is well-established that children in developing countries exhibit increased intestinal permeability when compared to their counterparts in industrialized nations [40]. Is this the result of repeated subclinical exposure to pathogens? If so, what is the true cost of this malady?

Many important questions remain in the fight against enteric disease among the world's poorest children. But perhaps for the first time, there exists a confluence of technology that will to bring to bear effective weapons. This effort will require continued application of energy and resources to benefit the majority of the world's children.

# References

1. Lacey SW (1995) Cholera: calamitous past, ominous future. Clin Infect Dis 20(5):1409–1419
2. Buechner JS, Constantine H, Gjelsvik A (2004) John Snow and the Broad Street pump: 150 years of epidemiology. Med Health R I 87(10):314–315
3. Von PM (1873) What We can Do against Cholera: Practical Instructions concerning what to Do to Prevent an Epidemic as Well as How to Guard against it during its Prevalence. Public Health Pap Rep 1:317–335
4. Macassa G, De Leon AP, Burstrom B (2006) The impact of water supply and sanitation on area differentials in the decline of diarrhoeal disease mortality among infants in Stockholm 1878–1925. Scand J Public Health 34(5):526–533
5. Farthing MJ (1988) History and rationale of oral rehydration and recent developments in formulating an optimal solution. Drugs 36(Suppl 4):80–90
6. Savarino SJ (2002) A legacy in 20th-century medicine: Robert Allan Phillips and the taming of cholera. Clin Infect Dis 35(6):713–720
7. Hirschhorn N, Kinzie JL, Sachar DB, Northrup RS, Taylor JO, Ahmad SZ et al (1968) Decrease in net stool output in cholera during intestinal perfusion with glucose-containing solutions. N Engl J Med 279(4):176–181
8. Farthing MJ (1994) Oral rehydration therapy. Pharmacol Ther 3:477–492
9. Black RE, Cousens S, Johnson HL, Lawn JE, Rudan I, Bassani DG et al (2010) Global, regional, and national causes of child mortality in 2008: a systematic analysis. Lancet 375(9730):1969–1987
10. Donnenberg MS (1995) Enteropathogenic Escherichia coli. In: Blaser MJ, Smith PD, Ravdin JI, Greenberg HB, Guerrant RL (eds) Infections of the Gastrointestinal Tract. Raven, New York, p 709–726
11. Esmaili A, Nazir SF, Borthakur A, Yu D, Turner JR, Saksena S et al (2009) Enteropathogenic Escherichia coli infection inhibits intestinal serotonin transporter function and expression. Gastroenterology 137(6):2074–2083
12. Royan SV, Jones RM, Koutsouris A, Roxas JL, Falzari K, Weflen AW et al (2010) Enteropathogenic E. coli non-LEE encoded effectors NleH1 and NleH2 attenuate NF-kappaB activation. Mol Microbiol 78(5):1232–1245
13. Weflen AW, Alto NM, Hecht GA (2009) Tight junctions and enteropathogenic E. coli. Ann N Y Acad Sci 1165:169–174
14. Caron E, Crepin VF, Simpson N, Knutton S, Garmendia J, Frankel G (2006) Subversion of actin dynamics by EPEC and EHEC. Curr Opin Microbiol 9(1):40–45
15. Dean P, Maresca M, Schuller S, Phillips AD, Kenny B (2006) Potent diarrheagenic mechanism mediated by the cooperative action of three enteropathogenic Escherichia coli-injected effector proteins. Proc Natl Acad Sci USA 103(6):1876–1881
16. Guerrant R, Schorling J, McAuliffe J, Souza Md (1992) Diarrhea as a cause and effect of malnutrition: Diarrhea prevents catch-up growth and malnutrition increases diarrhea frequency and duration. AJTMH 47(Suppl 1):28–35
17. Petri WA Jr, Miller M, Binder HJ, Levine MM, Dillingham R, Guerrant RL (2008) Enteric infections, diarrhea, and their impact on function and development. J Clin Invest 118(4):1277–1290
18. Guerrant RL, Oria RB, Moore SR, Oria MO, Lima AA (2008) Malnutrition as an enteric infectious disease with long-term effects on child development. Nutr Rev 66(9):487–505
19. Pitzer VE, Patel MM, Lopman BA, Viboud C, Parashar UD, Grenfell BT (2011) Modeling rotavirus strain dynamics in developed countries to understand the potential impact of vaccination on genotype distributions. Proc Natl Acad Sci USA
20. Grimwood K, Lambert SB, Milne RJ (2010) Rotavirus infections and vaccines: burden of illness and potential impact of vaccination. Paediatr Drugs 12(4):235–256
21. Telmesani AM (2010) Oral rehydration salts, zinc supplement and rota virus vaccine in the management of childhood acute diarrhea. J Family Community Med 17(2):79–82
22. Dutta P, Mitra U, Dutta S, Naik TN, Rajendran K, Chatterjee MK (2011) Zinc, vitamin A, and micronutrient supplementation in children with

diarrhea: a randomized controlled clinical trial of combination therapy versus monotherapy. J Pediatr 159(4):633–637

23. Frohna JG (2011) Oral rehydration solution with zinc and prebiotics decreases duration of acute diarrhea in children. J Pediatr 159(1):166–167

24. Bajait C, Thawani V (2011) Role of zinc in pediatric diarrhea. Indian J Pharmacol 43(3):232–235

25. Walker CL, Black RE (2010) Zinc for the treatment of diarrhoea: effect on diarrhoea morbidity, mortality and incidence of future episodes. Int J Epidemiol 39(Suppl 1):i63–i69

26. Alam DS, Yunus M, El AS, Chowdury HR, Larson CP, Sack DA et al (2011) Zinc treatment for 5 or 10 days is equally efficacious in preventing diarrhea in the subsequent 3 months among Bangladeshi children. J Nutr 141(2):312–315

27. Dalgic N, Sancar M, Bayraktar B, Pullu M, Hasim O (2011) Probiotic, zinc and lactose-free formula in children with rotavirus diarrhea: are they effective? Pediatr Int 53(5):677–682

28. Yakoob MY, Theodoratou E, Jabeen A, Imdad A, Eisele TP, Ferguson J et al (2011) Preventive zinc supplementation in developing countries: impact on mortality and morbidity due to diarrhea, pneumonia and malaria. BMC Public Health 11(Suppl 3):S23

29. Patel AB, Mamtani M, Badhoniya N, Kulkarni H (2011) What zinc supplementation does and does not achieve in diarrhea prevention: a systematic review and meta-analysis. BMC Infect Dis 11:122

30. Wadhwa N, Natchu UC, Sommerfelt H, Strand TA, Kapoor V, Saini S et al (2011) ORS containing zinc does not reduce duration or stool volume of acute diarrhea in hospitalized children. J Pediatr Gastroenterol Nutr 53(2):161–167

31. Gitanjali B, Weerasuriya K (2011) The curious case of zinc for diarrhea: unavailable, unprescribed, and unused. J Pharmacol Pharmacother 2(4):225–229

32. Oria RB, Patrick PD, Zhang H, Lorntz B, de Castro Costa CM, Brito GA et al (2005) APOE4 protects the cognitive development in children with heavy diarrhea burdens in Northeast Brazil. Pediatr Res 57(2):310–316

33. CHOICE Study Group (2001) Multicenter, randomized, double-blind clinical trial to evaluate the efficacy and safety of a reduced osmolarity oral rehydration salts solution in children with acute watery diarrhea. Pediatrics 107(4):613–618

34. Suh JS, Hahn WH, Cho BS (2010) Recent Advances of Oral Rehydration Therapy (ORT). Electrolyte Blood Press 8(2):82–86

35. Alam NH, Raqib R, Ashraf H, Qadri F, Ahmed S, Zasloff M et al (2011) L-isoleucine-supplemented oral rehydration solution in the treatment of acute diarrhoea in children: a randomized controlled trial. J Health Popul Nutr 29(3):183–190

36. Ramakrishna BS, Subramanian V, Mohan V, Sebastian BK, Young GP, Farthing MJ et al (2008) A randomized controlled trial of glucose versus amylase resistant starch hypo-osmolar oral rehydration solution for adult acute dehydrating diarrhea. PLoS One 3(2):e1587

37. Fischer Walker CL, Friberg IK, Binkin N, Young M, Walker N, Fontaine O et al (2011) Scaling up diarrhea prevention and treatment interventions: a Lives Saved Tool analysis. PLoS Med 8(3):e1000428

38. Bryce J, Terreri N, Victora CG, Mason E, Daelmans B, Bhutta ZA et al (2006) Countdown to 2015: tracking intervention coverage for child survival. Lancet 368(9541):1067–1076

39. Valentiner-Branth P, Steinsland H, Fischer TK, Perch M, Scheutz F, Dias F et al (2003) Cohort study of Guinean children: incidence, pathogenicity, conferred protection, and attributable risk for enteropathogens during the first 2 years of life. J Clin Microbiol 41(9):4238–4245

40. Lunn PG, Northrop-Clewes CA, Downes RM (1991) Intestinal permeability, mucosal injury, and growth faltering in Gambian infants. Lancet 338(8772):907–910

# Building Confidence in Vaccines

Jennifer C. Smith, Mary Appleton
and Noni E. MacDonald

**6**

## Abstract

Despite significant efforts by governments, organizations and individuals to maintain public trust in vaccines, concerns persist and threaten to undermine the effectiveness of immunization programs. Vaccine advocates have traditionally focused on education based on evidence to address vaccine concerns and hesitancy. However, being informed of the facts about immunization does not always translate into support for immunization. While many are persuaded by scientific evidence, others are more influenced by cognitive shortcuts, beliefs, societal pressure and the media, with the latter group more likely to hesitate over immunization.

Understanding evidence from the behaviour sciences opens new doors to better support individual decision-making about immunization. Drawing on heuristics, this overview explores how individuals find, process and utilize vaccine information and the role health care professionals and society can play in vaccine decision-making.

Traditional, evidence-based approaches aimed at staunching the erosion of public confidence in vaccines are proving inadequate and expensive. Enhancing public confidence in vaccines will be complex, necessitating a much wider range of strategies than currently used. Success will require a shift in how the public, health care professionals and media are informed and educated about vaccine benefits, risks and safety; considerable introspection and change in current academic and vaccine decision-making practices; development of proactive strategies to broadly address current and potential future concerns, as well as targeted interventions such as programs to address pain with immunization. This overview outlines ten such opportunities for change to improve vaccine confidence.

N. E. MacDonald (✉) · J. C. Smith · M. Appleton
Pediatric Infectious Diseases, Dalhousie University,
Canadian Center for Vaccinology, IWK Health Center,
Halifax, Canada
e-mail: noni.macdonald@dal.ca

J. C. Smith
e-mail: jaurora777@yahoo.ca

M. Appleton
e-mail: mary.appleton@iwk.nshealth.ca

N. Curtis et al. (eds.), *Hot Topics in Infection and Immunity in Children IX,*
Advances in Experimental Medicine and Biology 764, DOI 10.1007/978-1-4614-4726-9_6,
© Springer Science+Business Media New York 2013

**Table 6.1** Influenza vaccination coverage among sample of 1931 health-care personnel in US in 2010–2011 by selected characteristics. (Adapted from [6])

| Work setting | Vaccinated (%) | Occupation | Vaccinated (%) | Age | Vaccinated (%) |
|---|---|---|---|---|---|
| Hospital | 71 | Physician or dentist | 84 | 18–29 | 56 |
| Long-term care | 64 | Nurse | 70 | 30–44 | 58 |
| Home health | 54 | Allied health professional | 64 | 45–59 | 69 |
| Other | 47 | Assistant/aide | 56 | ≥60 | 74 |

## 6.1 Introduction

Immunization is heralded as one of the most important public health achievements of the last century and without question, vaccines have saved millions of lives and increased life expectancy. However, despite robust scientific evidence showing vaccines to be effective and safe, public doubts persist as they have since the first vaccine efforts by Jenner over two centuries ago. Other factors beyond the availability of evidence are clearly at play [1], undermining immunization programs and the potential for control of vaccine-preventable diseases. This chapter expands on a previous article that outlined several factors undermining confidence in vaccines and that suggested several strategies for how vaccination advocates might increase support and uptake of vaccines [2].

Key topics explored include perception of vaccine risk, the influence of heuristics and beliefs, the impact of the disease control cycle on vaccine uptake, and a discussion of several other factors that influence vaccine confidence. Current strategies to address vaccine concerns are examined followed by an exploration of several different and new opportunities for staunching the erosion of public confidence in vaccines.

## 6.2 The Vaccine Confidence Deficit

Many factors are involved in the decision to immunize. If the public was fully supportive of vaccines, immunization rates would be well over 95 %. Such is not the case. Lack of confidence or doubt in the safety and effectiveness of vaccines often results in the decision not to immunize. Unfortunately, there are no half measures with

immunization, a person is either immunized or not. Not immunizing, even if this occurs in only a small segment of the population, can undermine the control of vaccine preventable diseases. Public trust in vaccines and immunization programs has to be built; it does not occur by chance and it is not a simple task.

Vaccine confidence is not just a problem for the general public. Even some health care providers remain unconvinced that vaccines are safe, effective, and necessary for health. Although trained to rely on scientific evidence for their decision-making, some also succumb to other influences and question the value of immunization [3, 4]. The persistent annual problem of only moderate uptake of yearly influenza vaccine among health care providers well illustrates this conundrum [5, 6]. Even the threat of an influenza pandemic in 2009 did not drive voluntary compliance rates for the H1N1 vaccine high enough among American health care providers to provide adequate workplace protection [5]. Only health care workplaces that mandated influenza vaccine tended to reach the recommended 90 % target for coverage among their healthcare providers [5]. As shown in Table 6.1, modified from the United States' Centers for Disease Control and Prevention report on influenza vaccination coverage among healthcare personnel for the 2010 influenza season [6], the overall influenza vaccine uptake among the 1931 health care providers sampled was 63.5 % but the rates varied widely by workplace setting, health care worker type, and age.

Vaccine uptake by the public remains suboptimal in many industrialized countries in spite of a wealth of educational materials and a plethora of campaigns and government-funded immunization programs that removed financial barriers to vaccines. With all of this support, one would have anticipated high uptake of vaccines supplied by governments. Not so. For example, despite easy access to professionally developed evidence-based educational materials on the risks of cervical cancer and the benefits and safety of the human papilloma virus (HPV) vaccine, a publicly-funded school-based HPV vaccine program for girls in British Columbia, Canada, reported only a 65 % uptake in 2008–2009 [7].

Unfortunately, to be truly effective in reducing the burden of circulating vaccine-preventable infections, high immunization rates are required across entire populations. For example, in the case of measles, >95 % of the population must be immune [8]. Outbreaks can occur if the virus is introduced by a visitor into populations <95 % immune, as has recently occurred in parts of Canada [9]. Garnering a behavior compliance rate of this magnitude across entire populations for different vaccine-preventable infections is a daunting task. Few, if any, public health measures require such a high rate of compliance for success. Nurturing confidence in vaccines is thus critical to achieve this goal [10]. Outbreaks in areas where vaccines are easily available but uptake is suboptimal have repeatedly shown just how crucial these high immunization rates are and not just for measles. The pertussis outbreaks in California in 2010–2011, again, illustrate the problem created by emerging pockets of low vaccine uptake [11].

Evidence and fact-based strategies in support of immunization have been used for decades with some success as most parents choose to have their children immunized and many health care providers receive the annual influenza vaccine. But some remain unconvinced and hesitate to immunize despite the evidence, professional recommendations, and easy access to vaccines. To achieve the high vaccine uptake rates required to control diseases such as measles, the unconvinced and vaccine-hesitant minority must also become convinced. The current vaccine promotion strategies are not sufficient to achieve the required vaccine uptake rates.

Among commonly cited reasons for hesitation to immunize are concerns about vaccine safety and the perception that immunization risks out-weigh the danger of vaccine-targeted diseases [12, 13]. A better understanding of how vaccine-related risks are perceived is crucial for the development of future programs for success. The expansion of vaccinology from health sciences thinking to include that of the behaviour sciences enhances our understanding of why and how vaccine concerns arise despite the obvious benefits of immunization.

### 6.2.1 Vaccine Perception of Risk: the Role of Heuristics

Many people take pride in making major decisions believing these are based upon conscious and methodical deliberations of available information and evidence. Behaviour scientists, however, have long known that humans are hardwired to deal with threatening situations with reflexive reactions called heuristics (i.e., the use of simple cognitive shortcuts to rapidly solve complex problems) [14, 15]. These automatic mental associations, already biased by previously accessed information and experience, subconsciously influence choices including decisions about health [16]. Whether a parent decides to immunize his or her child or a health care provider opts for annual influenza immunization, automatic associations unconsciously influence the decision-making process. While at a conscious level we may think we are deliberating options and feel undecided, on an unconscious level the decision has already been made. Table 6.2 provides some examples of heuristics and the role heuristics play in vaccine decision-making [2, 14, 16].

One way to think of heuristics is to liken the amalgamation of these reflexive mental associations to a selective brain filter (Fig. 6.1). Beliefs and emotions are on the inside and facts on the outside. While evidence that bolsters preconceived convictions easily passes through the filter, facts contrary to previously held beliefs are met with resistance. Building on the analogy, evidence that can be reshaped in accordance with already held beliefs eventually enters. However, facts that remain contrary to convictions are rejected and may, in fact, serve to strengthen

**Table 6.2** Heuristics and Vaccines. (Adapted from [2, 14, 16])

| Heuristic | Simplified definition | Resulting behaviour or belief |
| --- | --- | --- |
| Anchoring | Start from a known value or belief i.e. anchor<br><br>Judge probability of a future event by what occurred in past | Healthcare worker sees a serious adverse event following immunization (AEFI) e.g. anaphylaxis following HPV vaccine and now believes AEFI following HPV vaccine to be more common than it is |
| Availability | Easily imagined or recalled and therefore judged as frequent or likely to occur | Serious vaccine preventable diseases not seen or recalled, but multiple sclerosis following HBV vaccine misinformation recently heard or seen in the media is believed as correct and common—hence fear of HBV vaccine (France) |
| Co-incidence dragon | After this therefore because of this | All adverse events that follow immunization must be due to the vaccine. The possibility of coincidence is ignored e.g. autism following MMR vaccine |
| Compression | Over estimate rare occurrence, under-estimate frequent. Misunderstanding of sample size | Incidence of Guillain-Barré syndrome following influenza immunization perceived by health care provider as more common and serious than influenza complications |
| Free Loading | Herd immunity will provide protection, no need to immunize myself or my child | Parent believe it is safe to not immunize their child because (1) all other children are immunized (2) the disease is gone. No need for their child to take on "risk" of vaccine |
| Compression | Over estimate rare occurrence, under-estimate frequent. Misunderstanding of sample size | Incidence of Guillain-Barré syndrome following influenza immunization perceived by health care provider as more common and serious than influenza complications |
| Omission Bias | Taking action is seen as more harmful than inaction, therefore do nothing | Not immunizing is seen as safer than immunizing. More health care providers took seasonal vaccine than the H1N1 vaccine in 2009–2010 i.e. not to take "new" H1N1 vaccine safer than taking it |
| Over confidence | Faith in own judgments. Believe they know their own health risks. Think nothing bad could happen to them | Have never had serious influenza and therefore don't need to be immunized. "I am healthy, wash my hands, no need for vaccine" |

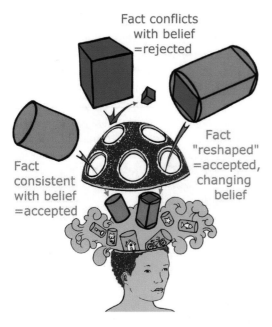

**Fig. 6.1** An illustration of heuristics at work filtering fact information. Original art work Kyla Francis with permission

preconceived contrary beliefs, thereby increasing the selectivity of the heuristic-driven filter. While the opinions derived from this process may feel rational, it is really beliefs that are driving the process and consequently biasing conclusions.

Take for example the impact of the *free loading* heuristic on decisions about vaccines (Table 6.2). *Free loading* refers to the belief that other people will assume the risk, so one does not have to put oneself at risk. A parent may choose not to have his/her child immunized because of the belief that sufficient numbers of other children in the community are immunized, thereby providing indirect protection for his/her child through herd immunity. Unfortunately, as has been shown in the pertussis outbreak in California [11], like-minded parents who do not immunize their children often cluster together so the immunization rate in their local under-immunized community is not sufficient for protection through herd immunity. *Free loading* does not offer adequate protection here. Non-immunized children can develop the vaccine-preventable disease when the disease is introduced into the community and even some of those who are immunized will succumb if they did not respond well to the vaccine [9]. They too were not able to benefit from herd immunity because the overall rate of immunization was not high enough in the locale.

*Omission bias*, another heuristic, notes that actions are more harmful than inactions. In this case, the risks of getting immunized—an active process—is perceived as carrying more risk than the vaccine-preventable disease—something obtained passively. This perception is incorrect. The H1N1 influenza vaccine saga illustrates this. In 2009, in the face of the H1N1 influenza pandemic, more health care providers were willing to get seasonal influenza vaccine than H1N1 vaccine [5] because the latter was new, untested and hence perceived to present a higher risk. Being immunized for H1N1 was viewed as more harmful than the consequences of not being immunized.

The *over confidence* heuristic refers to placing great faith in one's own judgments. For example, believing that being healthy, eating well, exercising often and carrying out good hand washing will protect oneself from vaccine-preventable diseases like influenza, thereby negating the need for annual immunization. Like the other heuristics in Table 6.2, this and the other automatic thought processes contribute to vaccine hesitancy and suboptimal immunization rates, leaving both the individual and populations at large susceptible to vaccine-preventable diseases.

Politics provides a strong exemplar of the great power of heuristics [17]. In an experiment, an ideology group was given mock news articles that contained either a misleading claim from a politician alone, or a misleading claim followed by a correction. The experiment revealed that subsequent corrections usually did not change the original misperceptions. The subjects remained convinced that the misinformation was correct because the misinformation that supported their beliefs went through their brain filter but the corrections did not. There are many instances

where working to correct misinformation have instead further reinforced erroneous beliefs. For example, some people continue to believe that US President Obama was not born in the US, in spite of a verified certificate showing his Hawaiian birthplace. They prefer to believe the misinformation and see the certificate as a cover up and government conspiracy (i.e., their belief has been reinforced by their misinterpretation of the evidence). Heuristics are powerful. The brain filter rejects facts that do not fit the belief. Beliefs begin early, are shaped through maturation, and are sustained through life. In order to avoid selective impermeability to positive vaccine practices, pro-immunization beliefs must be actively shaped and the process started early to maximize better outcomes.

### 6.2.1.1   The Internet, Heuristics and Perception of Vaccine Risks

Internet search technology also reinforces heuristics. Previous Internet inquiries shape future searches, since Internet search engines are preset to recognize patterns. Google™, currently the most utilized Internet search engine, personalizes search results based on previous browsing habits [18]. Combine this with the uneven quality of the health advice that parents can find through Google™ [19] and the problem for vaccines becomes clear. For example, one search for vaccine information that leads to an anti-vaccine site triggers the search engine to return to these websites in the next search for vaccine information. This confirms automatic associations in the user's mind between immunization and immunization concerns as highlighted on these anti-vaccine websites. Given that 70–80 % of households in industrialized countries have access to electronically organized and personalized information on the Internet, the Internet has become a major factor influencing human behaviour and decision-making.

The powerful impact of viewing of anti-vaccine websites and heuristics is shown in a German study where even 5–10 min spent viewing an anti-vaccine website had a significant negative impact on vaccine perceptions and decisions to immunize [20]. Another key observation from this study was that alarming anecdotal cases and testimonials viewed on anti-vaccine websites affected decisions. They easily passed through the brain filter to reinforce negative vaccine beliefs, while fact-based information on the pro-vaccine sites had minimal impact and was repelled by the brain filter.

The arguments against immunization posted on anti-vaccine web sites are also influenced by heuristics, exemplifying serious lapses in reasoning and logic [21]. For example, the natural human desire to find order and predictability in random events can result in the false assumption that events related only by time (temporally) must also be related causally. An example of this *co-incident dragon* heuristic (Table 6.2) is the assumed association between vaccines and autism. Given that autistic features emerge around the time of the immunization at 18 months to two years, causality between these two events has been erroneously assumed. Unfortunately, correcting for false assumptions is not easy. Despite the retraction of the controversial paper by Wakefield that purported a link between the measles, mumps and rubella (MMR) vaccine and autism, that was shown to be both fraudulent and in parts unethical [22, 23], and that Wakefield's medical license to practice medicine in the United Kingdom has been revoked, there are parents who remain convinced of the link between autism and the MMR vaccine [24]. They believe the attempts to discredit Wakefield are a conspiracy (see anti-vaccine website: http://www.naturalnews.com/028101_The_Lancet_Dr_Wake-field.html).

Reshaping of ambiguous and/or poor quality data to fit preconceived hypotheses is another source of erroneous reasoning on anti-vaccine websites [21]. The hypothesis that vaccine-preventable diseases are no longer a threat because of better hygiene and nutrition rather than immunization illustrates this point [21] as does the false assumption that all spontaneous reports of adverse events following immunization must be secondary to vaccines [21]. While many adverse events following immunization are reported to programs such as Vaccine Adverse Event Report-

ing System (VAERS) in the United States [25], many such events are subsequently proven to be unrelated to immunization [26]. However, the association has already been made in minds of many (i.e. the *co-incidence dragon* heuristic). Furthermore, the lack of formal feedback to the health care provider or parent who reported the adverse event, following formal causality assessment to determine if there is a relationship, does nothing to correct any misperceptions. Nor are these individual causality assessment outcomes readily accessible or made easily understandable for the public who have concerns about a specific event. There are no national or international programs that systematically provide a rapid summary of the formal causality assessments to help rectify public misperceptions about specific serious adverse events following immunization. Also, the causality assessment reports provided to vaccine program authorities are often written in very technical language. Even if the public had access to these documents they might not be easily understood.

The public does, however, have Internet access to academic publications and reports by august bodies on vaccine adverse events. Unfortunately, these documents also may not be well understood and conclusions may be lifted out of context and misinterpreted, further feeding negative vaccine beliefs. Public reaction to the August 2011 Institute of Medicine report "Adverse Effects of Vaccines" [26] illustrates this point. This report was written with an academic lens and thus assumes an understanding of medical and scientific terms plus a background in vaccinology. This report provides a rigorous review of the evidence for and against causality from a long list of serious vaccine adverse events following immunization that will be useful for health care providers and vaccine researchers. The lay public, however, may easily misinterpret the findings, as has been seen on Internet blogs following the release of the report. Some have taken parts of the well publicized report to support pre-existing anti-vaccine beliefs, overlooking reference to the very rare serious adverse events and the severity of vaccine-preventable diseases.

Thus even robust immunization data may be intentionally and sometimes even unintentionally misconstrued by anti-vaccine advocates, further undermining trust in vaccines. Some anti-vaccine websites fuel conspiracy theories (see above), suggesting that governments, pharmaceutical companies and even health care experts purposefully suppress evidence that immunizations are dangerous [21]. Heuristics mean that once these beliefs are set, shifting them with facts alone becomes very difficult and almost impossible. Anti-vaccine tales of horror (*availability* heuristic, see Table 6.2) are recalled instead of the evidence that vaccines are safe, effective and important for health.

### 6.2.2  Other Factors Contributing to the Vaccine Confidence Deficit

Other factors known to influence immunization acceptance include vaccination experience, the disease cycle stage, population-related factors, and the role of experts.

Given heuristics, it not surprising that older vaccines tend to be more trusted than newer ones; that is, the acceptability and safety of older vaccines has already been established, reinforced and is easily recalled (*anchoring* and *availability* heuristics—Table 6.2). In the case of new immunizations, trust must be earned and may not necessarily be transferred from previous vaccines. The earlier example of parents being hesitant about the human papilloma virus vaccine in British Columbia [7], noted above, with only a 65 % uptake rate, contrasts sharply with the high uptake rates for the older well established immunizations for children such as vaccines for measles, mumps, rubella and hepatitis B in the same population. While the *omission bias* heuristic may markedly sway their choices, other factors such as how others in the community are responding may also influence decisions. If parents know that many are refusing then more will.

Vaccine acceptance is also influenced by the disease cycle. A vaccine that prevents a prevalent disease with serious consequences that cannot otherwise be mitigated is highly valued [25]. However, when the disease disappears due to high immunization rates, the vaccine may no longer be seen as important and concerns about vaccine safety may overtake fear of the disease. The automatic assumption to immunize, in order to decrease the threat of disease, is eroded by increasing uncertainty about the value and safety of immunization. Less disease risk leads to more concern about vaccine safety (*anchoring* heuristic, see Table 6.2).

Automatic associations that influence vaccine benefit-to-risk assessments also vary among and within populations, by country and culture, and according to the background experience, environment and knowledge of the decision-maker (e.g., the variation within health-care providers for influenza vaccine as shown in Table 6.1). Context is key.

The role of experts is another factor that can influence immunization decisions. Trust in health care workers is critical to translating advice into action [10], confirming the importance of providers' knowledge of and commitment to immunization. Nurses in particular, because they administer the majority of immunizations, play a significant role as vaccine advisors. The understanding and beliefs held by nurses about the necessity, safety and effectiveness of vaccines not only affects their own immunization behaviour, but also the vaccine practices of others [3, 27]. When health care providers doubt the merit of immunization so do the patients they advise, leading to a negative impact on immunization programs.

In summary, vaccine confidence is indeed complicated and complex with many factors playing into the assessment by the public and health care providers of vaccine benefit, vaccine risks and disease risks. Thus it is not surprising that a 'one size fits all' evidence centered approach to vaccine education and promotion is insufficient to foster and maintain trust. Rather, an immunization program that is more comprehensive, adaptable and responsive is much needed.

## 6.3 Current Evidence-based Strategies to Deal with Vaccine Concerns

The public health community is well aware that more effort is needed to better address current vaccine safety concerns overall and to meet the needs of different subgroups [11]. While many information materials have been developed for parents these have often been very detailed and complex, many reading more like vaccine package inserts. Hence, a variety of more targeted vaccine information and training materials have been developed by governments and professional groups to better meet the needs of the general public for clear information on risks and benefits. One good example is the quick reference immunization communication tool designed by the Centre for Disease Control in British Columbia, Canada. Written in plain language, this guide provides immunizers with evidence to easily answer common parental vaccine questions and to help explain complex topics [28].

While using evidence and clear language to better answer vaccine-related queries is an important step forward, it does not address the complex behavioral factors noted above. Evidence may or may not get through the heuristic brain filter (Fig. 6.1). Easily understandable well packaged evidence is more likely to pass through and influence decisions but many other factors may prevent even this from happening, as noted above.

New vaccine concerns and questions have arisen for decades and will continue to arise, which means that ever more research will be needed to find evidence that proves or disproves each new allegation. This leaves many vaccine advocates frustrated and puts immunization experts continually on the defensive as they try to respond to these ever shifting concerns while still trying to retain public confidence. Public concerns deserve to be addressed. However, to tackle these unsubstantiated worries one by one through rigorous scientific research requires a substantial commitment of time, money and resources that might be better utilized address-

**Table 6.3** Ten novel approaches to consider for enhancing vaccine confidence

| |
| --- |
| 1. Exploiting heuristics to benefit vaccines in communication and social media |
| 2. Development of broad awareness of rigorous vaccine safety system |
| 3. Ensure clarity of language in all vaccine communications: public and academic |
| 4. Vaccine science: appropriateness, clarity and quality |
| 5. More open transparent decision making for vaccine approval, programs and policy |
| 6. Employ Strategies to Reduce Vaccine-Related Pain |
| 7. Enhance vaccinology education for health care providers, especially physicians and nurses |
| 8. Proactively educate children on vaccine necessity, benefits and safety |
| 9. Media vaccinology education |
| 10. Facilitate changes in vaccine decision making behaviour with a multi pronged approach |

ing other serious health-related issues. Even if evidence-based answers could be provided for all such questions in a very timely fashion (which is almost impossible as it often takes several years to study the question and find the evidence that refutes or proves a new allegation), public mistrust in vaccines would likely persist. Evidence does not necessarily translate into trust. As noted above, negative vaccine anecdotes are much more likely to be recalled than evidence. Hence, a proactive rather than reactive strategy and one that embraces the behavioral understanding of vaccine confidence is required to nurture and foster public support of immunization programs.

## 6.4 Novel Approaches to Enhance Vaccine Confidence

Given that current strategies have not garnered the confidence in vaccines that is needed and based upon a better understanding of factors that influence vaccine decisions, a number of novel approaches are proposed for consideration in Table 6.3. None of these are mutually exclusive or 'golden arrows' that will nail the target. Some may garner early wins such as pain control, while others like child education will take much time before they bear fruit. All are likely to interact synergistically.

### 6.4.1   Exploiting Heuristics for Communications and Social Media

Given that everyone uses heuristics to make decisions, albeit not at a conscious level, vaccine advocates need to understand and exploit heuristics to garner public support for immunization. Two potential heuristics to exploit are *anchoring* and *availability* (Table 6.2). *Anchoring* involves judging the probability of a future event based on what has already occurred, while *availability* involves judging an event as frequent or likely to occur when the event can easily be imagined or recalled. Informing the public of outbreaks of vaccine-preventable diseases in graphic detail can evoke "anchoring" and "availability" heuristics. The clustering of polio cases in Tajikistan in 2010 serves as a useful example. The large outbreak proved that polio can re-emerge, even in regions previously certified as polio-free, if vaccination rates fall below 90 % and asymptomatic cases are imported [29]. In the absence of a disease cure, polio paralysis can be devastating, resulting in death and disability. The image of a child on crutches is heart wrenching to see and readily recalled (*availability*). Hence, describing these unfortunate cases and showing the extent of the outbreak, explaining how the outbreak occurred in a designated polio-free region due to declining immunization rates and emphasizing the potential for similar outbreaks in other certified polio-free regions if immunization rates fall, can make polio real for parents (i.e., easy to recall as it was for parents in the 1950's and 60's because they probably knew someone who had polio). This helps cultivate the belief that vaccines are necessary and important for health. Similarly, the painful complications from the mumps outbreaks in young adults in Canada and the United States, where mumps had previously been controlled [30], provide another excellent

example. A description of the mumps orchitis pain as like having two basketballs on fire provides vivid unforgettable imagery (www.gov. ns.ca/hpp/images/testicular.jpg).

Raising public awareness, especially among parents, of these outbreaks of vaccine preventable diseases and reminding them of the consequences, reinforces *anchoring* and *availability* heuristics. This further nurtures the belief that vaccines are important for health. Parents are then more likely to recall these examples of vaccine-preventable morbidity and mortality when their child is due to be immunized. To craft these messages well takes skills not often present among public health professionals but more commonly found in advertising and marketing experts. Vaccinologists need to develop collaborations with a much broader range of experts than those in health care.

Vaccine advocates themselves should also be educated about historical rates of vaccine-preventable diseases, both locally and globally, and how these rates have waxed and waned with the success and failure of immunization programs [31–33]. Figure 6.2 demonstrates the impact of vaccine on *Haemophilus influenzae b* disease in Canada [31, 32]. Regularly sharing this type of information with the public supports the importance of immunization—again reinforcing *anchoring* and *availability* heuristics for a positive belief in vaccines. It is also worth highlighting where disease resurgences have occurred because vaccination rates dropped as this helps to further refute misinformation that these diseases disappeared due to factors other than immunization [33]. Outbreaks contradict this—vaccines are needed and are powerful. Not being immunized can be tragic. That is the belief that needs to get *anchored*.

In addition, vaccine advocates need to be well aware of the background rates of serious illnesses with unknown and little understood etiologies, such as Guillain Barré Syndrome and multiple sclerosis [34]. As new vaccines are introduced, causality may be inappropriately attributed. Information on background rates can help immunizers and their patients address the flawed reasoning of the *co-incidence dragon* heuristic (i.e., "after it therefore because of it"), as well as the *com-*

**Fig. 6.2** *Haemophilus influenzae* type b (Hib) Disease reported cases 1979–2003 in Canada. (Adapted from [33])

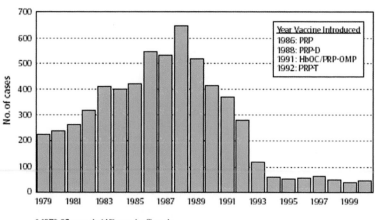

* 1979-85: reported Hib meningitis only
1986-2000: all invasive forms (meningitis and septicemia)

*pression* heuristic (i.e., believing that rare little-understood events are more common than serious disease-related adverse events) (Table 6.2). Many adverse events following immunization, when examined in a formal causality process, are not due to the vaccine but are just co-incidence [35]. As noted above, without transparency about the causality assessment process and a solid feedback loop to those who reported the adverse event following immunization on the findings of the causality assessment, mistrust in vaccine safety will persist among health care workers and the public alike as the *co-incidence dragon* reinforces the belief and is left unchallenged. Communication about the outcomes of causality assessments needs to be well done and well targeted to reach those most involved. The *co-incidence dragon* can be slain if there is only a temporal but not a causal association.

Anti-vaccine advocates well recognize the power of social media, very effectively utilizing the Internet-enabled technologies to garner support for their beliefs. While pro-vaccine websites tend to stick to facts, anti-vaccine Internet sites adopt alternative techniques, including the use of frightening anecdotal stories that easily slide through the brain filter to enhance anti-vaccine beliefs. As described above, based on heuristics, people are more likely to be persuaded by stories and associations rather than stark facts when facing complex decisions. Governments and public health organizations must take this into consid-

eration when designing pro-vaccine websites. They need to better utilize strategies that appeal to heuristics, such as storytelling and anecdotes, in addition to providing evidence. Stories about the terrible outcomes resulting from failure to immunize (e.g., death in children not immunized with *Haemophilus influenzae b* vaccines) [33] helps reinforce and *anchor* the importance of immunization. Fortunately, despite anti-vaccine advocates' efforts to generate an impression of widespread social support for the anti-vaccine stance, most parents continue to support childhood immunization. Vaccination advocates must ensure that parents know this so they will become less susceptible to the anti-vaccine pitch that immunization refusal is the norm in their community or region.

While vaccine advocates have long used printed materials and more recently websites, forays into new social media are now increasing, especially among young parents. Social media has become a normative communication tool for millions. Quality research is needed to ascertain how best to exploit new media; that is, what works well for enhancing confidence in vaccines and what does not. Facebook™, Twitter™, MySpace™, Friendster™, Linkedin™ and YouTube™ can be utilized for communication about vaccines but this takes appropriate skills to do well. Clever use of humour, well crafted messages and respect for the audience requires finesse, but the pay off in audience reach can be astounding.

Dedicated resources are needed to better develop the communication tools used by many parents today as well as future tools that will be utilized by upcoming generations. Pamphlets and television advertisements are not enough.

## 6.5  Develop Broad Awareness of Vaccine Safety System

Recognizing the importance of heuristics in decision-making, parents and health care providers must be continually reminded of the rigorous safety evaluation of all vaccines [36]. Doing so can help bolster trust in immunizations as well as help filter out vaccine safety misinformation. Unfortunately, most industrialized countries have invested little in educating health care providers and the public about the strength and reliability of their vaccine safety systems that are indeed more fulsome and rigorous than the drug safety systems. The first supplement of the National Vaccine Program Office of the US Department of Health and Human Services, outlining vaccine safety throughout the product cycle, was only published in 2011 [37]. Of note, an article in this supplement outlines how the Vaccine Safety Datalink system can be used to quickly check on purported adverse events linked to a vaccine, to determine if there is sufficient evidence to stimulate a more in-depth assessment [38] but there is no information on whether this is then reported back to the health care workers or parents who initiated the reports.

The Cuban government long ago recognized the importance of garnering community support for vaccines. Despite the fact that polio was eradicated from Cuba over 30 years ago, a national annual polio immunization program continues to flourish. Involving thousands of mothers and children, under the auspices of the Cuban Federation of Women, the neighborhood Committees for the Revolution and Public Health, this annual public show of support for immunization reinforces the community's trust in vaccines [36]. The Cuban national vaccine education program, starting with school-aged children and continuing into adulthood, further bolsters national confidence in vac-

cines. Interestingly, Cuba has not suffered from the vaccine safety angst seen elsewhere. Indeed, other countries can learn much from Cuba, where vaccine education is an integral part of citizenship and trust in immunization is the norm.

## 6.6  Clarity of Language

Academic and overly-scientific language can unintentionally subvert efforts to educate the public about vaccines. While scientific journals and technical academic reports used to be the purview only of those trained to read and interpret them, the widespread availability of these materials via the Internet, as noted above, has created the potential for misuse and misunderstanding, especially by lay readers including the media.

Academic jargon can easily obscure the intended scientific meaning, confusing untrained journalists and parents, and fostering misinterpretation by anti-vaccine advocates [39]. This can lead to anti-vaccine websites using misinformation in support of their beliefs. The Internet provides ready access to a multitude of scientific reports, including those from august bodies such as the Institute of Medicine in the United States and the National Institute for Health and Clinical Excellence in the United Kingdom and to a wide array of open access academic journals. Thus, greater care is now needed to ensure that the conclusions, implications, and executive summaries in academic articles and reports are easy to understand and are not obfuscated by jargon [39].

Unfortunately, the recent trend has been away from clear plain language in many academic articles. Technical reports are also becoming longer and more complex often with hundreds of references, making clarity of language in the executive summary and the final conclusions even more crucial as these will be the most read sections. Few people will read an exhaustive report in its entirety. The vaccine recommendations from the national immunization technical advisory groups (NITAGs) like the United States Advisory Committee on Immunization Practices (ACIP) exemplify this point. Even experienced users may struggle to find and interpret the major recommendations made by these advisory committees [40]. Ambiguous, complicated language and academic jargon are confusing for many, including many health care providers. Clarity in communications must become the norm whether they be academic articles, technical reports or advisory committee recommendations. As Einstein noted, *"Make everything as simple as possible, but not simpler".*

## 6.7 Vaccine Science: Appropriateness, Clarity and Quality

Polland and colleagues have also emphasized that great care must be used when research tools such as meta-analyses and evidence-based medicine techniques are applied to vaccines [41, 42]. Evidence-based analyses value outcomes such as infection rates much more than proxy outcomes such as immunogenicity. For rare diseases such as meningococcal infections or rabies, this would mean vaccine studies would require huge populations followed over several years for randomized controlled trials, an expensive and untenable position. Emphasizing such well-used research tools could lead to an undervaluing of vaccines studies compared to other drugs where more classic end points are used. This can lead to the erroneous implication that vaccine science is somehow less rigorous, thus less valuable and less credible. Similarly, assessing the findings from tools like meta-analyses needs to be used with care for vaccines, given their potential for undermining immunization program communications.

Journal editors also need to be reminded of the critical role vaccines play in population health. Not only must they critically assess the language of submitted articles for clarity and ease of understanding but also take great care in assessing the scientific quality of submissions. Poor quality science, when published in esteemed journals, has the potential to undermine public and professional confidence. Because high rates of vaccination are needed to control disease, even a small shift in confidence because of a well publicized academic article, such as occurred with the Wakefield article noted above, can lead to deaths. Much damage had already been done by the time that article was withdrawn years later.

While not suggesting that editors should decline to publish articles that cast doubt on vaccine benefits or highlight risks, they must bear in mind and be cautious of ecological and epidemiological studies that purport to show vaccine associations with diseases of unknown etiologies, such as multiple sclerosis, as the observed association may well be due to chance alone and not be causally related.

## 6.8 More Open Transparent Decision Making

Many anti-vaccine websites abound with allegations that governments conspire to suppress and even distort negative information about vaccines. These allegations are hard to refute when discussions and decisions about public funding of vaccines and programs, vaccine licensure, and vaccine safety assessments are not done openly. Although there has been some movement towards more open and transparent processes by some NITAGs and drug licensing bodies (i.e., ACIP in the US) [40], other organizations have lagged behind. As well, oft times, conflicts of interest in decision-making remains shielded from public scrutiny. Anti-vaccine websites harp on this flaw. This is not trivial. Previous failures to bring medication safety information to the public's attention in a timely fashion as occurred with Vioxx$^{TM}$, has led to an erosion in the public's confidence in drugs and the pharmaceutical industry in general [43]. This experience has only added fuel to the belief of many that vaccine problems are also being covered up. This is heuristics at work, augmenting public doubt in vaccines. Transparency is crucial.

Moving to a model of more robust disclosure will not be easy. NITAGs and drug licensing bodies will require a better understanding of the public's concerns about vaccine decision-making and then an open mind when considering how best to address these transparency concerns. Overly defensive responses to calls for more public scrutiny need to be avoided. Instead, organizations must learn to address public inquiries in a sensitive manner. Commitment to an open and transparent vaccine system can help enhance the public's confidence in vaccines and undermine anti-vaccine conspiracy allegations.

## 6.9 Strategies to Reduce Vaccine-related Pain

Pain from immunization has been long neglected as a cause for vaccine hesitancy, overshadowed by other vaccine-related concerns. However, pain with immunization is a significant source of distress for children, parents and health care providers—a concern expressed by 44 % of parents in one survey [44]. Fear of needles and anxiety about procedures can negatively impact vaccination and health-seeking behaviors [45]. In 2010, an evidence-based Canadian clinical practice guideline intended to reduce pain related to childhood immunization was published [45]. Designed by an interdisciplinary group of experts in pain and in vaccinology and approved by several professional societies, the guideline has been met with considerable interest by both immunizers and parents in Canada. Reducing pain with immunization is a step in the right direction to decrease vaccine hesitancy and increase vaccine uptake. Ignoring this for so long has been folly. However, the effectiveness of these approaches across populations and settings still requires thoughtful evaluation. Getting pain control right is important.

## 6.10 Vaccine Education for Physicians and Nurses

Formal education in medical and nursing schools has not typically provided sufficient information and training to prepare physicians and nurses to answer the vaccine-related questions from patients that often arise in their practices [46]. Low immunization uptake rates for annual influenza vaccine among health care providers, as noted above, demonstrates that physicians and nurses have not all been convinced during their training and while out in practice that immuniza-

tion is necessary to reduce the burden of diseases such as influenza. Education must be improved. In addition to enhancement of vaccine education in medical and nursing school, standardized competencies for those who immunize, such as the Immunization Competencies published by the Public Health Agency of Canada [47], could further reinforce the importance for health care providers of immunizing themselves and their patients. Given the important role health care providers play in patient vaccine decision making [12, 27], the training of these professionals needs urgent attention by curriculum planners and much pressure from vaccine programs and governments. Having a health care work force undereducated about vaccines critically undermines vaccine program success.

## 6.11 Vaccine Education for Children

Building confidence in vaccines should start early. A basic explanation of vaccines, including how they work and how vaccine safety is monitored, could be part of the health curriculum for elementary students and then reinforced in secondary school. Immunization is such an important public health measure that its significance could be incorporated into history and science classes as well. This would begin the process of building trust and shaping positive vaccine beliefs at an early age, a powerful technique for shaping adult decision making. There are a variety of teaching approaches that have been shown to make vaccine information meaningful to young students including drawings, word puzzles and stories of real people who have experienced vaccine-preventable diseases [48]. Figure 6.3 shows an example of a student's poster drawn after participating in such a program.

## 6.12 Vaccine Education for the Media

While health care providers play a major role in parental decisions to immunize their children [12, 27], what appears in the media, both tradi-

**Fig. 6.3** Immunization: protecting you, your family and your community. Ryan McGee winner 2008 National Immunization Poster Context. Canada. Public Health Agency of Canada. With permission

tional print and broadcast media as well as social media, is also influential as noted above [20]. Heuristics are at work here. Media stories that reinforce beliefs are well heard but those that disagree are filtered out. If there are no preconceived beliefs about vaccines, media stories may shape these beliefs, pro or con, depending on the story.

Unfortunately, vaccines are variably portrayed in the formal media. A 2011 report of a 10 year survey of vaccine-related articles in U.S. daily newspapers, with circulations equal to or over 50,000, found that 37 % of the articles suggested that vaccines were unsafe [49]. Over 60 % of all the vaccine articles were written by newspaper or wire service staff, many of whom likely only had a rudimentary background in science and probably no background in vaccine science. Recognizing the importance of journalists in shaping public opinion, many organizations that advocate for vaccines distribute news releases and backgrounders for the media, hoping journalists will simply use their vaccine facts in their articles. However, journalists are trained to question traditional sources and to present 'the other side' of the story which sometimes means the anti-vaccine side. As traditional media outlets experience more and more staff cuts, there are now fewer journalists who focus solely on science or health issues. Even fewer have substantive enough health or science expertise to fully brief themselves on the issues presented. Hence, it is

increasingly important to simplify the knowledge translation of vaccine science and to make the major findings in academic articles crystal clear, not only in the covering news release but also in the document's conclusions as noted above.

Ideally, journalists could attend educational workshops and seminars on vaccines, such as the European Media Workshop organized by the Fondation Mérieux in Annecy, France, April 2009, aptly titled "Unjustified scare or reasonable skepticism" [50]. Similar educational workshops for journalists have been offered at the biannual Canadian Immunization Conferences. While the importance of these education opportunities for media has not been formally assessed, this may be a promising avenue that leads to enhanced confidence in vaccines in the media and, by extension, the public. It is interesting to note that one of the most prominent health reporters in Canada who writes for the largest national daily is well educated about vaccines and is a staunch supporter of vaccines [51].

Given that annual national immunization weeks as well as regional and national immunization conferences are becoming more common throughout the world, journalists could be encouraged to not only report on the vaccine events but also to participate in short courses on vaccine science (online or in person) designed specifically for them. Other tools such as online briefings and backgrounders could also be made readily available to journalists including explanations of basic vaccinology terms and concepts relevant to the study—all provided in lay language. Expert spokespeople should also use plain language, and be clear about the implications of reports and articles. As outlined above, obfuscation is an open door to misunderstandings.

Working with social media, especially bloggers, presents an enormous challenge as there are so many of them. Blogs should be monitored and those with large followings responded to wherever possible if vaccine misinformation is being promulgated. While this takes time and effort, as politicians have well learned, you ignore bloggers at your peril.

## 6.13  Facilitate Changes in Behaviour

Most beliefs and behaviours are deep rooted, difficult to change, and well established by adulthood. Modifying beliefs and behaviors is thus, a complex process. There is no one way to build confidence in vaccines among adults that leads to high vaccine uptake. As noted above, sowing pro-vaccine belief seeds in childhood may be key but this will take at least a generation before accruing the benefits. What can be done in the interim? For adults, multiple approaches will be required, including many outlined above. Positive public vaccine beliefs will need to be actively shaped and reinforced.

### 6.13.1  How can this be Accomplished?

One model to consider is the World Health Organization Communication-for-Behavioural (WHO COMBI) program, a comprehensive, people-centered approach used to facilitate behavioural change at individual, family and societal levels [52]. Although typically applied to diseases in developing countries, the WHO COMBI strategy may be amenable to immunization programs nationally and globally. Drawing on the Cuban example noted above, comprehensive national immunization strategies that effectively influence behaviour can be highly successful. The poten-

tial of a COMBI like program for immunization deserves serious exploration.

As well, national vaccine programs need to involve a broader range of disciplines, expanding beyond traditional players with health backgrounds, to include contributions from the behaviour sciences, marketing, and communication. These disciplines would bring a wealth of new expertise and knowledge to the development of immunization programs and the field of vaccinology. Strategies for building vaccine confidence locally can be much stronger if they draw on local capacity to build robust immunization programs. Engaging community members in finding solutions to the vaccine confidence deficit problem will be fundamental work here as the literature has well shown that top down solutions often fail [53].

## 6.14  Conclusion

Fact-based communications and effective programs have achieved the high rates of immunization needed to control many vaccine-preventable diseases. These strategies need to continue. However, while many people have been convinced by scientific evidence, the vaccine-hesitant minority have not. Given that very high immunization rates are required for disease control across populations, ignoring the vaccine hesitant is not an option. The vaccine hesitant more often rely on beliefs and automatic assumptions and are not influenced by evidence. As stated by Goethe, a German polymath, over 150 years ago, *"Belief is not the beginning but the end of all knowledge"*. He continued, *"We are so constituted that we believe the most incredible things; and, once they are engraved upon the memory, woe to him who would endeavour to erase them"*.

A proactive, multidisciplinary, multifaceted approach capable of reshaping immunization beliefs is needed to address the vaccine confidence deficit. Devising such a strategy will not be easy. Vaccine advocates need to collaborate with experts in the behaviour and social sciences, in marketing and advertising, as well as members of the community to bring more depth and breadth to immunization programs. Milestone evaluations will be needed to ensure that the devised programs achieve the stepwise desired goals. Fortunately, there are some successful models to examine and build upon including the Cuban national immunization program and the WHO COMBI strategy for dengue fever control.

Vaccines are a critical public health strategy for population health and well being. The trust, support and confidence of health care providers and the general public in vaccine programs is essential for success. Given the high costs of failure, including health care costs and needless morbidity and mortality from vaccine-preventable diseases, immunization advocates need to aggressively address the vaccine confidence deficit at its roots by actively working to shape vaccine beliefs.

# References

1. Callereus T (2010) Perceptions of vaccine safety in a global context. Acta Paediatrica 99:166–171
2. MacDonald NE, Smith J, Appleton M (2012) Risk perception, risk management and safety assessment: What can governments do to increase public confidence in their vaccine system? Biologicals 40(5):384–388
3. Leitmeyer K, Buchholz U, Kramer M, Schenkel K, Stahlhut H, Köllstadt M, Haas W, Meyer C (2006) Influenza immunization in German health care providers: effects and findings after two rounds of a nationwide awareness campaign. Vaccine 24:7003–7008
4. Caban-Martinez AJ, Lee DJ, Davila EP, LeBlanc WG, Arheart KL, McCollister KE, Christ SL, Clarke T, Fleming LE (2010) Sustained low influenza immunization rates in US healthcare workers. Prev Med 50:210–212
5. Centers for Disease Control and Prevention (2010) Interim Results: Influenza A (H1N1) 2009 monovalent and seasonal influenza immunization coverage among health-care personnel—United States, August 2009–January 2010. MMWR 59(12):357–362
6. Centers for Disease Control and Prevention (2011) Influenza vaccination coverage among health-care personnel. United States, 2010–11 Influenza season. MMWR 60(32):1073–1077
7. Ogilvie G, Anderson M, Marra F, McNeil S, Pielak K, Dawar M, McIvor M, Ehlen T, Dobson S, Money D, Patrick DM, Naus M (2010) A population-based evaluation of a publicly funded, school-based HPV vaccine program in British Columbia, Can-

ada: parental factors associated with HPV vaccine receipt. PLoS Med 7(5):1000270
8. Andre FE, Booy R, Bock HL, Clemens J, Datta SK, John TJ, Lee BW, Lolekha S, Peltola H, Ruff TA, Santosham M, Schmitt HJ (2008) Vaccination greatly reduces disease, disability, death and inequity worldwide. Bull World Health Organ 86:140–146
9. Public Health Agency of Canada (2011) Vaccine-preventable diseases measles. http://www.phac-aspc.gc.ca/im/vpd-mev/measles-eng.php. Accessed 17 Aug 2011
10. Black S, Rappuoli R (2010) A crisis of public confidence in vaccines. Science translational medicine 2(61):61mr1. doi:10.1126/scitranslmed.3001738
11. California Department of Public Health (2011) Pertussis report January 7, 2011. http://www.cdph.ca.gov/programs/immunize/Documents/Pertussis-Report2011-01-07.pdf
12. Benin AL, Wisler-Scher DJ, Colson E, Shapiro ED, Holmboe ES (2006) Qualitative analysis of mothers' decision-making about vaccines for infants: the importance of trust. Pediatrics 117(5):1532–1541
13. Freed GL, Clark SJ, Butchart AT, Singer DC, Davis MM (2010) Parental vaccine safety concerns in 2009. Pediatrics 125(4):654–659
14. Ball LK, Evans G, Bostrom A (1998) Risky business: challenges in vaccine risk communication. Pediatrics 101:453–458
15. Lau AY, Coiera EW (2007) Do people experience cognitive biases while searching for information? J Am Med Inform Assoc 14(5):599–608
16. Galdi S, Arcuri L, Gawronski B (2008) Automatic mental associations predict future choices of undecided decision-makers. Science 321(5892):1100–1102
17. Nyhan B, Reifler J (2010) When corrections fail: the persistence of political misperceptions. Polit Behav 32:303–330
18. Google Privacy Center (2011) http://www.google.ca/intl/en/privacy/faq.html. Accessed 22 Aug 2011
19. Scullard P, Peacock C, Davies P (2010) Googling children's health: reliability of medical advice on the internet. Arch Dis Child. doi:10.1136/adc.2009.168856
20. Betsch C, Renkewitz F, Betsch T, Ulshöfer C (2010) The influence of vaccine-critical websites on perceiving immunization risks. J Health Psychol 15:446–455
21. Jacobson RM, Targonski PV, Poland GA (2007) A taxonomy of reasoning flaws in the anti-vaccine movement. Vaccine 25(16):3146–3152
22. Godlee F, Smith J, Marcovitch H (2011) Wakefield's article linking MMR vaccine and autism was fraudulent. BMJ 342:7452
23. Deer B (2011) How the case against the MMR vaccine was fixed. BMJ 342:5347. doi:10.1136/bmj.c5347
24. Hilton S, Petticrew M, Hunt K (2007) Parents' champions vs. vested interests: who do parents

believe about MMR? A qualitative study. BMC Public Health 7:42. doi:10.1186/1471-2458-7-42

25. Chen RT, Rastogi SC, Mullen JR, Hayes SW, Cochi SL, Donlon JA, Wassilak SG (1994) The Vaccine Adverse Event Reporting System (VAERS). Vaccine 12:542–550

26. Institute of Medicine (2011) Adverse effects of vaccines: evidence and causality. August 2011. http://www.iom.edu/Reports/2011/Adverse-Effects-of-Vaccines-Evidence-and-Causality.aspx correct referencing? Accessed Aug 2011

27. Zhang J, While AE, Norman IJ (2010) Knowledge and attitudes regarding influenza immunization among nurses: a research review. Vaccine 28(44):7207–7214

28. British Columbia Centre for Disease Control (2011) Immunization communication tool for immunizers. http://www.bccdc.ca/NR/rdonlyres/DADA3304–7590-48AC-8D2C-65D54ADFC77E/0/CDC_IC_Tool.pdf. Accessed 22 Aug 2011

29. MacDonald NE, Hebert PC (2010) Polio outbreak in Tajikistan is cause for alarm. CMAJ 182(10):1013

30. MacDonald N, Hatchette T, Elkout L, Sarwal S (2011) Mumps is back: why is mumps eradication not working? Adv Exp Med Biol 697:197–220

31. National Advisory Committee on Immunization (2006) Haemophilus Vaccine. 7th Edition Canadian Immunization Guide. Pub Health Agency Canada 172–178

32. Scheifele DW, Bettinger JA, Halperin SA, Law B, Bortolussi R (2008) Ongoing control of Haemophilus influenzae Type B infections in Canadian children, 2004–2007. Pediatr Infect Dis J 27:755–757

33. Centers for Disease Control and Prevention (2009) Invasive Haemophilus influenzae Type B disease in five young children—Minnesota, 2008. MMWR 58:58–60

34. Black S, Eskola J, Siegrist C, Halsey N, MacDonald NE, Law B, Miller E, Andrews N, Stowe J, Salmon D, Vannice K, Izurieta HS, Akhtar A, Gold M, Oselka G, Zuber P, Pfeifer D, Vellozzi C (2009) Importance of background rates of disease in assessment of vaccine safety during mass immunization with pandemic H1N1 influenza vaccines. Lancet 374(9707):2115–2122

35. Collet JP, MacDonald N, Cashman N, Pless R (2000) Monitoring signals for vaccine safety: the assessment of individual adverse event reports by an expert advisory committee. Advisory Committee on Causality Assessment. Bull World Health Organ 78(2):178–185

36. MacDonald NE, Pickering L (2009) Canadian paediatric society, infectious diseases and immunization committee. Canada's eight-step vaccine safety program: vaccine literacy. Paediatr Child Health 14(9):605–608

37. Salmon D, Pavia A, Gellin B (2011) Vaccine safety throughout the product life cycle. Pediatrics 127(Suppl 1):1–137

38. Baggs J, Gee J, Lewis E, Fowler G, Benson P, Lieu T, Naleway A, Klein NP, Baxter R, Belongia E, Glanz J, Hambidge SJ, Jacobsen SJ, Jackson L, Nordin J, Weintraub E (2011) The Vaccine Safety Datalink: a model for monitoring immunization safety. Pediatrics 127(Suppl 1):45–53

39. MacDonald NE, Picard A (2009) A plea for clear language on vaccine safety. CMAJ 180(7):697–698

40. Smith JC (2010) The structure, role and procedures of the U.S. Advisory Committee on Immunization Practices (ACIP). Vaccine 28S:68–75

41. Jacobson RM, Targonski PV, Poland GA (2007) Meta-analyses in vaccinology. Vaccine 25(16):3153–3159

42. Jacobson RM, Targonski PV, Poland GA (2007) Why is evidence-based medicine so harsh on vaccines? An exploration of the method and its natural biases. Vaccine 20/25(16):3165–3169

43. Faunce T, Townsend R, McEwan A (2010) The Vioxx pharmaceutical scandal: Peterson v Merke Sharpe & Dohme (Aust) Pty Ltd (2010) 184 FCR 1. Law Med 18(1):38–49

44. Kennedy A et al (2011) Vaccine attitudes, concerns, and information sources reported by parents of young children: results from the 2009 HealthStyles survey. Pediatrics 127(Suppl):92–99

45. Taddio A, Appleton M, Bortolussi R, Chambers C, Dubey V, Halperin S, Hanrahan A, Ipp M, Lockett D, MacDonald N, Midmer D, Mousmanis P, Palda V, Pielak K, Riddell RP, Rieder M, Scott J, Shah V (2010) Reducing the pain of childhood immunization: an evidence-based clinical practice guideline (summary). CMAJ 182(18):1989–1995

46. Pelly L, Pierrynowski MacDougall D, Halperin B, Strang R, Bowles S, Baxendale M, McNeil S (2010) The VaxEd project: an assessment of immunization education in canadian health professional programs. BMC Med Educ 2010, 10, 86. Published online 2010 November 26. 10.1186/1472-6920-10-86 Copyright ©2010 Pelly et al. licensee BioMed Central Ltd

47. Immunization Competencies for Health Professionals (2011) Prepared by the Professional Education Working Group of the Canadian Immunization Committee. Approved by the Communicable Disease Control Expert Group and the Pan-Canadian Public Health Network. Published by the Public Health Agency of Canada Centre for Immunization and Respiratory Infectious Diseases November 2008. http://www.phac-aspc.gc.ca/im/ic-ci-eng.php. Accessed 23 Aug 2011

48. Teachers Kit, National Immunization Poster Contest (2011) Canadian Coalition for Immunization Awareness & Promotion and the Public Health Agency of Canada. http://www.immunize.cpha.ca/en/events/imm-poster-contest.aspx. Accessed 19 Jul 2011

49. Hussain H, Omer SB, Manganello JA, Kromm EE, Carter TC, Kan L, Stokley S, Halsey NA, Salmon DA (2011) Immunization safety in US print media, 1995–2005. Pediatrics 127(Suppl 1):100–106

50. Fondation Mérieux (2011) European Media Workshop on Vaccine Safety: Unjustified scare or reasonable scepticism. 20–21 April 2009, Les Pensières, Annecy (France) http://www.fondation-merieux.org/European-Media-Workshop-on-Vaccine,2321.html. Accessed 31 Aug 2011

51. Picard A (2011) The return of measles: where did we go wrong? Globe & Mail. June 8 2011 http://www.theglobeandmail.com/life/health/new-health/andre-picard/the-return-of-measles-where-did-we-go-wrong/article2052432/. Accessed 31 Aug 2011

52. World Health Organization (2011) Mobilizing for Action. Communication-for-Behavioural-Impact (COMBI) http://www.k4health.org/system/files/COMBI.pdf. Accessed 31 Aug 2011

53. Lasker RD, Guidy JA (2009) Engaging the community in decision making. Case studies tracking participation, voice and influence. McFarland and Company 2009, Jefferson, North Carolina USA, 2009

# Chronic Recurrent Multifocal Osteomyelitis

**7**

## Marion R. Roderick and Athimalaipet V. Ramanan

### Abstract

Chronic recurrent multifocal osteomyelitis (CRMO) is an autoinflammatory bone disease occurring primarily in children and adolescents. Episodes of systemic inflammation occur due to immune dysregulation without autoantibodies, pathogens or antigen-specific T cells. CRMO is characterised by the insidious onset of pain with swelling and tenderness over the affected bones. Clavicular involvement was the classical description; however, the metaphyses and epiphyses of long bones are frequently affected. Lesions may occur in any bone, including vertebrae. Characteristic imaging includes bone oedema, lytic areas, periosteal reaction and soft tissue reaction. Biopsies from affected areas display polymorphonuclear leucocytes with osteoclasts and necrosis in the early stages. Subsequently, lymphocytes and plasma cells predominate followed by fibrosis and signs of reactive new bone forming around the inflammation. Diagnosis is facilitated by the use of STIR MRI scanning, potentially obviating the need for biopsy and unnecessary long-term antibiotics due to incorrect diagnosis. Treatment options include non-steroidal anti-inflammatory drugs and bisphosphonates. Biologics have been tried in resistant cases with promising initial results. Gene identification has not proved easy although research in this area continues. Early descriptions of the disease suggested a benign course; however, longer-term follow up shows that it can cause significant morbidity and longer term disability.

Although it has always been thought of as very rare, the prevalence is likely to be vastly underestimated due to poor recognition of the disease.

A. V. Ramanan (✉) · M. R. Roderick
Bristol Royal Hospital for Children, Bristol, UK
e-mail: Athimalaipet.Ramanan@UHBristol.nhs.uk

A. V. Ramanan
Royal National Hospital for Rheumatic Diseases,
Bath, UK

M. R. Roderick
e-mail: rodericks1000@hotmail.com

## 7.1 Introduction

Chronic recurrent multifocal osteomyelitis (CRMO) is an inflammatory bone disease occurring primarily in children and adolescents. It

N. Curtis et al. (eds.), *Hot Topics in Infection and Immunity in Children IX*,
Advances in Experimental Medicine and Biology 764, DOI 10.1007/978-1-4614-4726-9_7,
© Springer Science+Business Media New York 2013

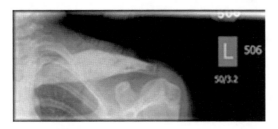

**Fig. 7.1** Plain radiograph showing the left clavicle expansion and sclerosis of entire shaft

was first described in 1972 by Giedion as "an unusual form of multifocal bone lesions with subacute and chronic symmetrical osteomyelitis" [1]. Subsequently, Björkstén called it CRMO, describing nine patients with fever, pain and radiological findings suggestive of osteomyelitis and described recurrent pustulosis palmaris and plantaris, which paralleled the exacerbations in bony lesions [2].

There are now around 300 cases of CRMO described in the literature, predominantly as case series. It is difficult to guess the true prevalence as it is likely that as a little-known disease it is underdiagnosed [3]. CRMO is related to Synovitis, Acne, Pustulosis, Hyperostosis, Osteitis (SAPHO) syndrome which may represent a similar disease process in adults [4].

CRMO has recently been classified as an autoinflammatory disorder (rather than autoimmune). These inflammatory conditions are characterised by episodes of systemic inflammation including serological signs of inflammation occurring in the absence of autoantibodies, pathogens or antigen-specific T cells [5–7]. They appear to be caused by a primary dysfunction of the innate immune system [8].

## 7.2 Clinical Features

The consistent feature of CRMO is the insidious onset of pain with swelling and tenderness localised over the affected bones. Lesions may occur in any bone, including vertebrae [9–17]. Involvement of the clavicle is classical however, the metaphyses and epiphyses of the femur, tibia or humerus are also frequently affected. Involve-

ment of the skin includes palmoplantar pustulosis, occurring in approximately 20 %, acneiform lesions or psoriasis [2, 15, 18–20]. Inflammatory bowel disease, particularly Crohn's-like lesions have been associated with CRMO and may represent "enteropathic CRMO" [21–25]. There is a female preponderance with twice as many girls affected and a median age of onset of around 10 years [3, 5, 26, 27].

Fever may be a feature during episodes of disease flare, and inflammatory markers such as erythrocyte sedimentation rate (ESR), C-reactive protein (CRP) level, leucocytes, and fibrinogen are commonly elevated. However, the increases are usually moderate and return to normal during quiescent periods. If these markers are very significantly elevated it favours an alternate cause and this should be investigated further. [5, 10, 16, 27, 28]

## 7.3 Radiology

Plain x-rays may be normal in the early stages of disease, but later will show the usual features of osteomyelitis, such as osteolysis, sclerosis, and new bone formation, or combinations of these. Different patients have very different radiological findings with some lesions showing periosteal reactions and lytic areas mimicking tumours [29] (Figs. 7.1 and 7.2).

Characteristic imaging findings include bone expansion, medullary lytic areas, periosteal reaction, bone oedema, and inflammatory soft tissue reaction. These features may help distinguish the diagnosis from other diseases [5].

Radionuclide bone scan are still used in some departments to identify silent lesions in other bones (as foci of increased uptake) and to monitor the response to therapy. However, whole body MRI using short tau inversion recovery (STIR) sequencing permits identification of the characteristic features of areas of bone marrow oedema (hyperintensity on STIR images). It is also more sensitive than radiographs or bone scans to identify silent lesions without radiation. In many cases this will help identify multifocal lesions

**Fig. 7.2** Bone scan demonstrating increased areas of uptake in the left clavicle and both proximal humeri

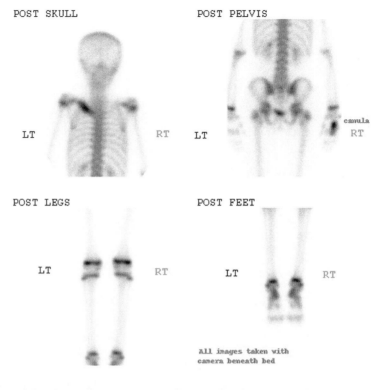

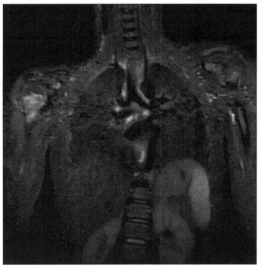

**Fig. 7.3** MRI stir coronal image showing high signal in right proximal humerus

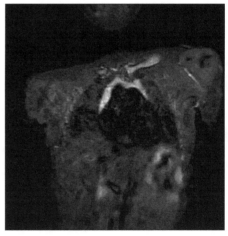

**Fig. 7.4** MRI stir coronal image showing high signal in the shaft of left clavicle

where only one has been noticed clinically and make the diagnosis plain [30] (Figs. 3 and 4).

If the clinical presentation, laboratory findings and radiology strongly suggest CRMO then, potentially, the need for a bone biopsy can be avoided [5, 27].

## 7.4 Diagnosis

Various diagnostic criteria have been suggested [9, 15, 16] but essentially CRMO remains a diagnosis of exclusion, with the association of positive and negative findings. Multifocal bone lesions are rare in childhood but can have a diverse aetiology including leukaemia, neuroblastoma, Langerhans cell histiocytosis and staphylococcal osteomyelitis.

CRMO is particularly difficult to diagnose when there is only one bony lesion; care must be taken to exclude bacterial osteomyelitis which may only be possible after biopsy, failure to respond to antibiotics and then a good response to non-steroidal anti-inflammatory drugs without antibiotics [15].

## 7.5 Pathophysiology

Bone biopsies from affected areas show non-specific inflammation. There is a spectrum of changes beginning with an acute inflammatory process with early lesions consisting of polymorphonuclear leucocytes accumulating in the marrow. At this stage there is also osteoclastic bone resorption and necrosis. Subsequently, lymphocytes and plasma cells become more common in the inflammatory infiltrates and some cases display granulomatous foci. Fibrosis occurs in the later stages and osteoblasts are seen with signs of reactive new bone formation around the inflammation which parallels that seen radiologically as osteosclerosis [31, 32]. Due to the wide variation of reparative changes seen, histological examination alone cannot differentiate CRMO from bacterial infection [9, 33–35].

## 7.6 Management

There is currently no way of identifying those patients that will have spontaneous resolution of symptoms and those who will have persistent disease. In all cases with active disease, the bone pain may be severe, resulting in reduced quality of life, disruption of childhood and potentially permanent bony deformities [36–39]. The aims of treatment are to minimise pain and maintain normal bone growth and function of adjacent joints [40]. There are no published randomised controlled trials of treatment in CRMO.

### 7.6.1 Nonsteroidal Anti-Inflammatory Drugs

Nonsteroidal anti-inflammatory drugs (NSAIDs) are first-line treatment and may be very effective; producing a clinical and radiological improvement in many patients. One case series found indomethacin provided a dramatic response, which was sustained throughout the four year follow-up [41]. Naproxen has also been used with good effect for both primary lesions and in relapse. [29, 42] In a review compiling published case reports of CRMO, overall 79 % of patients had a good response to a variety of NSAIDs [15].

None of these studies were randomised controlled trials; however, as Geschick points out, if duration of symptoms is compared with the historical controls first described by Björkstén et al. [2], there is a marked difference in duration in favour of NSAIDs [42, 43].

If there is a failure to respond to one NSAID it is worth trying one or two others, as the response is recognised to vary between patients [9, 26–28]. It is postulated that NSAIDs act by inhibition of the cyclooxygenase pathway and thereby inhibiting prostaglandin synthesis [29].

### 7.6.2 Steroids

For those patients who fail to respond to NSAIDs, steroid therapy with intravenous methylprednisolone, oral prednisolone or hydrocortisone has been used with good effect. However, the resolution of pain may be transient with symptoms recurring once steroids are stopped [9, 29, 44, 45].

### 7.6.3 Disease Modifying Agents

Methotrexate and azathioprine are generally considered to be ineffective [9].

### 7.6.4 Pamidronate

Evidence is building for the use of the bisphosphonate pamidronate in CRMO when NSAIDs have failed to control symptoms. It is generally very effective in reducing pain, sometimes dramatically; with improvement of functional outcomes and evidence of reduction in lesion size as shown by a variety of imaging methods [39, 46–48]. It has also been shown to resolve vertebral hyperintensities (seen on MRI) either partially or completely, with improvement in vertebral modelling and height [14, 49].

Pamidronate was the first bisphosphonate to be used in CRMO as there was already experience of its use in a paediatric population in osteogenesis imperfecta [50]. The same three-day paediatric protocols were used. Neridronate (an aminobisphosphonate) has also been used in CRMO and appears to have similar effects to pamidronate [51].

The mechanism of action remains uncertain. Bisphosphonates are known to inhibit osteoclasts [52] (which are seen in biopsies of lesions), and may reduce lesion expansion by action at this site. Using bisphosphonates as anti-cytokine agents in inflammatory conditions is becoming more widespread [53] with effects on the release tumour necrosis factor-$\alpha$ and interleukin-6 from T-lymphocytes [54–55]; Minor flu-like symptoms and a mild increase in pain for 24 h are common with pamidronate infusion and are found most frequently after the first infusion. Subsequent infusions tend to be better tolerated [56–57]. Although the rapidity of resolution of pain in patients whose pain has been resistant to NSAIDs lends weight to the efficacy of bisphosphonates for treatment of CRMO [46]. Randomised controlled trials would be helpful to explore the efficacy and potential adverse effects of this relatively recently described use of bisphosphonates.

## 7.6.5   Interferon

There are some case reports of good results using interferon but they have been superseded by the use of biological agents in CRMO resistant to NSAIDs, bisphosphonates or steroids [58, 59].

## 7.6.6   Biologics

Initial reports are being published on the use of biological agents in CRMO. However, it must be remembered that the patients treated with biologics represent a subset of patients with disease that is resistant to NSAIDs, pamidronate and, in some cases, steroids.

Infliximab is a chimeric monoclonal IgG1k antibody with action against tumour necrosis factor-alpha (TNF-$\alpha$) and has been used in a handful of patients in varying doses with mostly positive results.

A large (89 patients) retrospective cohort study found two patients had been treated successfully with infliximab [9]. Another child with CRMO followed by Crohn's disease also had a good response to infliximab with resolution of pain and decrease in size of the bony lesion [22]. A further case report using a lower dose of 3 mg/kg at the same intervals led to resolution of pain at the third dose, and continuing for the 2 years of follow-up. In this instance, bone scans were used to demonstrate radiological improvement [60].

With less promising results, a case report of two children receiving Infliximab found that although both patients initially showed improvement of symptoms, one child had to increase treatment to every 4 weeks due to a recurrence of symptoms with azathioprine added before symptoms were fully controlled; the second child stopped infliximab due to a possible fungal skin infection [61]. Using a different dosage, a child with CRMO resistant to all other treatments was treated with infliximab (5 mg/kg) every 4 weeks for 12 months and then at 8 weekly intervals. This resulted in remission of symptoms and allowed the cessation of NSAIDs and steroids [62].

There is one published report of two cases where etanercept (a recombinant, humanised TNF receptor which antagonises soluble TNF) was used in combination with methotrexate and led to clinical remission [63].

The role of TNF in the progression of CRMO is unclear. Whilst the case reports of efficacy of TNF blockade suggest some involvement, the mouse model of CRMO has not implicated TNF as the major factor [64].

Anakinra (recombinant IL-1ra) was used in one patient with CRMO having first demonstrated marked elevation of interleukin 1ra. Using a daily dose of 2 mg/kg, there was a good response initially that persisted for 17 months before further symptoms developed and, after 3 years, necessitated switching therapy to adalimumab (humanized IgG1 anti-TNF-α mAb) [61].

There is more evidence for use of biologics in patients with SAPHO, in whom an increase in TNF-α has been well described [65–70].

## 7.7 Genetics

There are several reports of family members being affected, including two pairs of monozygotic twins, suggesting a genetic susceptibility [9, 18, 71].

A mouse with a spontaneous mutation on chromosome 18 following an autosomal recessive inheritance was found to have similar abnormalities to human CRMO [72]. Two human disorders of bone have been localised to human chromosome 18q [73, 74] and therefore this seems like a possible area for a susceptibility gene. A significant association of CRMO with a rare allele of marker D18S60 microsatellite polymorphism suggests the existence of a susceptibility gene on chromosome 18 which may contribute to the aetiology of CRMO [71].

A missense mutation in PSTPIP2 in the chronic multifocal osteomyelitis mouse [75] led to this being investigated as a candidate gene; however, a series screening ten patients with CRMO for mutations in the genes PSTPIP1 and PSTPIP2 found no mutations [9].

From studies to date it is not possible to infer a monogenic inheritance pattern.

## 7.8 Outcomes

The course of CRMO is very variable with symptoms remitting and relapsing. The disease was initially thought to resolve spontaneously, going into remission particularly during puberty. However, as follow-up studies grow in length, it is apparent that many patients continue to have long-term impairments, such as bony deformity, disability and chronic pain [10, 11, 37]. Some patients evolve into an atypical spondyloarthropathy [20].

## 7.9 Conclusion

It is now 30 years since CRMO was first described but recently our understanding of its nature has changed. Rather than being an auto-immune disorder as first thought, it is now thought to be one of the autoinflammatory disorders, where systemic inflammation occurs due to a dysregulation of the immune system. Although it has always been thought of as very rare, the prevalence is likely to be vastly underestimated due to poor recognition of the disease. Diagnosis may be made easier by the advent of STIR MRI scanning, potentially obviating the need for biopsy and avoiding unnecessary long-term antibiotic treatment due to incorrect diagnosis. Treatment options are increasing with an expanding evidence base for the use of bisphosphonates. It is early days for the use of biologics but initial results appear promising in resistant cases. Gene identification has not proved as easy as first thought when a mouse model was identified, although research in this area continues.

The early descriptions of CRMO suggest a benign course; however, longer-term follow up is showing that the disease can cause significant morbidity and longer term disability.

## References

1. Giedion A, Holthusen W, Masel LF, Vischer D (1972) Subacute and chronic "symmetrical" osteomyelitis. Ann Radiol 15(3):329–342 (Paris)
2. Björkstén B, Gustavson KH, Eriksson B, Lindholm A, Nordström S (1978) Chronic recurrent multifocal osteomyelitis and pustulosis palmoplantaris. J Pediatr 93(2):227–231
3. Iyer RS, Thapa MM, Chew FS (2011) Chronic recurrent multifocal osteomyelitis: review. AJR Am J Roentgenol 196:87–91
4. Chamot AM, Benhamou CL, Kahn MF, Beraneck L, Kaplan G, Prost A (1987) Acne-pustulosis-hyperos-

tosis-osteitis syndrome. Results of a national survey. 85 cases. Rev Rhum Mal Osteoartic 54(3):187–196

5. Gikas PD, Islam L, Aston W, Tirabosco R, Saifuddin A, Briggs TW, Cannon SR, O'Donnell P, Jacobs B, Flanagan AM (2009) Nonbacterial osteitis: a clinical, histopathological, and imaging study with a proposal for protocol-based management of patients with this diagnosis. J Orthop Sci 14(5):505–516

6. Schedel J, Bach B, Kümmerle-Deschner JB, Kötter I (2011) Autoinflammatory syndromes/ fever syndromes. Hautarzt 62(5):389–401

7. Ferguson PJ, El-Shanti HI (2007) Autoinflammatory bone disorders. Curr Opin Rheumatol 19(5):492–498

8. Galeazzi M, Gasbarrini G, Ghirardello A, Grandemange S, Hoffman HM, Manna R, Podswiadek M, Punzi L, Sebastiani GD, Touitou I, Doria A (2006) Autoinflammatory syndromes. Clin Exp Rheumatol 24:79–85

9. Jansson A, Renner ED, Ramser J, Mayer A, Haban M, Meindl A, Grote V, Diebold J, Jansson V, Schneider K, Belohradsky BH (2007) Classification of non-bacterial osteitis: retrospective study of clinical, immunological and genetic aspects in 89 patients. Rheumatology 46:154–160 (Oxford)

10. Catalano-Pons C, Comte A, Wipff J, Quartier P, Faye A, Gendrel D, Duquesne A, Cimaz R, Job-Deslandre C (2008) Clinical outcome in children with chronic recurrent multifocal osteomyelitis. Rheumatology 47(9):1397–1399 (Oxford)

11. Huber AM, Lam PY, Duffy CM, Yeung RS, Ditchfield M, Laxer D, Cole WG, Kerr Graham H, Allen RC, Laxer RM (2002) Chronic recurrent multifocal osteomyelitis: clinical outcomes after more than five years of follow-up. J Pediatr 141(2):198–203

12. Coinde E, David L, Cottalorda J, Allard D, Bost M, Lucht F, Stéphan JL (2001) Chronic recurrent multifocal osteomyelitis in children: report of 17 cases. Arch Pediatr 8(6):577–583

13. Khanna G, Sato TS, Ferguson P (2009) Imaging of chronic recurrent multifocal osteomyelitis. Radiographics 29(4):1159–1177

14. Hospach T, Langendoerfer M, von Kalle T, Maier J, Dannecker GE (2010) Spinal involvement in chronic recurrent multifocal osteomyelitis (CRMO) in childhood and effect of pamidronate. Eur J Pediatr 169(9):1105–1111

15. Schultz C, Holterhus PM, Seidel A, Jonas S, Barthel M, Kruse K, Bucsky P (1999) Chronic recurrent multifocal osteomyelitis in children. Pediatr Infect Dis J 18:1008–1013

16. King SM, Laxer RM, Manson D, Gold R (1987) Chronic recurrent multifocal osteomyelitis: a noninfectious inflammatory process. Pediatr Infect Dis J 6:907–911

17. Jurik AG, Helmig O, Ternowitz T, Møller BN (1988) Chronic recurrent multifocal osteomyelitis: a follow-up study. J Pediatr Orthop 8:49–58

18. Paller AS, Pachman L, Rich K, Esterly NB, Gonzalez-Crussi F (1985) Pustulosis palmaris et plantaris: its association with chronic regional multifocal osteomyelitis. J Am Acad Dermatol 12:927–930

19. Van Howe RS, Starshak RJ, Chusid MJ (1989) Chronic, recurrent multifocal osteomyelitis. Case report and review of the literature. Clin Pediatr 28(2):54–59 (Phila)

20. Huaux JP, Esselinckx W, Rombouts JJ, Maldague B, Malghem J, Devogelaer JP, Nagant de Deuxchaisnes C (1988) Pustulotic arthroosteitis and chronic recurrent multifocal osteomyelitis in children: report of three cases. J Rheumatol 15:95–100

21. Bousvaros A, Marcon M, Treem W, Waters P, Issenman R, Couper R, Burnell R, Rosenberg A, Rabinovich E, Kirschner BS (1999) Chronic recurrent multifocal osteomyelitis associated with chronic inflammatory bowel disease in children. Dig Dis Sci 44:2500–2507

22. Carpenter E, Jackson MA, Friesen CA, Scarbrough M, Roberts CC (2004) Crohn's-associated chronic recurrent multifocal osteomyelitis responsive to infliximab. J Pediatr 144(4):541–544

23. Bognar M, Blake W, Agudelo C (1998) Chronic recurrent multifocal osteomyelitis associated with Crohn's disease. Am J Med Sci 315:133–135

24. Schilling F, Märker-Hermann E (2003) Chronic recurrent multifocal osteomyelitis in association with chronic inflammatory bowel disease: enteropathic CRMO. Z Rheumatol 62(6):527–538

25. Bazrafshan A, Zanjani KS (2000) Chronic recurrent multifocal osteomyelitis associated with ulcerative colitis: a case report. J Pediatr Surg 35:1520–1522

26. Carr AJ, Cole WG, Robertson DM, Chow CW (1993) Chronic multifocal osteomyelitis. J Bone Joint Surg Br 75:582–591

27. Wipff J, Adamsbaum C, Kahan A, Job-Deslandre C (2011) Chronic recurrent multifocal osteomyelitis. Joint Bone Spine 78(6):555–560

28. Job-Deslandre C, Krebs S, Kahan A (2001) Chronic recurrent multifocal osteomyelitis: five-year outcomes in 14 pediatric cases. Joint Bone Spine 68:245–251

29. Girschick HJ, Krauspe R, Tschammler A, Huppertz HI (1998) Chronic recurrent osteomyelitis with clavicular involvement in children: diagnostic value of different imaging techniques and therapy with non-steroidal anti-inflammatory drugs. Eur J Pediatr 157:28–33

30. Fritz J, Tzaribatchev N, Claussen CD, Carrino JA, Horger MS (2009) Chronic recurrent multifocal osteomyelitis: comparison of whole-body MR imaging with radiography and correlation with clinical and laboratory data. Radiology 252(3):842–851

31. Bjorkstén B, Boquist L (1980) J Histopathological aspects of chronic recurrent multifocal osteomyelitis. Bone Joint Surg Br 62(3):376–380

32. Probst FP, Bjorksten B, Gustavson K-H (1978) Radiological aspect of chronic recurrent multifocal osteomyelitis. Ann Radiol 21:115–112 (Paris)

33. Chow LT, Griffith JF, Kumta SM, Leung PC (1999) Chronic recurrent multifocal osteomyelitis: a great

clinical and radiologic mimic in need of recognition by the pathologist. APMIS 107(4):369–379

34. Girschick HJ, Huppertz HI, Harmsen D, Krauspe R, Müller-Hermelink HK, Papadopoulos T (1999) Chronic recurrent multifocal osteomyelitis in children: diagnostic value of histopathology and microbial testing. Hum Pathol 30(1):59–65

35. Mortensson W, Edeburn G, Fries M, Nilsson R (1988) Chronic recurrent multifocal osteomyelitis in children. A roentgenologic and scintigraphic investigation. Acta Radiol 29(5):565–570

36. Yu L, Kasser JR, O'Rourke E, Kozakewich H (1989) Chronic recurrent multifocal osteomyelitis. Association with vertebra plana. J Bone Joint Surg Am 71:105–112

37. Duffy CM, Lam PY, Ditchfield M, Allen R, Graham HK (2002) Chronic recurrent multifocal osteomyelitis: review of orthopaedic complications at maturity. J Pediatr Orthop 22(4):501–505

38. Manson D, Wilmot DM, King S, Laxer RM (1989) Physeal involvement in chronic recurrent multifocal osteomyelitis. Pediatr Radiol 20:76–79

39. Miettunen PM, Wei X, Kaura D, Reslan WA, Aguirre AN, Kellner JD (2009) Dramatic pain relief and resolution of bone inflammation following pamidronate in 9 pediatric patients with persistent chronic recurrent multifocal osteomyelitis (CRMO). Pediatr Rheumatol Online J 7:2

40. Chiu CK, Singh VA (2009) Chronic recurrent multifocal osteomyelitis of the first metatarsal bone: a case report. J Orthop Surg 17(1):119–122 (Hong Kong)

41. Abril JC, Ramirez A (2007) Successful treatment of chronic recurrent multifocal osteomyelitis with indomethacin: a preliminary report of five cases. J Pediatr Orthop 27(5):587–591

42. Girschick HJ, Raab P, Surbaum S, Trusen A, Kirschner S, Schneider P et al (2005) Chronic non-bacterial osteomyelitis in children. Ann Rheum Dis 64:279–285

43. Handrick W, Hormann D, Voppmann A, Schille R, Reichardt P, Trobs RB et al (1998) Chronic recurrent multifocal osteomyelitis—report of eight patients. Pediatr Surg Int 14:195–198

44. Holden W, David J (2005) Chronic recurrent multifocal osteomyelitis: two cases of sacral disease responsive to corticosteroids. Clin Infect Dis 40:616–619

45. Ishikawa-Nakayama K, Sugiyama E, Sawazaki S, Taki H, Kobayashi M, Koizumi F, Furuta I (2000) Chronic recurrent multifocal osteomyelitis showing marked improvement with corticosteroid treatment. J Rheumatol 27:1318–1319

46. Simm PJ, Allen RC, Zacharin MR (2008) Bisphosphonate treatment in chronic recurrent multifocal osteomyelitis. J Pediatr 152(4):571–575

47. Compeyrot-Lacassagne S, Rosenberg AM, Babyn P, Laxer RM (2007) Pamidronate treatment of chronic noninfectious inflammatory lesions of the mandible in children. J Rheumatol 34(7):1585–1589

48. Yamazaki Y, Satoh C, Ishikawa M, Notani K, Nomura K, Kitagawa Y (2007) Remarkable response of juvenile diffuse sclerosing osteomyelitis of mandible to pamidronate. Oral Surg Oral Med Oral Pathol Oral Radiol Endod 104(1):67–71

49. Gleeson H, Wiltshire E, Briody J, Hall J, Chaitow J, Sillence D, Cowell C, Munns C (2008) Childhood chronic recurrent multifocal osteomyelitis: pamidronate therapy decreases pain and improves vertebral shape. J Rheumatol 35(4):707–712

50. Glorieux FH, Bishop NJ, Plotkin H, Chabot G, Lanoue G, Travers R (1998) Cyclic administration of pamidronate in children with severe osteogenesis imperfecta. N Engl J Med 339(14):947–952

51. De Cunto A, Maschio M, Lepore L, Zennaro F (2009) A case of chronic recurrent multifocal osteomyelitis successfully treated with neridronate. J Pediatr 154(1):154–155

52. Rogers MJ, Gordon S, Benford HL et al (2000) Cellular and molecular mechanisms of action of bisphosphonates. Cancer 88(12 Suppl): 2961–2978

53. Toussirot E, Wendling D (2007) Antiinflammatory treatment with bisphosphonates in ankylosing spondylitis. Curr Opin Rheumatol 19(4):340–345

54. Dicuonzo G, Vincenzi B, Santini D, Avvisati G, Rocci L, Battistoni F, Gavasci M, Borzomati D, Coppola R, Tonini G (2003) Fever after zoledronic acid administration is due to increase in TNF-alpha and IL-6. J Interferon Cytokine Res 23(11):649–654

55. Hewitt RE, Lissina A, Green AE, Slay ES, Price DA, Sewell AK (2005) The bisphosphonate acute phase response: rapid and copious production of proinflammatory cytokines by peripheral blood gd T cells in response to aminobisphosphonates is inhibited by statins. Clin Exp Immunol 139(1):101–111

56. Gallacher SJ, Ralston SH, Patel U, Boyle IT (1989) Side-effects of pamidronate. Lancet 2:42–43

57. Adami S, Bhalla AK, Dorizzi R, Montesanti F, Rosini S, Salvagno G, Lo Cascio V (1987) The acute-phase response after bisphosphonate administration. Calcif Tissue Int 41:326–331

58. Gallagher KT, Roberts RL, MacFarlane JA, Stiehm ER (1997) Treatment of chronic recurrent multifocal osteomyelitis with interferon gamma. J Pediatr 131(3):470–472

59. Andersson R (1995) Effective treatment with interferon-alpha in chronic recurrent multifocal osteomyelitis. J Interferon Cytokine Res 15(10):837–838

60. Marangoni RG, Halpern AS (2010) Chronic recurrent multifocal osteomyelitis primarily affecting the spine treated with anti-TNF therapy. Spine (Phila Pa 1976) 35(7):E253–256

61. Eleftheriou D, Gerschman T, Sebire N, Woo P, Pilkington CA, Brogan PA (2010) Biologic therapy in refractory chronic non-bacterial osteomyelitis of childhood. Rheumatology 49(8):1505–1512 (Oxford)

62. Deutschmann A, Mache CJ, Bodo K, Zebedin D, Ring E (2005) Successful treatment of chronic recurrent multifocal osteomyelitis with tumor necrosis factor-alpha blockage. Pediatrics 116(5):1231–1233

63. Eisenstein EM, Syverson GD, Vora SS, Williams CB (2011) Combination therapy with methotrexate and

etanercept for refractory chronic recurrent multifocal osteomyelitis. J Rheumatol 38(4):782–783

64. Hentunen TA, Choi SJ, Boyce BF, Dallas MR, Dallas SL, Shen-Ong GL, Roodman GD (2000) A murine model of inflammatory bone disease. Bone 26:183–188

65. Wagner AD, Andresen J, Jendro MC, Hulsemann JL, Zeidler H (2002) Sustained response to tumor necrosis factor alpha-blocking agents in two patients with SAPHO syndrome. Arthritis Rheum 46:1965–1968

66. Hurtado-Nedelec M, Chollet-Martin S, Nicaise-Roland P, Grootenboer-Mignot S, Ruimy R, Meyer O, Hayem G (2008) Characterization of the immune response in the synovitis, acne, pustulosis, hyperostosis, osteitis (SAPHO) syndrome. Rheumatology 47:1160–1167

67. Olivieri I, Padula A, Ciancio G, Salvarani C, Niccoli L, Cantini F (2002) Successful treatment of SAPHO syndrome with infliximab: report of two cases. Ann Rheum Dis 61:375–376

68. Widmer M, Weishaupt D, Brühlmann P, Michel BA, Forster A (2003) Infliximab in the treatment of SAPHO syndrome: clinical experience and MRI response (Abstract). Ann Rheum Dis 62(Suppl. I):250–251

69. Asmussen KH (2003) Successful treatment of SAPHO syndrome with infliximab. A case report. Arthritis Rheum 48(Suppl.1):S621

70. Kyriazis NC, Tachoula AV, Sfontouris CI (2004) Successful treatment of refractory SAPHO syndrome with infliximab. Ann Rheum Dis 63(Suppl. I):388–389

71. Golla A, Jansson A, Ramser J, Hellebrand H, Zahn R, Meitinger T, Belohradsky BH, Meindl A (2002) Chronic recurrent multifocal osteomyelitis (CRMO): evidence for a susceptibility gene located on chromosome 18q21.3–18q22. Eur J Hum Genet 10(3):217–221

72. Byrd L, Grossmann M, Potter M, Shen-Ong GL (1991) Chronic multifocal osteomyelitis, a new recessive mutation on chromosome 18 of the mouse. Genomics 11(4):794–798

73. Hughes AE, Shearman AM, Weber JL, Barr RJ, Wallace RG, Osterberg PH, Nevin NC, Mollan RA (1994) Genetic linkage of familial expansile osteolysis to chromosome 18q. Hum Mol Genet 3:359–361

74. Cody JD, Singer FR, Roodman GD, Otterund B, Lewis TB, Leppert M, Leach RJ (1997) Genetic linkage of Paget disease of the bone to chromosome 18q. Am J Hum Genet 61:1117–1123

75. Ferguson PJ, Bing X, Vasef MA, Ochoa LA, Mahgoub A, Waldschmidt TJ, Tygrett LT, Schlueter AJ, El-Shanti H (2006) A missense mutation in pstpip2 is associated with the murine autoinflammatory disorder chronic multifocal osteomyelitis. Bone 38(1):41–47

# Recognition and Treatment of Chlamydial Infections from Birth to Adolescence

**8**

Toni Darville

## Abstract

The "silent epidemic" of *Chlamydia trachomatis* threatens to cause reproductive damage and infertility in many of the 50 million women who acquire it each year. Female reproductive tract infection has more recently been linked to stillbirth and premature delivery. Innate immune cells and mediators appear to be the primary players in pathogenesis, with neutrophils playing a prominent role in disease development. Although adaptive antibody and CD4 T cell responses appear primarily protective, these responses are inefficient. Infections are frequently chronic as a result, and when infection is diagnosed and treated with appropriate antibiotics, repeated infection is the rule. The lack of acute symptoms in many infected individuals contributes to the high prevalence of chlamydial infection. Although chronic sequelae are relatively rare in men, and many women sustain infection without developing pelvic inflammatory disease or chronic sequelae, the extremely high prevalence of chlamydial infection leads to significant morbidity and healthcare costs. A vaccine is urgently needed to prevent infection, but given the difficulties of inducing a CD4 T cell memory response that can home quickly to the genital tract, induction of sterilizing immunity may not be possible. A vaccine that prevents disease by lowering bacterial burden and dampening production of tissue-damaging responses may be possible. Until an efficacious vaccine is developed, screening and treatment programs appear to be the best method of disease prevention.

## 8.1 Introduction

T. Darville (✉)
Division of Pediatric Infectious Diseases, College of Medicine, University of Pittsburgh Medical Center, Pittsburgh, PA, USA
e-mail: darvilletonil@chp.edu

*Chlamydia trachomatis* is an obligate, intracellular, nonmotile, Gram-negative bacterium recognized as one of the most common sexually transmitted agents in the world. Chlamydial genital infection primarily affects sexually

N. Curtis et al. (eds.), *Hot Topics in Infection and Immunity in Children IX*,
Advances in Experimental Medicine and Biology 764, DOI 10.1007/978-1-4614-4726-9_8,
© Springer Science+Business Media New York 2013

active adolescents and young adults. Large-scale screening programs routinely detect infection rates of 5–10 % in young adults (19–25 years of age) [1, 2], and 10–20 % or greater in sexually active adolescents 15–19 years of age [3]. Most infected persons do not have symptoms, thus they do not seek medical care, and their infections go undetected. Consequently, screening is necessary to identify and treat this infection. The large reservoir of unrecognized infected individuals helps sustain transmission of this organism. Among men, urethritis is the most common illness resulting from C. trachomatis infection. Complications (e.g., epididymitis) affect a minority of infected men and rarely result in sequelae. Women bear the brunt of disease due to infection, for if left untreated, infection can ascend from the cervix to infect the uterus and Fallopian tubes to cause pelvic inflammatory disease (PID). Inflammation of the Fallopian tubes can lead to subsequent scar formation, and tubal occlusion. Tubal obstruction can lead to ectopic pregnancy, subfertility and infertility. In addition, an infected pregnant woman can transmit the organism to her newborn at the time of delivery, potentially resulting in neonatal conjunctivitis and/or afebrile pneumonia.

## 8.2   The Pathogen

Chlamydiae are obligate intracellular parasites that have been classified under the order Chlamydiales with their own family and genus (Chlamydiaceae, *Chlamydia*). Chlamydiae are small in size (0.25–0.8 µm in diameter) compared with typical bacteria such as *Escherichia coli* (1.0 µm) and have small chromosomes ranging from 1.0 to 1.2 megabases in size. They are Gram-negative in architecture and composition, with an outer membrane containing lipopolysaccharide (LPS), which is truncated, and not very endotoxic and a cytoplasmic membrane. While the classic bacterial cell wall component peptidoglycan has not been confirmed by isolation and identification, chlamydiae possess all the genes needed for its synthesis and are susceptible to β-lactam antibiotics [4]. Although chlamydiae contain DNA,

RNA, and ribosomes, during growth and replication these obligate intracellular bacteria parasitize their host epithelial cell for nutrients and are auxotrophic for several amino acids and three of the four nucleoside triphosphates; the demand for host cell adenosine triphosphate (ATP) has led to their designation as an "energy parasite." *Chlamydia trachomatis* encodes an abundant protein called the major outer membrane protein (MOMP or OmpA) that is surface exposed and is the major determinant of serologic classification. Almost all strains of C. trachomatis harbor a plasmid, which confers the virulence properties of enhanced attachment/uptake and activation of the innate immune receptor, Toll-like receptor 2 (TLR2) [5].

### 8.2.1   Chlamydial Developmental Cycle

The biphasic developmental cycle of chlamydiae is unique among microorganisms and involves two highly specialized morphologic forms, both of which are required for infection and disease to occur: the infectious, extracellular form called an elementary body (EB) and the noninfectious but metabolically active intracellular form called a reticulate body (RB). The EB contains extensive disulfide cross-links both within and between outer membrane proteins giving it an almost spore like structure that is stable outside of the cell. The small infectious EB is inactive metabolically. Infection is initiated by attachment of EBs to the apical surfaces of epithelial cells of the conjunctiva, urogenital, or respiratory tracts, followed by receptor-mediated endocytosis. The EBs quickly modify their early endosomal membrane to exit the endosomal pathway, thereby avoiding fusion with lysosomes, and traffic on microtubules to the peri-Golgi/nuclear hof region. The EB-containing endosomes of C. trachomatis then fuse homotypically with one another to form their one nascent microcolony called an inclusion. The EBs then transform into RBs, the chromosome becomes relaxed and transcriptionally active, and metabolic growth and binary fission ensue to generate progeny. Chlamydiae-directed modification of their inclusion membrane permits interception of trans-Golgi vesicles for transfer of sphingomyelin and glycerolphospholipids to the inclusion membrane, which can expand to accommodate some 200–1,000 progeny; this strategy of acquiring host cell markers for the inclusion also provides some degree of camouflage for the chlamydiae within. After 48–72 h, multiplication ceases and nucleoid condensation occurs as the RBs transform to new infectious EBs. The EBs are then released from the cell, allowing for infection of new host cells to occur.

The biphasic and relatively prolonged developmental cycle of chlamydiae are survival advantages. Antibiotic treatment or the host immune response must be able to kill both extracellular non-replicating infectious EBs as well as intracellular replicating RBs hidden within their protective vacuole if they are to rid the host of infection. Thus, antibiotic treatment requires multiple-dose regimens for 7–14 days. Single-dose azithromycin treats genital *C. trachomatis* infection effectively because it has a half-life in host cells of 5–7 days. The ability to cause prolonged, often subclinical infection is a major characteristic of chlamydiae.

## 8.2.2 Classification

*Chlamydia trachomatis* has been divided into subgroups based on antigenic variation in the major outer membrane proteins (serovars) and clinical expression. Microimmunofluorescence and monoclonal antibody testing have shown that there are more than 18 serovars of *C. trachomatis* with several distinctive clinical patterns of disease: trachoma is caused by serovars A, B, Ba, and C; oculogenital and neonatal disease by serovars B, Da, Ga, Ia, and D-K; and lymphogranuloma venereum (LGV) by serovars L1, L2, L2a, and L3. LGV infections are more invasive, as these serovars can replicate in macrophages, whereas replication of the other serovars of *C. trachomatis* is confined to mucosal epithelial cells.

## 8.3 Pathogenesis

### 8.3.1 Immunopathogenesis

In the realm of infectious diseases, it has often been observed that an overly aggressive inflammatory host response can be more problematic than the infection that initiated it. This is certainly true in the case of chlamydial infection, where the pathology that leads to the serious morbidities of chronic pelvic pain, ectopic pregnancy and infertility after female genital tract infection, is the result of the host inflammatory response. The *cellular paradigm* of chlamydia pathogenesis [6] states that the host response to chlamydiae is initiated and sustained by epithelial cells that are the primary targets of chlamydial infection. Infected host epithelial cells act as first responders, initiating and propagating immune responses through recognition of various chlamydial ligands via pathogen recognition receptors. They secrete chemokines that recruit inflammatory leukocytes to the site of infection, as well as cytokines that induce and augment the cellular inflammatory response and these mediators induce direct damage to the tissues. Unfortunately, this response is frequently ineffective at resolving infection, and ongoing stimulation of the host cells and bystander cells leads to continued release of tissue-damaging mediators. Since reinfection with chlamydiae is a frequent occurrence, repeated inflammatory responses may lead to repeated insult to the tissues, and promote further scarring. Since the host cell response to bacteria is the inciting inflammatory event, increased and prolonged bacterial burden correlates directly with disease development. Pathogen-specific and environmental factors that promote infection and bacterial survival lead to enhanced disease.

*In vivo* immunological studies in animal models and immune-epidemiological studies in humans indicate that resolution of infection can occur with minimal to no disease provided that the correct responses are induced in the right amount. Recognizing that the immune response to this organism leads to tissue damage, it is important to delineate the specific host responses involved in disease promotion both for rational vaccine design and the discovery of biomarkers to monitor the effectiveness of candidate therapeutics and vaccines. Limited studies have been conducted in women investigating human genetic functional polymorphisms related to innate immune molecules. Investigation of polymorphisms in the gene for the innate immune receptor, TLR2 revealed single nucleotide polymorphisms associated with protection against tubal disease following *C. trachomatis* infection [7]. Interestingly, studies in the mouse model of female chlamydial genital tract infection have revealed activation of TLR2 is a key mechanism for induction of oviduct pathology [8]. Chlamydial induced activation of TLR2 leads to enhanced neutrophil influx, cytokine and protease production and prolongs neutrophil longevity [9]. Multiple reports indicate an important role for neutrophils and their products in oviduct tissue damage [10–13].

### 8.3.2  Immunoprotection

The adaptive immune response that occurs with chlamydial infection may lead to natural resolution of infection over time. However, the chronicity of infection in women indicates the suboptimal nature of the response. The natural course of *C. trachomatis* infection was described in a study of Columbian women followed for a 5-year period [14]. Eighty-two women found to be positive for *C. trachomatis* at the start of the study were studied at 6-month intervals. Most of the women (57.3 %) were >30 years of age (70.7 % were >25 years of age). Infection was classified as persistent if the same serotype was found at follow-up visits. Women who had taken antibiotics effective against *C. trachomatis* while infected were excluded. All study women reported 1–2 lifetime sex partners (82.9 % reported a single lifelong sex partner), thus the potential for repeated infection from an untreated male sex partner was high. Approximately 46 % of the infections were persistent at 1 year, 18 % at 2 years, and 6 % at 4 years of follow-up as determined by PCR of cervical scrape samples. Thus,

in nearly half of this female cohort, an adaptive immune response effective in eradicating their infection or in preventing repeat infection did not develop for up to 1 year.

The adaptive response is also suboptimal with respect to protection from reinfection. The high frequency of repeat infections found in clinic-based studies has led some authors to recommend screening female adolescents for *Chlamydia* as frequently as every 6 months [15]. However, there is a strong inverse relationship between age and susceptibility to chlamydial infection even when corrected for frequency of sexual contact, suggesting effective adaptive immunity eventually develops. Lymphoproliferative responses, but not serum antibody titers increase with age [16]. Data from humans point to MHC Class II–restricted CD4+ T cells of the Th1 phenotype as being critical to recovery from chlamydial infection as well as having a role in protection from disease [17, 18]. In a cohort of female commercial sex workers with HIV, susceptibility to chlamydial PID increased as numbers of CD4+ T cells decreased. Furthermore, in a prospective cohort study of commercial sex workers in Nairobi at high risk of exposure, production of IFN-$\gamma$ by peripheral blood mononuclear cells stimulated with chlamydial antigen strongly correlated with protection against incident *C. trachomatis* infection [19].

Data from the mouse model of genital tract infection reveal that chlamydia-specific B cells and antibody effectively lower the bacterial burden upon challenge, thereby partially protecting the oviduct from infection and disease [20, 21]. Surface proteins including MOMP are likely the principal targets of neutralizing antibodies. Anti-chlamydial antibodies are not sufficient to protect humans from reinfection. In female sex workers levels of chlamydial EB-specific IgA and IgG detected in endocervical mucus and plasma were not significantly associated with a decreased risk of infection [19]. Although antibody may not play a primary role in protection from reinfection, it may help control the shedding of organisms and protect against upper tract disease. One study reported the prevalence of mucosal IgA antibodies was inversely related to the quantity of *C. trachomatis* shed from the human endo-cervix [22], and another found the presence of serum IgA and IgG antibodies reduced the risk for ascending infection among women undergoing therapeutic abortion [23].

## 8.4  Epidemiology

*Chlamydia trachomatis* is the most common bacterial sexually transmitted infection, with an estimated 92 million cases occurring globally each year, including more than four million in sexually active adolescents and adults in the United States [24]. In reports from other parts of the world, the prevalence ranges from 28.5 % among female sex workers in Dakar [25], to 5.7 % among pregnant women in Thailand [26], and 0.8 % overall among women seen in private gynecology practices in Paris and 5.2 % for those under the age of 21 years [27].

In the US, substantial racial/ethnic disparities are present in the prevalence of both chlamydial and gonococcal infections. One large study of US female military recruits found a chlamydial prevalence of 9 % that was maintained over four consecutive years [28]. Young age, black race, home-of-record from the south, more than one sex partner, a new sex partner, lack of condom use, and a history of having a sexually transmitted disease were correlates of chlamydia infection.

Urine screening for chlamydial infection in Louisiana public schools revealed the overall prevalence of *C. trachomatis* was 6.5 %, with rates among girls more than twice that of boys (9.7 % vs. 4.0 %). The highest prevalence for boys occurred in 12th grade (8.9 %), whereas the highest prevalence for girls occurred in 10th grade (15.8 %) [29]. The high prevalence rates in this cohort are in contrast to a rate of 0.9 % for a cohort of 1,114 patients aged 15–24 years in two pediatric private practices in suburban North Carolina. In sexually active participants, prevalence was 2.1 %; in sexually active females, 2.7 %; and in sexually active males, 0.9 %. Most participants were female (63 %), white (87 %), and from highly educated families (64 % of their mothers graduated from college) [30].

Identification of persons with the highest risk of infection should enhance cost-effectiveness of screening and treatment programs. However, in a recent study among 3,202 sexually active adolescent females attending middle school health centers in Baltimore, MD chlamydial infection was found in 771 first visits (24.1 %) and 299 repeat visits (13.9 %); 29.1 % had at least one positive test result [15]. Unfortunately, independent predictors of chlamydial infection—reason for clinic visit, clinic type, prior sexually transmitted diseases, multiple or new partners, or inconsistent condom use-failed to identify a subset of adolescent females with the majority of infections.

## 8.5  Clinical Manifestations

Most studies report that 25 % of men and 70 % of women infected with *C. trachomatis* are asymptomatic or minimally symptomatic. The National Longitudinal Study of Adolescent Health Study collected data prospectively from 14,322 US adolescents and followed them into adulthood [1]. Of the participants that tested positive for chlamydial infection, 95 % did not report symptoms in the 24 h preceding specimen collection. Among men with chlamydial infection, the prevalence of urethral discharge and dysuria were only 3.3 % and 1.9 %, respectively. Among women with chlamydial infection, the prevalence of vaginal discharge and dysuria were 0.3 % and 4.2 %, respectively. Among the small number of young men reporting urethral discharge ($n = 17$), the prevalence of chlamydial infection was high (38.5 %), whereas the prevalence of chlamydial infection was only 0.9 % among women reporting vaginal discharge ($n = 98$). Of note, 6.0 % of the women reporting dysuria ($n = 232$) had chlamydial infection [1].

### 8.5.1  Infections in Males

When symptomatic, males frequently complain of dysuria or note a clear or mucopurulent urethral discharge at least 7–14 days following contact with an infected partner [31]. The discharge may be so slight as to be demonstrable only after penile stripping and then only in the morning. Some patients may deny the presence of discharge but may note stained underwear in the morning resulting from scant discharge overnight. The primary complications of chlamydial urethritis in men are (1) epididymitis; (2) sexually reactive arthritis, including Reiter's syndrome; and (3) transmission to women. *Chlamydia trachomatis* and *N. gonorrhoeae* are the most frequent causes of epididymitis in men under age 35; urethritis also is usually present.

Although asymptomatic rectal carriage of *C. trachomatis* occurs in both infants [32] and adults, [33] *C. trachomatis* is a fairly common cause of proctitis and proctocolitis in men who have sex with men [34]. If the infection is due to a lymphogranuloma venereum strain, a severe proctocolitis can develop. Approximately 1 % of men with nongonococcal urethritis develop acute aseptic arthritis of presumed immune-mediated etiology [35]. One third of patients have the full complex of Reiter syndrome (arthritis, nonbacterial urethritis, and conjunctivitis); most such patients carry the histocompatibility antigen HLA-B27 [36].

## 8.5.2 Infections in Females

In women, chlamydial infections may cause PID, tubal infertility, chronic pelvic pain, and ectopic pregnancy. Chlamydial infection may also be linked to cervical cancer [37]. Chlamydial and gonococcal infections may increase susceptibility to and transmission of HIV in both men and women [38].

Symptoms in females include mild abdominal pain, intermittent bleeding, vaginal discharge, or dysuria-pyuria syndrome. The cervix can appear normal or exhibit edema, erythema, friability, or mucopurulent discharge. In prepubertal girls, vaginitis can occur secondary to infection of transitional cell epithelium by *C. trachomatis*. In contrast, the squamous epithelium of the adult vagina is not susceptible to chlamydiae, and vaginal discharge generally reflects endocervical infection.

Pelvic inflammatory disease is a sexually transmitted infection that ascends from the vagina and cervix to involve the uterus, ovaries, and peritoneal tissues as well as the fallopian tubes. Lower abdominal pain, usually bilateral, is the most common presenting symptom. Pain may be associated with an abnormal vaginal discharge, abnormal uterine bleeding, dysuria, dyspareunia, nausea, vomiting, fever, or other constitutional symptoms. It may also be present in a subclinical form that lacks the typical acute symptoms, but continues to lead to the associated long-term sequelae of infertility and ectopic pregnancy [39]. The most important causative organisms are *C. trachomatis* and *N. gonorrhoeae*; one or both of these agents cause well over half of cases. Other microorganisms implicated in PID include organisms found in the abnormal vaginal flora of women with bacterial vaginosis, such as bacteroides species, anaerobic cocci, *Mycoplasma hominis*, and *Ureaplasma urealyticum*. *Escherichia coli* and other enteric organisms have also been found.

A recent review of 24 studies examined PID diagnosis and sequelae after untreated chlamydial infection [40]. In one study, eighteen of 109 (16.5 %) asymptomatic adolescent women infected with *C. trachomatis* followed for 2 months or more became symptomatic, but only 2 (1.8 %) developed clinical PID [41]. On average in high-risk settings, 2–5 % of untreated females developed PID within the ~2-week period between testing positive for *C. trachomatis* and returning for treatment [42, 43]. However, the rate of progression to PID in the general, asymptomatic population followed up for longer periods appeared to be low [44]. The best data regarding the risk of PID over a long period of time is probably from the prevention of pelvic infection (POPI) trial conducted in the UK. Women in the control group had their vaginal swab stored. When analyzed 12 months later, 7 of 75 (9.5 %) women positive for chlamydial infection at baseline had developed PID [45].

The spectrum of PID associated with *C. trachomatis* infection ranges from acute, severe disease with perihepatitis and ascites (Fitz-Hugh-Curtis syndrome), to asymptomatic or "silent" disease. When women with chlamydial salpingitis are compared to women with gonococcal or with nongonococcal-nonchlamydial salpingitis, they are more likely to experience a chronic, subacute course with a longer duration of abdominal pain before seeking medical care. Yet, they have as much or more tubal inflammation at laparoscopy [46]. In several studies, repeated chlamydial infection was associated with PID and other reproductive sequelae, although it was difficult to determine whether the risk per infection increased with each recurrent episode [17, 47]. No prospective studies have directly assessed the risk of infertility after untreated *C. trachomatis* infection. However, according to the largest studies, after symptomatic PID of any cause has occurred, up to 18 % of women may develop infertility [48].

### 8.5.3 Infections in Neonates

Neonatal infection generally is acquired during passage through an infected birth canal. Prospective studies of infants born to women with a chlamydial infection of the cervix have shown a 50–75 % risk that the infant will acquire *C. trachomatis* infection at one or more anatomic sites [49–51]. In exposed infants, risk of conjunctivitis is 20–50 % and risk of pneumonia is 5–20 %.

Ophthalmia neonatorum is the major clinical manifestation of neonatal chlamydial infection [52]. The usual incubation period is 5–14 days after birth, but symptoms can occur earlier after premature rupture of membranes or as late as 6 weeks after birth [53]. Typically, the most common presenting symptom described is a watery ocular discharge which becomes progressively more purulent (95 %), followed by swelling of the eyelids (73 %) and conjunctival erythema (65 %) [52]. The majority of chlamydial conjunctivitis resolves spontaneously during the first few months of life. However, if the condition is untreated, a chronic conjunctivitis can develop and persist for weeks or months. Although conjunctivitis may be quite severe, corneal ulceration, scars and pannus formation are rare and recovery is usual without visual impairment. Mild or subclinical infection can persist in some cases for years [54]. The neovascularization of

the cornea resulting from repeated infection in classic trachoma does not occur with neonatal disease.

Afebrile pneumonia caused by *C. trachomatis* in infancy occurs characteristically between 3 and 12 weeks of age, but may sometimes present later [55–57]. Characteristically, the infant has been symptomatic for three or more weeks before presentation. Most infants are only moderately ill and are afebrile. Symptoms of nasal obstruction and a pertussis-like, non-productive staccato cough gradually worsen over a week or more. Physical findings include tachypnea and rales but not wheezing. About 50 % of the affected infants have a history or evidence of conjunctivitis; a similar proportion has middle ear abnormalities [58]. Laboratory findings may include hyperinflation with symmetric interstitial infiltrates on chest radiography, peripheral eosinophilia ($>400$ cells/mm$^3$), and increased levels of serum immunoglobulins.

## 8.6 Laboratory Diagnosis

A positive laboratory test for *C. trachomatis* can be utilized for patient education and increases both compliance with drug therapy and the likelihood of referral of sexual partners. Although the development of tissue cell culture methods in the 1960s for detecting *C. trachomatis* was a major advance, the availability of nonculture tests has dramatically increased the availability and decreased the cost of laboratory detection. Definitive diagnosis of chlamydial infection, as would be required in a medicolegal setting (i.e., suspected sexual abuse or rape), requires isolation of *C. trachomatis* in cell culture or a positive nucleic acid amplification test (NAAT) confirmed by a second NAAT that targets a different sequence [59].

### 8.6.1 Diagnostic Specimens

Many screening tests for *C. trachomatis* require appropriately handled samples containing columnar epithelium from mucosal sites (e.g., endocervix, urethra, or conjunctiva) rather than exudate; the adequacy of specimens should be verified by periodic cytologic evaluations. The discomfort caused by obtaining a urethral swab in males has precluded its widespread use in asymptomatic men. A dipstick test for leukocyte esterase performed on the first portion of a voided urine is a cost-effective and moderately sensitive screen (47–58 %) for detection of chlamydial infection in asymptomatic young males [60]. When feasible, urine NAAT provides a much more sensitive and equally noninvasive method of detecting *Chlamydia*.

## 8.6.2 Cell Culture

Use of chlamydial transport media containing antibiotics maximizes recovery and reduces the likelihood of culture overgrowth by other bacteria. Swabs used to obtain a specimen should have plastic or metal shafts, as soluble components from wooden shafts can have a toxic effect on cell cultures. Storage at −4 °C or maintenance at −70 °C is required if inoculation within 24 h is not possible. Cycloheximide-treated McCoy or HeLa cell lines are used most frequently to isolate *C. trachomatis.* Centrifugation techniques appear to enhance absorption of chlamydiae to cells. Intracytoplasmic inclusions can be detected at 48–72 h with species-specific immunofluorescent monoclonal antibodies for *C. trachomatis* and Giemsa or iodine stains. Generally, a higher isolation rate using cell culture is found in symptomatic patients than asymptomatic ones.

## 8.6.3 Nonculture Tests for *C. trachomatis*

NAATs amplify nucleic acid sequences that are specific for the organism being detected, and can detect as little as a single copy of target DNA or RNA. These tests have a higher sensitivity than all other tests, while retaining high specificity when cross contamination is being avoided [61]. NAATs have FDA approval for cervical swabs from women, urethral swabs from men, and urine from men and women [62, 63]. Similar to other

nonculture tests, NAATs do not require viable organisms. NAATs detect *C. trachomatis* in urine or in self-administered vaginal swab specimens with sensitivity comparable to clinician obtained urogenital swab specimens, which makes noninvasive testing for chlamydial infections possible on individual as well as pooled specimens from a single patient [60, 64]. Multiple studies have determined that NAAT of self-obtained vaginal swabs are an acceptable, simple and sensitive diagnostic sample for the detection of *C. trachomatis,* as well as the sexually transmitted disease pathogens, *N. gonorrhoea* and *Trichomonas vaginalis* [65, 66]. Data suggest that NAATs are equivalent to or better than culture for the detection of *C. trachomatis* in the conjunctiva and nasopharynx of infants [67], and are currently being used in evaluation of newborns with conjunctivitis or pneumonia.

Enzyme immunoassays (EIAs) use enzyme-labeled chlamydial-specific antibodies to detect chlamydial LPS. The enzyme converts a colorless substrate into a colored product, which is detected by a spectrophotometer. EIAs are less sensitive than culture and NAATs; especially when using samples that contain few organisms (asymptomatic infections). EIAs lack specificity and will detect *C. pneumoniae* in respiratory specimens.

## 8.6.4 Serology

Antibodies to *Chlamydia spp.* are best detected with a microimmunofluorescent (MIF) assay, but these assays are not widely available. Serologic screening is of very little value in uncomplicated genital infections but may be useful for population studies. The MIF assay is species-specific and sensitive but is available only at a limited number of clinical laboratories.

## 8.7 Treatment

The most widely used treatments for uncomplicated oculogenital infections caused by *C. trachomatis* in nonpregnant adolescents and adults is doxycycline for 7 days or azithromycin in a single dose [68]. In populations in which compliance with treatment is poor, azithromycin may be more cost-effective because it provides single-dose, directly observed therapy. Doxycycline costs less than azithromycin, and it has been used extensively for a longer period. Ofloxacin is similar in efficacy to doxycycline and azithromycin, but it is more expensive and offers no advantage with regard to dosage regimen. Erythromycin is less efficacious than both azithromycin and doxycycline, and gastrointestinal side effects discourage adherence to treatment.

Sex partners should be evaluated, tested, and treated if they had sexual contact with the patient during the 60 days preceding either diagnosis of *Chlamydia* or onset of symptoms in the patient. The most recent sex partner should be treated even if the time of the last sexual contact was >60 days before diagnosis of the index case. Patients do not need to be retested for *Chlamydia* after completing treatment with doxycycline or azithromycin unless symptoms persist or reinfection is suspected. A test of cure may be considered 3 weeks after completion of treatment with erythromycin. Testing at <3 weeks after completion of therapy to identify cases that did not respond to therapy may not be valid [69].

Azithromycin is currently recommended as first choice to treat *C. trachomatis* in pregnant women with amoxicillin as alternative [70]. Doxycycline and ofloxacin are contraindicated in pregnant women. Azithromycin is widely prescribed during pregnancy and lactation. Although it is excreted in breast milk, the dose delivered to the infant is quite low and not likely clinically significant [71]. Based on available data, the benefits of human milk feeding outweigh the risks of infant exposure to the small amounts of azithromycin transmitted through breastfeeding.

The optimal treatment for neonatal chlamydial conjunctivitis or pneumonia is uncertain. Erythromycin is recommended in many guidelines and has been the most widely-used antibiotic for neonatal chlamydial infection despite its association with infantile hypertrophic pyloric stenosis (particularly in the first 2 weeks of life). Azithromycin is an alternative but has not been well studied in this setting and it is also uncertain whether there is a decreased risk of infantile hypertrophic pyloric stenosis with this macrolide.

## 8.8  Complications and Sequelae

*Chlamydia trachomatis* has been implicated as a pathogen in 8–54 % of women who have PID and has been associated with the long-term consequences of tubal infertility (17 %), ectopic pregnancy (10 %), or chronic pelvic pain (17 %) [72–74]. *C. trachomatis* infection has been associated with spontaneous abortion, though not consistently, and stillbirth [75–77]. Additionally, *C. trachomatis* infection has been associated with chorioamnionitis, premature rupture of membranes and preterm delivery. A 2.6 and 3-fold increased risk of preterm delivery were determined in women with positive serology detected at 17 weeks gestation or diagnosed with cervical infection at 24 weeks gestation, respectively [76, 78]. Evidence grows with the use of NAATs that preterm delivery is associated with *C. trachomatis* infection [79–81].

In males, epididymitis, prostatitis, and reactive arthritis are the most common sequelae. Furthermore, untreated or incorrectly treated chlamydial conjunctivitis may result in chronic conjunctivitis that can develop alone or as part of Reiter syndrome. Given the high prevalence of chlamydial infections, complications due to this pathogen account for serious morbidity and economic cost.

## 8.9  Prevention

Because chlamydial infections usually are not associated with overt symptoms, prevention of infection and screening of asymptomatic high-risk patients is the most effective means of preventing disease and sequelae. Behavioral interventions (i.e., delaying intercourse, decreasing the number of sex partners, and use of barrier contraception) should be pursued aggressively. High-risk patients who should be routinely tested for *Chlamydia* include women with mucopurulent cervicitis, sexually active women less than

20 years old, and older women with more than one sex partner during the last 3 months or inconsistent use of barrier contraception while in a nonmonogamous relationship [70]. Because of the frequency of repeated chlamydial infections within the first several months following treatment of an initial infection, [15] more frequent (e.g., every 6 months) screening of asymptomatic sexually active adolescents may be necessary. Clinicians and health-care agencies should consider advising all women with chlamydial infection to be rescreened 3–4 months after treatment. Providers are also strongly encouraged to rescreen all women treated for chlamydial infection whenever they next present for care within the following 12 months, regardless of whether the patient believes that her sex partners were treated. Screening and treatment programs have resulted in reduced rates of complications [82, 83].

In a recent study conducted in the UK, seven of 74 women randomized to deferred chlamydial testing and treatment developed clinical PID over a 12 month follow-up period (9.5 %) compared to one of 63 (1.6 %) treated at the time of enrollment [45]. Although this difference was not significant ($p=0.07$) the annual incidence of PID (38 out of 2,377 women; 1.6 %) was less than the 3 % used in the sample size calculations, and thus the study was underpowered. In addition, participants were advised to be screened independently, and the one in five who acted on this advice had a high prevalence of chlamydial infection. Importantly, most cases of PID over 12 months occurred in women who were negative for chlamydia at baseline, indicating an importance for incident infection and the need to focus on testing those at higher risk, such as women with a new sexual partner.

In areas where screening and treatment programs have been established, rates of infection in screened populations have risen, causing speculation that early case identification and treatment interferes with development of immunity [84]. Others have insisted that the detection of increased rates of infection reflect a greater awareness of the infection, which has led to more testing being done, and still a larger number of positive tests [85]. Until a vaccine is developed to combat chlamydial disease, screening and treatment of infected persons remains the most logical mechanism of disease prevention.

To prevent maternal postnatal complications and chlamydial infections among infants, pregnant women should be screened for *Chlamydia* during the third trimester to permit completion of treatment before delivery. Ocular prophylaxis with topical erythromycin or tetracycline has reduced the incidence of gonococcal ophthalmia but does not appear to be effective against *C. trachomatis* [86]. Infants born to infected mothers are at high risk for infection; however, prophylatic antibiotic treatment is not indicated, and the efficacy of such treatment is unknown. Infants should be monitored and treated appropriately if symptoms develop.

## 8.10    Future Directions

During the extracellular EB stage, antibodies can act to inhibit infection. However, since the replicating RB form resides within the intracellular inclusion, bacterial killing at this stage requires a cell-mediated immune response with the primary effectors being IFN-γ secreting CD4 T cells. Thus, an ideal *C. trachomatis* vaccine should induce both local antibodies to prevent infection by EBs, and a strong Th1 response to limit infection once initiated. Efforts to develop a *C. trachomatis* vaccine have concentrated primarily on the use of recombinant *Chlamydia* antigens with immune adjuvants [87, 88]. The use of a purified native preparation of MOMP combined with Th1-inducing adjuvants induced significant resistance in mice, but sterilizing immunity was not achieved [89]. Stimulation of long-term mucosal immunity in the genital tract is a challenge; persons are susceptible to reinfection with *C. trachomatis* after a brief period of immunity because memory cells are not retained in the genital tract. It is unclear whether all genital infections could be prevented or whether only more invasive disease, such as salpingitis, might be preventable using vaccine technology. Markers for protection from upper genital tract infection and/or disease in the female will be necessary if vaccine candidates are to be tested in humans.

Although current antibiotic treatment is highly successful when administered, most persons infected with *C. trachomatis* are asymptomatic and thus go undiagnosed and untreated. Although rates of infection have increased in certain areas where widespread screening and treatment programs have been in place [84], complications from infection have decreased [82, 83], indicating the utility of such programs. Public health officials should pursue such strategies in parallel with the ongoing research for effective vaccines.

**Acknowledgements**   Toni Darville is supported by United States National Institutes of Health grants (AI054624 and U19 AI084024).

# References

1. Miller WC, Ford CA, Morris M, Handcock MS, Schmitz JL, Hobbs MM et al (2004) Prevalence of Chlamydial and gonococcal infections among young adults in the United States. JAMA 291:2229–2236
2. LaMontagne DS, Fenton KA, Randall S, Anderson S, Carter P (2004) Establishing the national Chlamydia screening programme in England: results from the first full year of screening. Sex Transm Infect 80:335–341
3. Ford CA, Pence BW, Miller WC, Resnick MD, Bearinger LH, Pettingell S et al (2005) Predicting adolescents' longitudinal risk for sexually transmitted infection: results from the National Longitudinal Study of Adolescent Health. Arch Pediatr Adolesc Med 159:657–664
4. Griffiths E, Gupta RS (2002) Protein signatures distinctive of Chlamydial species: horizontal transfers of cell wall biosynthesis genes glmU from archaea to Chlamydiae and murA between Chlamydiae and Streptomyces. Microbiology 148:2541–2549
5. O'Connell CM, Abdelrahman YM, Green E, Darville HK, Saira K, Smith B et al (2011) TLR2 activation by Chlamydia trachomatis is plasmid-dependent and plasmid-responsive chromosomal loci are coordinately regulated in response to glucose limitation by *C. trachomatis* but not by *C. muridarum*. Infect Immun 79(3):1044–1056
6. Stephens RS (2003) The cellular paradigm of Chlamydial pathogenesis. Trends Microbiol 11:44–51
7. Karimi O, Ouburg S, de Vries HJ, Pena AS, Pleijster J, Land JA et al (2009) TLR2 haplotypes in the susceptibility to and severity of Chlamydia trachomatis infections in Dutch women. Drugs Today (Barc) 45(Suppl B):67–74
8. Darville T, O'Neill JM, Andrews CW Jr, Nagarajan UM, Stahl L, Ojcius DM (2003) Toll-like receptor-2, but not toll-like receptor-4, is essential for development of oviduct pathology in Chlamydial genital tract infection. J Immunol 171:6187–6197
9. Frazer LC, O'Connell CM, Andrews CW Jr, Zurenski MA, Darville T (2011) Enhanced neutrophil longevity and recruitment contribute to the severity of oviduct pathology during C. muridarum infection. Infect Immun 79(10):4029–4041
10. Shah AA, Schripsema JH, Imtiaz MT, Sigar IM, Kasimos J, Matos PG et al (2005) Histopathologic changes related to fibrotic oviduct occlusion after genital tract infection of mice with Chlamydia muridarum. Sex Transm Dis 32:49–56
11. Darville T, Andrews CW Jr, Laffoon KK, Shymasani W, Kishen LR, Rank RG (1997) Mouse strain-dependent variation in the course and outcome of Chlamydial genital tract infection is associated with differences in host response. Infect Immun 65:3065–3073
12. Ramsey KH, Sigar IM, Schripsema JH, Shaba N, Cohoon KP (2005) Expression of matrix metalloproteinases subsequent to urogenital Chlamydia muridarum infection of mice. Infect Immun 73:6962–6973
13. Imtiaz MT, Distelhorst JT, Schripsema JH, Sigar IM, Kasimos JN, Lacy SR et al (2007) A role for matrix metalloproteinase-9 in pathogenesis of urogenital Chlamydia muridarum infection in mice. Microbes Infect
14. Molano M, Meijer CJ, Weiderpass E, Arslan A, Posso H, Franceschi S et al (2005) The natural course of Chlamydia trachomatis infection in asymptomatic Colombian women: a 5-year follow-up study. J Infect Dis 191:907–916
15. Burstein GR, Gaydos CA, Diener-West M, Howell MR, Zenilman JM, Quinn TC (1998) Incident Chlamydia trachomatis infections among inner-city adolescent females (see comments). JAMA 280:521–526
16. Arno JN, Katz BP, McBride R, Carty GA, Batteiger BE, Caine VA et al (1994) Age and clinical immunity to infections with Chlamydia trachomatis. Sex Transm Dis 21:47–52
17. Kimani J, Maclean IW, Bwayo JJ, MacDonald K, Oyugi J, Maitha GM et al (1996) Risk factors for Chlamydia trachomatis pelvic inflammatory disease among sex workers in Nairobi, Kenya. J Infect Dis 173:1437–1444
18. Brunham RC, Kimani J, Bwayo J, Maitha G, Maclean I, Yang C et al (1996) The epidemiology of Chlamydia trachomatis within a sexually transmitted diseases core group. J Infect Dis 173:950–956
19. Cohen CR, Koochesfahani KM, Meier AS, Shen C, Karunakaran K, Ondondo B et al (2005) Immunoepidemiologic profile of Chlamydia trachomatis infection: importance of heat-shock protein 60 and interferon- gamma. J Infect Dis 192:591–599
20. Su H, Feilzer K, Caldwell HD, Morrison RP (1997) Chlamydia trachomatis genital tract infection of

antibody-deficient gene knockout mice. Infect Immun 65:1993–1999

21. Morrison SG, Morrison RP (2005) A predominant role for antibody in acquired immunity to Chlamydial genital tract reinfection. J Immunol 175:7536-7542

22. Brunham RC, Kuo CC, Cles L, Holmes KK (1983) Correlation of host immune response with quantitative recovery of Chlamydia trachomatis from the human endocervix. Infect Immun 39:1491–1494

23. Brunham RC, Peeling R, Maclean I, McDowell J, Persson K, Osser S (1987) Postabortal Chlamydia trachomatis salpingitis: correlating risk with antigen-specific serological responses and with neutralization. J Infect Dis 155:749–755

24. World Health Organization (WHO) (2001) Global prevalence and incidence of selected curable sexually transmitted infections: overview and estimates. WHO, Geneva

25. Sturm-Ramirez K, Brumblay H, Diop K, Gueye-Ndiaye A, Sankale JL, Thior I et al (2000) Molecular epidemiology of genital Chlamydia trachomatis infection in high-risk women in Senegal, West Africa. J Clin Microbiol 38:138–145

26. Kilmarx PH, Black CM, Limpakarnjanarat K, Shaffer N, Yanpaisarn S, Chaisilwattana P et al (1998) Rapid assessment of sexually transmitted diseases in a sentinel population in Thailand: prevalence of Chlamydial infection, gonorrhoea, and syphilis among pregnant women–1996. Sex Transm Infect 74:189–193

27. Warszawski J, Meyer L, Weber P (1999) Criteria for selective screening of cervical Chlamydia trachomatis infection in women attending private gynecology practices. Eur J Obstet Gynecol Reprod Biol 86:5–10

28. Gaydos CA, Howell MR, Quinn TC, McKee KT Jr, Gaydos JC (2003) Sustained high prevalence of Chlamydia trachomatis infections in female army recruits. Sex Transm Dis 30:539–544

29. Cohen DA, Nsuami M, Etame RB, Tropez-Sims S, Abdalian S, Farley TA et al (1998) A school-based Chlamydia control program using DNA amplification technology. Pediatrics 101:e1

30. Best D, Ford CA, Miller WC (2001) Prevalence of Chlamydia trachomatis and Neisseria gonorrhoeae infection in pediatric private practice. Pediatrics 108:E103

31. Stamm WE, Koutsky LA, Benedetti JK, Jourden JL, Brunham RC, Holmes KK (1984) Chlamydia trachomatis urethral infections in men. Prevalence, risk factors, and clinical manifestations. AnnIntern Med 100:47–51

32. Schachter J, Grossman M, Holt J, Sweet R, Spector S (1979) Infection with Chlamydia trachomatis: involvement of multiple anatomic sites in neonates. J Infect Dis 139:232–234

33. Jones RB, Rabinovitch RA, Katz BP, Batteiger BE, Quinn TS, Terho P et al (1985) Chlamydia trachomatis in the pharynx and rectum of heterosexual

patients at risk for genital infection. AnnIntern Med 102:757–762

34. Quinn TC, Goodell SE, Mkrtichian E, Schuffler MD, Wang SP, Stamm WE et al (1981) Chlamydia trachomatis proctitis. N Eng J Med 305:195–200

35. Rahman MU, Cantwell R, Johnson CC, Hodinka RL, Schumacher HR, Hudson AP (1992) Inapparent genital infection with Chlamydia trachomatis and its potential role in the genesis of Reiters syndrome. DNA Cell Biol 11:215–219

36. Keat A, Thomas BJ, Taylor Robinson D (1983) Chlamydial infection in the aetiology of arthritis. Br Med Bull 39:168–174

37. Koskela P, Anttila T, Bjorge T, Brunsvig A, Dillner J, Hakama M et al (2000) Chlamydia trachomatis infection as a risk factor for invasive cervical cancer. Int J Cancer 85:35–39

38. Plummer FA, Simonsen JN, Cameron DW, Ndinya-Achola JO, Kreiss JK, Gakinya MN et al (1991) Cofactors in male-female sexual transmission of human immunodeficiency virus type 1. J Infect Dis 163:233–239

39. Paavonen J, Vesterinen E, Mardh PA (1982) Infertility as a sequela of Chlamydial pelvic inflammatory disease. Scand J Infect Dis Suppl 32:73–76

40. Haggerty CL, Gottlieb SL, Taylor BD, Low N, Xu F, Ness RB (2010) Risk of sequelae after Chlamydia trachomatis genital infection in women. J Infect Dis 201(Suppl 2):S134–55

41. Rahm VA, Gnarpe H, Odlind V (1988) Chlamydia trachomatis among sexually active teenage girls. Lack of correlation between Chlamydial infection, history of the patient and clinical signs of infection. Br J Obstet Gynaecol 95:916–919

42. Bachmann LH, Richey CM, Waites K, Schwebke JR, Hook III EW (1999) Patterns of Chlamydia trachomatis testing and follow-up at a University Hospital Medical Center. Sex Transm Dis 26:496–499

43. Geisler WM, Wang C, Morrison SG, Black CM, Bandea CI, Hook III EW (2008) The natural history of untreated Chlamydia trachomatis infection in the interval between screening and returning for treatment. Sex Transm Dis 35:119–123

44. Morre SA, van den Brule AJ, Rozendaal L, Boeke AJ, Voorhorst FJ, de Blok S et al (2002) The natural course of asymptomatic Chlamydia trachomatis infections: 45 % clearance and no development of clinical PID after one-year follow-up. Int J STD AIDS 13(Suppl 2):12–18

45. Oakeshott P, Kerry S, Aghaizu A, Atherton H, Hay S, Taylor-Robinson D et al (2010) Randomised controlled trial of screening for Chlamydia trachomatis to prevent pelvic inflammatory disease: the POPI (prevention of pelvic infection) trial. BMJ 340:c1642

46. Svensson L, Westrom L, Ripa KT, Mardh PA (1980) Differences in some clinical and laboratory parameters in acute salpingitis related to culture and serologic findings. Am J Obstet Gynecol 138(7 Pt 2):1017–1021

47. Hillis SD, Owens LM, Marchbanks PA, Amsterdam LE, Mac Kenzie WR (1997) Recurrent Chlamydial infections increase the risks of hospitalization for ectopic pregnancy and pelvic inflammatory disease. Am J Obstet Gynecol 176:103–107

48. Ness RB, Smith KJ, Chang CC, Schisterman EF, Bass DC (2006) Prediction of pelvic inflammatory disease among young, single, sexually active women. Sex Transm Dis 33:137–142

49. Schachter J, Grossman M, Sweet RL, Holt J, Jordan C, Bishop E (1986). Prospective study of perinatal transmission of Chlamydia trachomatis. JAMA 255:3374–3377

50. Datta P, Laga M, Plummer FA, Ndinya-Achola JO, Piot P, Maitha G et al (1988) Infection and disease after perinatal exposure to Chlamydia trachomatis in Nairobi, Kenya. J Infect Dis 158:524–528

51. Alexander ER, Harrison HR (1983) Role of Chlamydia trachomatis in perinatal infection. Rev Infect Dis 5:713–719

52. Rours IG, Hammerschlag MR, Ott A, De Faber TJ, Verbrugh HA, de Groot R et al (2008) Chlamydia trachomatis as a cause of neonatal conjunctivitis in Dutch infants. Pediatrics 121:e321–326

53. Chandler JW, Alexander ER, Pheiffer TA, Wang SP, Holmes KK, English M (1977) Ophthalmia neonatorum associated with maternal Chlamydial infections. Trans Am Acad Ophthalmol Otola 83:302–308

54. Persson K, Ronnerstam R, Svanberg L, Pohla MA (1983) Neonatal Chlamydial eye infection: an epidemiological and clinical study. Br J Ophthalmol 67:700–704

55. Beem MO, Saxon EM (1977) Respiratory-tract colonization and a distinctive pneumonia syndrome in infants infected with Chlamydia trachomatis. N Engl J Med 296:306–310

56. Schachter J, Lum L, Gooding CA, Ostler B (1975) Pneumonitis following inclusion blennorrhea. J Pediatr 87:779–780

57. Rours GI, Hammerschlag MR, Van Doornum GJ, Hop WC, de Groot R, Willemse HF et al (2009) Chlamydia trachomatis respiratory infection in Dutch infants. Arch Dis Child 94:705–707

58. Tipple MA, Beem MO, Saxon EM (1979) Clinical characteristics of the afebrile pneumonia associated with Chlamydia trachomatis infection in infants less than 6 months of age. Pediatrics 63:192–197

59. Johnson RE, Newhall WJ, Papp JR, Knapp JS, Black CM, Gift TL et al (2002) Screening tests to detect Chlamydia trachomatis and Neisseria gonorrhoeae infections–2002. MMWR Recomm Rep 51:1–38

60. Blake DR, Lemay CA, Gaydos CA, Quinn TC (2005) Performance of urine leukocyte esterase in asymptomatic male youth: another look with nucleic acid amplification testing as the gold standard for Chlamydia detection. J Adolesc Health 36:337–341

61. Black CM (1997) Current methods of laboratory diagnosis of Chlamydia trachomatis infections. Clin Microbiol Rev 10:160–184

62. Schachter J, Stamm WE, Quinn TC, Andrews WW, Burczak JD, Lee HH (1994) Ligase chain reaction to detect Chlamydia trachomatis infection of the cervix. J Clin Microbiol 32:2540–2543

63. Jaschek G, Gaydos CA, Welsh LE, Quinn TC (1993) Direct detection of Chlamydia trachomatis in urine specimens from symptomatic and asymptomatic men by using a rapid polymerase chain reaction assay. J Clin Microbiol 31:1209–1212

64. Rours GI, Verkooyen RP, Willemse HF, van der Zwaan EA, van Belkum A, de Groot R et al (2005) Use of pooled urine samples and automated DNA isolation to achieve improved sensitivity and cost-effectiveness of large-scale testing for Chlamydia trachomatis in pregnant women. J Clin Microbiol 43:4684–4690

65. Shafer MA, Moncada J, Boyer CB, Betsinger K, Flinn SD, Schachter J (2003) Comparing first-void urine specimens, self-collected vaginal swabs, and endocervical specimens to detect Chlamydia trachomatis and Neisseria gonorrhoeae by a nucleic acid amplification test. J Clin Microbiol 41:4395–4399

66. Knox J, Tabrizi SN, Miller P, Petoumenos K, Law M, Chen S et al (2002) Evaluation of self-collected samples in contrast to practitioner-collected samples for detection of Chlamydia trachomatis, Neisseria gonorrhoeae, and Trichomonas vaginalis by polymerase chain reaction among women living in remote areas. Sex Transm Dis 29:647–654

67. Hammerschlag MR, Roblin PM, Gelling M, Tsumura N, Jule JE, Kutlin A (1997) Use of polymerase chain reaction for the detection of Chlamydia trachomatis in ocular and nasopharyngeal specimens from infants with conjunctivitis. Pediatr Infect Dis J 16:293–297

68. Martin DH, Mroczkowski TF, Dalu ZA, McCarty J, Jones RB, Hopkins SJ et al (1992) A controlled trial of a single dose of azithromycin for the treatment of Chlamydial urethritis and cervicitis. N Engl J Med 327:921–925

69. Gaydos CA, Crotchfelt KA, Howell MR, Kralian S, Hauptman P, Quinn TC (1998) Molecular amplification assays to detect Chlamydial infections in urine specimens from high school female students and to monitor the persistence of Chlamydial DNA after therapy. J Infect Dis 177:417–424

70. Workowski KA, Berman SM (2006) Sexually transmitted diseases treatment guidelines, 2006. MMWR Recomm Rep 55:1–94

71. Kelsey JJ, Moser LR, Jennings JC, Munger MA (1994) Presence of azithromycin breast milk concentrations: a case report. Am J Obstet Gynecol 170:1375–1376

72. Cates W Jr, Wasserheit JN (1991) Genital Chlamydial infections: Epidemiology and reproductive sequelae. Am J Obstet Gynecol 164(Suppl):1771–1781

73. Stamm WE, Holmes KK, Mardh PA, Sparling PF, Wiesner PJ (1999) Chlamydia trachomatis infections of the adult. Sexually transmitted diseases. Mcgraw-hill book company, New York, p 407–422

74. Westrom L, Joesoef R, Reynolds G, Hagdu A, Thompson SE (1992) Pelvic inflammatory disease and fertility. A cohort study of 1,844 women with laparoscopically verified disease and 657 control women with normal laparoscopic results. Sex Transm Dis 19:185–192

75. Gencay M, Koskiniemi M, Ammala P, Fellman V, Narvanen A, Wahlstrom T et al (2000) Chlamydia trachomatis seropositivity is associated both with stillbirth and preterm delivery. APMIS 108:584–588

76. Hollegaard S, Vogel I, Thorsen P, Jensen IP, Mordhorst CH, Jeune B (2007) Chlamydia trachomatis C-complex serovars are a risk factor for preterm birth. In Vivo 21:107–112

77. Fejgin MD, Cohen I, Horvat-Kohlmann M, Charles AG, Luzon A, Samra Z (1997) Chlamydia trachomatis infection during pregnancy: can it cause an intrauterine infection? Isr J Med Sci 33:98–102

78. Andrews WW, Goldenberg RL, Mercer B, Iams J, Meis P, Moawad A et al (2000) The Preterm Prediction Study: association of second-trimester genitourinary Chlamydia infection with subsequent spontaneous preterm birth. Am J Obstet Gynecol 183:662–668

79. Kovacs L, Nagy E, Berbik I, Meszaros G, Deak J, Nyari T (1998) The frequency and the role of Chlamydia trachomatis infection in premature labor. Int J Gynaecol Obstet 62:47–54

80. Rastogi S, Das B, Salhan S, Mittal A (2003) Effect of treatment for Chlamydia trachomatis during pregnancy. Int J Gynaecol Obstet 80:129–137

81. Blas MM, Canchihuaman FA, Alva IE, Hawes SE (2007) Pregnancy outcomes in women infected with Chlamydia trachomatis: a population-based cohort study in Washington State. Sex Transm Infect 83:314–378

82. Moss NJ, Ahrens K, Kent CK, Klausner JD (2006) The decline in clinical sequelae of genital Chlamydia trachomatis infection supports current control strategies. J Infect Dis 193:1336–1338 (author reply 8–9)

83. Brunham RC, Pourbohloul B, Mak S, White R, Rekart ML (2006) Reply to Hagdu and to Moss et al. J Infect Dis 193:1338–1339

84. Brunham RC, Pourbohloul B, Mak S, White R, Rekart ML (2005) The Unexpected Impact of a Chlamydia trachomatis Infection Control Program on Susceptibility to Reinfection. J Infect Dis 192:1836–1844

85. Vickers DM, Osgood ND (2010) Current crisis or artifact of surveillance: insights into rebound Chlamydia rates from dynamic modelling. BMC Infect Dis 10:70

86. Hammerschlag MR, Cummings C, Roblin PM, Williams TH, Delke I (1989) Efficacy of neonatal ocular prophylaxis for the prevention of Chlamydial and gonococcal conjunctivitis. N Eng J Med 320:769–772

87. Cong Y, Jupelli M, Guentzel MN, Zhong G, Murthy AK, Arulanandam BP (2007) Intranasal immunization with Chlamydial protease-like activity factor and CpG deoxynucleotides enhances protective immunity against genital Chlamydia muridarum infection. Vaccine 25:3773–3780

88. Ifere GO, He Q, Igietseme JU, Ananaba GA, Lyn D, Lubitz W et al (2007) Immunogenicity and protection against genital Chlamydia infection and its complications by a multisubunit candidate vaccine. J Microbiol Immunol Infect 40:188–200

89. Pal S, Peterson EM, Rappuoli R, Ratti G, De La Maza LM (2006) Immunization with the Chlamydia trachomatis major outer membrane protein, using adjuvants developed for human vaccines, can induce partial protection in a mouse model against a genital challenge. Vaccine 24:766–775

# Treatment of Resistant Bacterial Infections in Children: Thinking Inside and Outside the Box

**9**

Gilat Livni and Shai Ashkenazi

### Abstract

Antimicrobial resistance of bacteria causing pediatric infections has become more common and complicated in recent years. Although formerly confined to hospital settings, multi-drug resistant bacteria now also cause community-acquired infections. Treatment of infections caused by resistant pathogens is difficult, necessitating thinking both inside and outside the box. Determination of the precise minimal inhibitory concentration (MIC) is often crucial for selecting the most appropriate antibiotics, their doses, and use of prolonged infusions. For some multiply-resistant bacteria, off-label use of antibiotics, sometimes with no evidence from controlled studies ("salvage therapy") is unavoidable.

## 9.1 Introduction

### 9.1.1 Antibiotic Resistance

Though recognized for decades, bacterial resistance to antimicrobial agents has increased dramatically over recent years [1–4], as witnessed by several illustrative phenomena. First is the appearance of acquired multi-drug resistant bacteria, both Gram-positive bacteria that are resistant to multiple agents, including glycopeptides [5], and Gram-negative bacteria resistant to multiple agents, including carbapenems [1, 2, 6]. Resistance is mediated by complex and diverse mechanisms, which mainly include drug inactivation, target modification, target bypass, decreased outer membrane permeability, and increased efflux [6, 7].

Bacteria have devised varied and efficient mechanisms for the transfer of resistance-encoding genes among other bacteria, including those of different species, using genetic recombination and mobile elements, such as plasmids, integrons, and transposons [6, 8]. The result is the rapid dissipation of resistant bacteria among departments, institutions, countries, and continents [1, 2, 6, 9].

In the past, antibiotic resistance, and particularly multiply-resistant bacteria, was limited to

S. Ashkenazi (✉)
Professor of Pediatrics, Sackler Faculty of Medicine, Schneider Children's Medical Center, Petach Tikva, Israel
e-mail: ashai@post.tau.ac.il

G. Livni
Sackler Faculty of Medicine, Department of Pediatrics A, Schneider Children's Medical Center of Israel, Petach Tikva, Israel
e-mail: gilat@orange.net.il

**Table 9.1** Stages of antibiotic therapy

1. Initial empiric antibiotics
   Based on:
   A. The suspected bacterial causes of the infection
   B. The expected *in vitro* antibiotic susceptibility of these pathogens
   C. Penetration to the site of infection
   D. Results of clinical studies
   E. Presence of contraindications or drug interactions
2. After 16–24 h: based on initial growth, but the final identification and antibiotic susceptibility are not yet available
3. After an additional 8–24 h: according to identification of the pathogen(s) and its antibiotic susceptibility

hospital settings, especially to intensive care units and to children with serious underlying health problems. The situation has changed, as demonstrated by increased rates in the last 10 years of community-acquired methicillin-resistant *Staphylococcus aureus* (CA-MRSA), which now accounts for 70 % of all CA-*S. aureus* infections and 74 % of all *S. aureus* pneumonia in Texas Children's Hospital, USA [10]. Community-acquired Gram-negative bacteria that are carbapenem-resistant have also been reported [11]. This increasing and widespread bacterial resistance is occurring in an era when the options for new antimicrobial agents seem very limited [3, 12].

### 9.1.2 Approach to Antibiotic Therapy

Antibiotic therapy can be divided into three phases (Table 1). The first is the empiric antibiotic regimen, which is initiated when bacterial infection is suspected. Selection of specific antimicrobial agents is based primarily on the expected pathogens and their antibiotic susceptibilities, as well as on several other factors that are listed in Table 1. In positive bacterial cultures, initial growth of pathogens (i.e., of Gram-positive cocci or Gram-negative bacilli) is usually observed after 16–24 h, though definitive identification of the bacteria and their antibiotic susceptibility takes longer. Modifications in the antibiotic regimen may be required at this intermediary stage. Finally, after an additional 8–24 h, conclusive

identification of the pathogen(s) and their antibiotic susceptibility is generally possible. At this stage, "fine-tuning" of the antibiotic regimen is required to target the antibiotics according to the results from the microbiological laboratory.

"Direct susceptibility testing" is sometimes used, particularly for severe infections, to more rapidly attain antibiotic susceptibility. According to this approach, susceptibility testing is performed on the clinical specimen directly, without waiting for growth of a pathogen. As an example, due to limited penetration of most antimicrobial agents into the cerebrospinal fluid (CSF), the precise minimal inhibitory concentration (MIC) is crucial for optimizing treatment of bacterial meningitis. Therefore, in many institutions, if lumbar puncture reveals a purulent CSF, especially with a positive Gram stain, E-tests are placed directly on the CSF sediment, yielding the MIC within less than 24 h.

## 9.2 Treating Resistant Bacterial Infections in Children

The discussion of all resistant bacteria in children and their treatment options is obviously an impossible mission. We will focus therefore on major resistant bacteria causing pediatric infections and emphasize the most recent and clinically applicable data.

### 9.2.1 Penicillin-resistant Streptococcus pneumoniae

#### 9.2.1.1 Pneumococcal Infections

*S. pneumoniae* is one of the most common pediatric pathogens, both in the community and in hospitalized children, causing mucosal (acute otitis media, mastoiditis, sinusitis, and pneumonia) and invasive (bacteremia, meningitis, septic arthritis, and osteomyelitis) infections [13]. More than 90 serotypes of *S. pneumoniae* have been described. Changes in the relative prevalence of the serotypes have occurred with and without relation to the conjugate vaccine use. For example, increased incidence of invasive pneumococcal disease caused by serotype 19A was reported in the United States after introducing the pneumococcal conjugate vaccine in 2000, but an increase was also detected in other locations prior to vaccine use [13, 14].

### 9.2.1.2  Antibiotic Resistance

Resistance of *S. pneumoniae* to penicillin is caused by mutations in the penicillin-binding protein (PBP)-encoding genes [7, 14, 15]. These genes can spread by clonal dissemination of resistant strains, as well as by homologous recombination among different strains, either within pneumococci or with the closely related viridians streptococci. The modified PBPs have reduced affinity to penicillin and all other β-lactams. In addition, penicillin-resistant pneumococci often show resistance to other antimicrobial agents, including erythromycin, azithromycin, trimethoprim-sulfamethoxsazole (TMP-SMX), and tetracyclines [13, 14]. Tolerance of pneumococci to penicillin, mediated by changes in the peptidoglycan structure, has been reported, but its clinical significance is not completely clear. Based on recent pharmacokinetic/pharmacodynamic data, the Clinical and Laboratory Standards Institute (CLSI) updated in 2011 the susceptibility breakpoints for pneumococci [16].

### 9.2.1.3  Nonmeningeal Infections

Since pneumococci present reduced susceptibility but usually not full resistance, the principle of treating these infections is that high doses of β-lactams increase the fraction of time the antibiotic concentration in the area of the infection is above the MIC (fT > MIC), leading to microbiological and clinical efficacy [7, 13]. For nonmeningeal infections, the updated CLSI breakpoints of *S. pneumoniae* to parenteral penicillin are ≤2, 4, and ≥8 μg/ml, for susceptible, intermediate, and resistant strains, respectively [16]. Since MIC ≥8 μg/ml is very uncommon, most nonmeningeal infections (i.e., pneumonia, bacteremia) caused by *S. pneumoniae* can be treated successfully with high dosages (200,000–400,000 units/kg/day) of parenteral penicillin. Parenteral 3rd-generation cephalosporin (ceftriaxone or cefotaxime) may sometimes be indicated. Because of the high bioavailability, the 2011 breakpoints for oral amoxicillin are the same as for parenteral penicillin [16]; for children who can take oral medication, high-dosage amoxicillin (80–90 mg/kg/d) is efficacious for most mucosal pneumococcal infections.

### 9.2.1.4  Meningeal Infections

Since the penetration of all β-lactams through the blood-brain barrier is very limited, the breakpoints for meningeal infections are much lower: ≤0.06, 0.12–1, and ≥2 μg/ml for parenteral penicillin, and ≤0.5, 1, and ≥2 μg/ml for 3rd-generation cephalosporins, for susceptible, intermediate, and resistant strains, respectively [16]. Determination of the MIC of the *S. pneumoniae* isolate causing meningitis is therefore crucial for optimal treatment: when the MIC to penicillin is < 0.1 μg/ml, high-dose parenteral penicillin is appropriate; when the MIC to penicillin is > 0.1 μg/ml but < 0.5 μg/ml to 3rd-generation cephalosporins, ceftriaxone or cefotaxime can be used; when the MIC to 3rd-generation cephalosporin is 1 μg/ml, vancomycin should be added to the ceftriaxone or cefotaxime; and when the MIC to 3rd-generation cephalosporins is ≥2 μg/ml, the addition of rifampin should be considered, especially if adjunctive steroid therapy is administered. For empiric treatment of bacterial meningitis in children, a 3rd-generation cephalosporin and vancomycin are recommended and the addition of rifampin should be considered if steroids are used.

### 9.2.2  Methicillin-resistant Staphylococcus aureus

Resistance of *S. aureus* to methicillin is mediated by mecA gene-encoding PBP2a with low affinity to methicillin, and in fact to all β-lactams (see below). Although methicillin is not currently in use, the term MRSA is still used to describe these strains [7, 10]. The expression of mecA genes can be constitutive or inducible; they are situated on mobile genetic elements called Staphylococcal Cassette Chromosome mec (SCCmec), which often contain genes coding for resistance to other antibiotics.

**Table 9.2** Treatment options for multi-drug resistant Gram-positive infections

| Antimicrobial agent | Comments |
|---|---|
| Vancomycin | The mainstay for resistant Gram-positive bacteria |
| Clindamycin | Mainly for CA-MRSA |
| Trimethoprim-sulfa-methoxazole | For SSTIs |
| Teicoplanin | Registered (not in the US) for SSTIs, bacteremia |
| Quinupristin-dalfopristin | A streptogramin, approved for SSTIs in >16 year olds |
| Tigecycline | Glycycline antibiotic, inferior to β-lactams |
| Linezolid | An oxazolidinone antibiotic, excellent oral bioavailability |
| Daptomycin | A cyclic lipopeptide, not for pneumonia |
| Ceftaroline | 5th-generation cephalosporin, approved for adults with SSTIs and community-acquired pneumonia |

*CA-MRSA* community-acquired methicillin-resistant *Staphylococcus aureus*
*SSTIs* skin and soft tissue infections

### 9.2.2.1 Community-acquired MRSA

For many years MRSA were limited to hospital settings, and particularly to intensive care units. However, from recent reports, mainly from the United States [10], rates of CA-MRSA infections in children seem to be rising steadily, with 10–22 % carriage rates documented in healthy children. Such high rates of CA-MRSA have not been reported in all locations. Particularly in the U.S. pediatric CA-MRSA has been shown to be particularly severe, causing syndromes like bacteremia with septic shock and purpura fulminans, necrotizing pneumonia and empyema, severe orbital cellulitis, multiple-site osteomyelitis with deep vein thrombosis, pyomyositis, and necrotizing fasciitis. A study from the US Children's Hospital showed that, compared to CA-methicillin-sensitive *S. aureus* infections, CA-MRSA infections result in higher numbers of days with fever (3.9 vs 1.8), of hospitalization (12 vs 9), and with positive blood cultures (3.4 vs 1.5) [17].

In contrast to nosocomial MRSA, most isolates of CA-MRSA are susceptible to non-β-lactam antibiotics, like clindamycin, TMP-SMX,

aminoglycosides, fluoroquinolones, and tetracycline [7, 10, 17], thus enabling broader treatment options. This is because most CA-MRSA is encoded by the relatively small type I SCCmec, which does not contain other resistant genes.

### 9.2.2.2 Treatment of MRSA

Guidelines for the treatment of MRSA were published in the United States [18] and in the United Kingdom [19]. We highlight first that for staphylococcal cutaneous abscess, incision and drainage comprises the primary management and is likely to be adequate alone. For example, in a double-blind, randomized, placebo-controlled study of 161 children with skin abscesses who underwent incision and drainage—no significant differences were noted between TMP-SMX and placebo after 10-day treatment or a 90-day follow-up call [20]. Systemic antibiotics are recommended for skin abscesses associated with the following conditions: severe local disease (multiple sites, rapidly progressed cellulitis, and septic phlebitis), systemic illness, comorbidities or immunosuppression, abscesses in difficult to drain areas (face, hands, and genitalia) and lack of response to incision and drainage alone [18].

The main antibiotic used for the treating of severe infections caused by MRSA is IV vancomycin. The recommended regimen for children with serious or invasive disease is 15 mg/kg/dose every 6 h, although data to support this dosing recommendation are limited [18]. In adults with serious infections (bacteremia, infective endocarditis, osteomyelitis, meningitis, pneumonia, and necrotizing fasciitis) due to MRSA, vancomycin trough concentrations of 15–20 µg/ml are recommended. The efficacy and safety of targeting similar concentrations in children are currently unclear [18].

Other treatment options exist (Table 9.2). Clindamycin is used mainly for CA-MRSA infections, because these strains are usually susceptible to this agent. It is recommended for children with osteomyelitis or septic arthritis who are stable, without ongoing bacteremia or intravascular infection. Initial treatment should be intravenous, with a dosage of 40 mg/kg/day divided into 3–4 doses; oral therapy can then follow. A review of the medical records of 39 children with CA-MRSA treated with clindamycin (4 of them received additional antibiotics) showed recovery in all patients, except one child with undrained pyomyositis and septic thrombophlebitis [17]. Since clindamycin-resistant CA-MRSA has been reported in some locations, local resistance patterns should be monitored. Usually a 3 to 4-week antibiotic course is recommended for septic arthritis and a 4 to 6-week course for osteomyelitis; however recent data from controlled studies show that for uncomplicated infections shorter courses are acceptable [21]. TMP-SMX can be used for uncomplicated skin and soft tissue infections (SSTIs), as can tetracycline in children older than 7 years. Teicoplanin is not registered in the United States, but is recommended in the United Kingdom for the treatment of SSTIs and bacteremia [19].

Linezolid is the first agent of a new class of antibiotics, the oxazolidinones, to be approved for use in adults and children [22, 23]. It is mostly bacteriostatic, and has broad antimicrobial activity against Gram-positive bacteria, including resistant streptococci, MRSA, methicillin-resistant *Staphylococcus epidermidis* (MRSE), and vancomycin-resistant enterococci (VRE). Because of its excellent bioavailability, the oral and intravenous dosages are similar. In addition to the common, yet mild, adverse events of diarrhea, vomiting, and headache, hematologic (mainly thrombocytopenia) and neurologic complications have been reported, mainly in adults [23]. Summarizing seven randomized controlled trials (six with vancomycin and one with a cephalosporin) and four uncontrolled trials, Garazzino and Tovo concluded that linezolid is safe and efficacious in children with serious Gram-positive infections. They recommended that it should be "reserved for children who are intolerant to or fail conventional agents" [22]. The recommended regimen is 10 mg/kg/dose, IV or PO, three times a day in children under the age of 11 years and twice a day in older children. In our pediatric department, we also have good experience with linezolid, regarding tolerability, safety, and clinical efficacy.

Daptomycin is a novel cyclic lipopeptide, whose activity is based on disruption of membrane functions of Gram-positive bacteria [23]. The antibacterial bactericidal activity includes resistant bacteria, such as MRSA and VRE, and resistant streptococci, with demonstrated *in vitro* synergy with aminoglycosides, β-lactams, and rifampin. Daptomycin is approved for the treatment of SSTIs, bacteremia, and right-sided endocarditis. Since it is inactivated by alveolar surfactant, it should not be used to treat pneumonia. This antibiotic is shown to cause muscle pain and weakness, with elevated creatine kinase levels. Pediatric data are very limited. A retrospective review from Dallas Children's Hospital (U.S.) described the experience of 16 children (15 with invasive S. aureus infections (14 MRSA) and one with VRE) treated with daptomycin [24]. The addition of daptomycin to "conventional" antibiotics resulted in clinical improvement in all but two patients and in bacteriologic cure in six of seven children with persistent bacteremia; no adverse effects were attributed to the daptomycin therapy [24].

Ceftaroline is the first "fifth-generation" cephalosporin (others are in the pipeline) approved by the FDA, in October 2010, for adults with SSTIs and community-acquired pneumonia [25, 26]. Though its activity is similar to that of ceftriaxone, ceftaroline was designed to have a high affinity to PBP2a, and is accordingly active against MRSA, VRE, and resistant pneumococci. The most common adverse events are diarrhea, nausea, and rash. Clinical studies in children are currently in the planning stages.

Dalfopristin/quinupristin is a combination streptogramin that is active against MRSA, MRSE, resistant pneumococci, and vanA-mediated resistant *E. faecium*; it is not active against resistant *E. faecalis* [23]. The agent is approved for children over age 16 years and adults with complicated SSTIs.

### 9.2.2.3 MRSA with Reduced Vancomycin Susceptibility

Several levels of reduced susceptibility of S. aureus to vancomycin have been defined.

**Vancomycin-resistant S. aureus (VRSA)** is defined as MIC $\geq 16$ µg/ml, and is usually mediated by vanA [16, 27]. Although very alarming, it has been reported in only a few (~7) patients. These strains typically appear after prolonged vancomycin use, and are often resistant to other antimicrobial agents as well.

**Vancomycin-intermediate *S. aureus* (VISA)** is defined as MIC of 4–8 μg/ml, and is probably caused by cell wall thickening with reduced levels of peptidoglycan cross linking; it is uncommon and associated with previous vancomycin use [7, 16, 28].

**Increased MIC ("vancomycin creep")** reflects the trend of increased MIC of *S. aureus* to vancomycin, although still within the susceptibility range (MIC ≤2 μg/ml) [16]. In adults, an increase in MIC from 0.5–2 μg/ml has been shown to double clinical failure rates of staphylococcal infections [28], and an increase in MIC from 1 to 2 μg/ml of staphylococcal bacteremia to double the 30-day mortality rate [29]. Consequently, the recommended dosages of vancomycin for MRSA were increased [18], although elevated levels of vancomycin MIC were not observed in a study of pediatric MRSA isolates [30].

## 9.2.3 Extended-spectrum β-lactamases

Since first described in Germany in 1983, extended-spectrum β-lactamases (ESBLs) have spread worldwide, with increasing frequency among Gram-negative isolates [3, 4, 7]. These enzymes are viewed as new β-lactamases, as they hydrolyze the oxyimino- cephalosporins and monobactams [31]. They are inhibited by β-lactamase inhibitors, a property that assists their identification in the bacteriologic laboratory. Currently, >500 types of ESBLs have been reported: >160 in the TEM class, >140 in the OXA family, >100 in the SHV family, and >80 in the CTX-M class [7, 31]. The Amp C β-lactamase, which is found predominantly in *Enterobacter* sp and is often inducible and chromosomally-encoded, also reduces permeability by porin channel loss in the outer membrane.

ESBLs are most frequently present in *Klebsiella pneumoniae*, *Klebsiella oxytoca*, and *Escherichia coli*, mediating their resistance to broad-spectrum cephalosporins [7, 31]; yet they have also been detected in other Gram-negative organisms, including typhoidal and non-typhoidal *Salmonella* sp, *Pseudomonas aeruginosa*, and *Proteus sp*. ESBL-producing organisms present mainly in pediatric, and especially neonatal, intensive care units.

Most ESBLs are encoded by large plasmids, and therefore can spread relatively easily among bacteria, also of different genera [3, 7, 32]. The plasmids can carry additional resistance genes, such as those mediating resistance to aminoglycosides and fluoroquinolones, leading to multi-drug resistant organisms with limited options for treatment [32].

Rates of Gram-negative bacteria that express ESBLs vary greatly by geographical region. Rates among *K. pneumoniae* tend to be highest. According to the European Antimicrobial Resistance Surveillance Network (EARS-Net), rates in 2008 were <5 % in Scandinavian countries, 25–50 % in Italy, Portugal, and Turkey, and >50 % in Greece. In the United Kingdom, the rate increased from 2 % in 2000 to 12 % in 2007 [32]. ESBL-producing organisms are usually associated with hospital-acquired infections, sometimes as outbreaks, and mainly in neonatal intensive care units. Previous exposure to antimicrobial agents, especially to ceftazidime, has been related to increased prevalence of ESBL-producing bacteria [7, 32]. These bacteria mostly affect children with comorbidities; most commonly causing bacteremia, septicemia, pneumonia, urinary tract infections (UTIs), wound infections, and ocular infections, with a greater risk of clinical treatment failure than for susceptible strains [32].

### 9.2.3.1 The Treatment of ESBL-producing Bacteria

Since ESBLs are inhibited by β-lactamase inhibitors, the combination of a β-lactam-β-lactamase inhibitor, such as piperacillin-tazobactam, is often used to treat infections caused by ESBL-producing bacteria; another possibility is an aminoglycoside, mainly amikacin [23]. A "thinking-outside-the box" approach uses prolonged β-lactam infusion to overcome the increased MIC of resistant Gram-negative organisms, ultimately optimizing the PK/PD of the available antibiotics

**Table 9.3** Treatment options for multi-drug resistant Gram-negative bacteria

| Antimicrobial agent | Comments |
| --- | --- |
| Ampicillin-sulbactam | Intrinsic activity against *A. baumani* |
| Fluoroquinolones | For ESBL-producing bacteria, *P. aeruginosa* |
| Colistin | For CRE, *A. baumani*, *P. aeruginosa* |
| Tigecycline | For CRE, *A. baumani*; inferior to β-lactams |

*ESBL* extended-spectrum β-lactamase
*CRE* carbapenem-resistant Enterobacteriaceae

[33]. Since the efficacy of β-lactams is primarily determined by the fraction of time in which the drug concentration exceeds the MIC (fT>MIC), prolonging the infusion time results in more consistent serum levels and increases the fT>MIC. Prolonged infusion consists of either extended or continuous infusion; extended infusion is defined as intermittent infusion lasting ≥3 h, and continuous infusion as a 24-h fixed rate administration [33]. The clinical benefit of prolonged infusion has been demonstrated in adults, using piperacillin-tazobactam or ceftazidime, whereas studies in children are pending [33].

As the ESBL-encoding plasmids sometimes carry additional resistance genes, the resistance pattern of these organisms should be monitored in each medical unit, and the empiric antibiotic determined accordingly. In some cases fluoroquinolones are required, although they are not generally approved for individuals >17 years (Table 9.3). ESBLs are not capable of hydrolyzing carbapenems, which are usually the last option for the treatment of the ESBL-producing Gram-negative bacteria.

## 9.2.4 Carbapenem Resistance

### 9.2.4.1 Carbapenem Use in Children

Carbapenems are the mainstay in the treatment of pediatric infections caused by multi-drug resistant Gram-negative bacteria; due to increasing resistance, carbapenems are often the last resort for these infections [3, 23, 32]. All four carbap-

enems—meropenem, imipenem, ertapenem, and doripenem—are administered parenterally, and their elimination is predominantly by glomerular filtration in the kidneys. They all have broad spectrum antimicrobial activity, though some differences exist among them [23]. Meropenem, which is the most active against Gram-negative bacteria, is most commonly used in neonates and children. Imipenem, which is more active against Gram-positive bacteria than is meropenem, is administered with cilastatin to avoid hydrolysis and to reduce nephrotoxicity. Seizures are an important adverse effect of imipenem, occurring in about 3 % of the treated patients who do not have central nervous system infections, and in rates reaching and exceeding 30 % in patients treated for bacterial meningitis [23]. Ertapenem demonstrates reduced activity against *Pseudomonas* isolates; in contrast, doripenem is most active against *P. aeruginosa*, but pediatric data are very limited. Carbapenems are not active against MRSA, *Stenotrophomonas maltophilia*, *Burkholderia cepacia*, *E. faecium*, *Mycoplasma* sp, and *Chlamydia* sp.

### 9.2.4.2 Resistance to Carbapenems

Resistance of Gram-negative bacteria to carbapenems is usually mediated by one of the following three mechanisms: impermeability due to porin channel loss in the outer membrane, alterations in PBPs, and carbapenemase-mediated hydrolysis [1, 3, 9]. Most common and alarming is the production of carbapemases [6, 9, 34]. The carbapenemases are divided into three classes (A, B, and D), and include *K. pneumoniae* carbapenemases (KPCs), several serine carbapenemases, and several metallo-β-lactamases [34]. KPC enzymes, classified as types 1–8, are no longer confined to *K. pneumoniae*, but have spread among other Gram-negative bacteria, including other *Klebsiella* sp, *E. coli*, *Salmonella enterica*, *Enterobacter* sp, *P. aeruginosa,* and others [1, 9, 34]. Outbreaks of KPC-producing bacteria have been reported in the United States, South America, Greece, Israel, India, and China [34].

Rapid dissemination of the New Delhi metallo-β-lactamase 1 (NDM-1) was recently reported [6, 11]. The $bla_{NDM-1}$-encoded gene was first detected in a Swedish patient who had been hospitalized in New Delhi and was colonized with *K. pneumoniae* and *E. coli* containing this enzyme. The resistant gene was carried on plasmids of varying size, which readily transferred *in vitro* between bacterial strains. Indeed, this new type of metallo-β-lactamase spread rapidly to the United Kingdom and Pakistan, and subsequently to all continents [9]. The bla$_{NDM-1}$-encoded mobile gene disseminated into diverse bacteria, including *Enterobacter* sp, *Proteus* sp, *Citrobacter* sp, and *A. baumani*. Infections caused by these resistant organisms comprised septicemia, pneumonia, urinary tract infections, SSTIs, and diarrhea, with many fatal cases [6, 9]. The observation that some of these infections caused by the resistant bacteria were community-acquired is very alarming [9].

### 9.2.4.3 Treatment

Treatment options for multi-drug resistant Gram-negative bacteria are listed in Table 9.3. Ampicillin-sulbactam has intrinsic activity against *A. baumani*, and is used for infections caused by multiply-resistant *A. baumani* [7]. The fluoroquinolones, which inhibit DNA replication by inhibiting the topoisomerase, have broad antimicrobial activity, mainly against Gram-negative organisms [23]. They are active against ESBL-producing bacteria and *P. aeruginosa* [23, 35]. Pediatric effectiveness has been shown for meningitis, infections in children with cystic fibrosis, complicated urinary tract infections, bacterial diarrhea, and chronic suppurative otitis media. The major drawback, however, is that the fluoroquinolones induced changes in the immature joint cartilage of all laboratory animals studied. In children, reversible arthralgia or arthritis has been documented, but no clear permanent arthropathy. The fluoroquinolones are therefore not approved for use in children, except for in unusual settings; ciprofloxacin, for example, is approved for use in children in many countries for limited indications. Fluoroquinolone use in children is only recommended when an alternative antimicrobial therapy is not feasible [23, 35].

Colistin, Polymixin E, is a large cationic polypeptide that binds to the bacterial anionic lipopolysaccharide and increases membrane permeability, ultimately leading to cell death. This polypeptide is active against most aerobic Gram-negative bacteria, including those resistant to other antibiotics, such as multi-drug resistant *A. baumani* and *P. aeruginosa* [36]. Colistin has concentration-dependent bactericidal activity, with a considerable post-antibiotic effect. Colistimethate, a prodrug that is hydrolyzed to the bioactive colistin, is the commercial intravenous formulation, which can be administered also by inhalation. While generally not absorbed from the gastrointestinal tract, it has been administered orally for bowel decontamination, as well as topically for bacterial skin infections. It does not cross the blood-brain barrier; case reports of intrathecal or intraventricular use for resistant Gram-negative bacteria have been described. Excretion is mainly by glomerular filtration. Because of the major concern of nephrotoxicity during colistin therapy, its IV use was abandoned in the 1980s. Recent data from adults, however, indicate that colistin-mediated nephrotoxicity is less prevalent than previously thought, occurring in 8–19 % of treated patients, mainly as tubular damage [34]. Neurotoxicity presented in about 7 % of the patients, and allergic reactions in about 2 %.

Colistin is typically used as "salvage therapy" for bacteria that are resistant to carbapenems and other antibiotics (Table 9.3).

Tamma and Lee reviewed colistin use in children [36]. Of 271 children without cystic fibrosis who were treated with colistin, 86.5 % were cured; nephrotoxicity was noted in 2.8 %, and neurotoxicity was not observed. A retrospective study evaluated colistin use for the treatment of septicemia in a tertiary neonatal intensive care unit during 2009 [37]. Eighteen critically ill neonates who received colistin and at least one additional antibiotic were identified. A favorable clinical response was observed in 76 %, and microbiological clearance in 81 %, with no major adverse events. The authors concluded that IV colistin appears to be safe and efficacious in neonates, including preterm babies [37].

Tigecycline is the first glycylcycline antibiotic, a new class of antimicrobial agents, that have been approved in many countries for the treatment of complicated SSTIs, complicated intra-abdominal infections, and community-acquired pneumonia in adults [38]. It is active *in vitro* against a wide-range of resistant bacteria, including MRSA, carbapenem-resistant Enterobacteriaceae, and *A. baumani*. The dosage in children above 8 years old is 1 mg/kg every 12 h. A recent systematic review and meta-analysis identified 15 randomized controlled trials that examined the efficacy of tigecycline. Compared with other antibiotic regimens, tigecycline therapy was associated with higher clinical failure, development of septic shock, microbiological failure, and overall mortality [38]. The authors concluded that "clinicians should avoid tigecycline monotherapy in the treatment of severe infections and reserve it as a last-resort drug" [38]. Accordingly, the US Food and Drug administration (FDA) issued a warning regarding the use of this drug.

## 9.3   Conclusions

Increasing rates of multi-drug resistant bacteria have recently been witnessed, with some organisms resistant to all currently available antimicrobial agents [2, 3, 6, 11]. In a world in which travel is very common, resistant bacteria spread rapidly [6]. In an era with very limited options for new antimicrobial agents [12], our ability to effectively treat bacterial infections in adults and in children is threatened [11]. Immediate action is needed, especially to decrease the spread of resistant bacteria. A recent study demonstrated the success of a nationally implemented broad infection control program in containing a nationwide outbreak of carbapenem-resistant *K. pneumoniae*; monthly numbers of new cases of resistant strains decreased by 79 %, as detected by active surveillance [39]. This evidence of the feasibility of controlling antibiotic resistance gives hope for the efficacy of antibiotics in the future.

## References

1.  Gupta N, Limbago BM, Patel JB, Kallen AJ (2011) Carbapenem-resistant Enterobacteriaceae: epidemiology and prevention. Clin Infect Dis 53(1):60–67
2.  Perez F, Hujer AM, Hujer KM, Decker BK, Rather ON, Bonomo RA (2007) Global challenge of multi-drug resistant *Acinetobacter baumani*. Antimicrobial Agents Chemother 51(10):3471–3484
3.  Paterson DL (2008) Impact of antibiotic resistance in Gram-negative bacilli on empirical and definitive antibiotic therapy. Clin Infect Dis 47(Suppl1):14–20
4.  Paterson DL, Rogers BA (2010) How soon is now? The urgent need for randomized controlled trials evaluating treatment for multi-drug resistant bacterial infections. Clin Infect Dis 51(11):1245–1247
5.  Sievent DM, Rudrick JT, Patel JB, McDonald LC, Wilkins MJ, Hageman JC (2008) Vancomycin-resistant *Staphylococcus aureus* in the Unites States, 2002–2006. Clin Infect Dis 46(5):668–674
6.  Kumarasamy KK, Toleman MA, Walsh TR, Bagaria J, Butt F, Balakrishnan R et al (2010) Emergence of a new antibiotic resistance mechanism in India, Pakistan and the UK: a molecular, biological and epidemiological study. Lancet Infect Dis 10(9):597–602
7.  Chen LF, Chopra T, Kaye KS (2009) Pathogens resistant to antimicrobial agents. Infect Dis Clin North Am 23(4):817–845
8.  Normark BH, Normark S (2002) Evolution and spread of antibiotic resistance. J intern Med 252(2):91–106
9.  Castanheira M, Mendes RE, Woosley LN, Jones RN (2011) Trends in carbapenemase-producing *Escherichia coli* and *Klebsiella* app from Europe and the Americas: report from the SENTRY antimicrobial surveillance programme (2007-09). J Antimicrobial Chemother 66(6):1409–1411
10. Carrillo-Marquez MA, Hulten KG, Hammerman W, Lamberth L, Mason EO, Kaplan SL (2011) *Staphylococcus aureus* pneumonia in children in the era of community-acquired methicillin-resistance at Texas Children's Hospital. Pediatr Infect Dis J 30(7):545–550
11. Nordmann P, Poirel L, Toleman MA, Walsh TR (2011) Does broad-spectrum β-lactam resistance due to NDM-1 herald the end of the antibiotic era for treatment of infections caused by Gram-negative bacteria? J Antimicrobial Chemother 66(4):689–692
12. British Society for Antimicrobial Chemotherapy (2011) The urgent need for new antimicrobial agents. J Antimicrobial Chemother 66(6):1939–1940
13. Dagan R, Greenberg D, Jacobs MR, Phillips BL (2009) Pneumococcal infections. In: Feigin RD, Cherry JD, Demmler-Harrison GJ, Kaplan SL (eds) Textbook of Pediatric Infectious Diseases, 6th edn. p 1288–1342
14. Hsu KK, Shea KM, Stevenson AE, Pelton SI (2010) Changing serotypes causing childhood invasive pneumococcal disease: Massachusetts, 2001–2007. Pediatr Infect Dis J 29(4):289–293
15. Markiewicz Z, Tomasz A (1989) Variations in penicillin-binding protein patterns of penicillin-resistant clinical isolates of pneumococci. J Clin Microbiol 27(3):405–410
16. Clinical and Laboratory Standards Institute (2011) Performance standards for antimicrobial susceptibility testing (Table 2G), Wayne

17. Martinez-Aguilar G, Hammerman WA, Mason EO, Kaplan SL (2011) Clindamycin treatment of invasive infections caused by community-acquired methicillin-resistant and methicillin-sensitive *Staphylococcus aureus* in children. Pediatr Infect Dis J 22(7):593–598

18. Liu C, Bayer A, Cosgrove SE, Daum RS, Fridkin SK, Gorwitz RJ et al (2011) Clinical practice guidelines by the Infectious Diseases Society of America for the treatment of methicillin-resistant *Staphylococcus aureus* infections in adults and children. Clin Infect Dis 52(3):285–292

19. Gemmel CG, Edwards DI, Fraise AP, Gould FK, Ridgway GL, Warren RE (2006) Guidelines for the prophylaxis and treatment of methicillin-resistant *Staphylococcus aureus* infections in the UK. J Antimicrobial Chemother 57(4):589–608

20. Markwell S, Peter J, Barenkamp S (2010) Randomized controlled trial of antibiotics in the management of community-acquired skin abscesses in the pediatric patients. Ann Emerg Med 55(5):401–407

21. Peltola H, Paakkonen M, Kallio MJT (2010) Short- versus long-term antimicrobial treatment for acute osteomyelitis of children: prospective, randomized trial on 131 culture-proven cases. Pediatr Infect Dis J 29(12):1123–1128

22. Garazzino S, Tovo PA (2011) Clinical experience with linezolid in infants and children. J Antimicrobial Chemother 66(Suppl 4):23–41

23. Chavez-Bueno S, Stull TL (2009) Antibacterial agents in children. Infect Dis Clin North Am 23(4):865–880

24. Ardura MI, Mejias A, Katz KS, Ravell P, McCracken GH, Sanchez PJ (2007) Daptomycin therapy for invasive Gram-positive bacterial infections in children. Pediatr Infect Dis J 26(12):1128–1132

25. The Pediatric Infectious Disease Newsletter (2011) Ceftaroline approved. Pediatr Infect Dis J 30(4):A7

26. Laudano JB (2011) Ceftaroline fosamil: a new broad-spectrum cephalosporin. J Antimicrobial Chemother 66(Suppl 3):11–18

27. Sievent DM, Rudrik JT, Patel JB (2008) Vancomycin-resistant *S. aureus* in the United States, 2002–2006. Clin Infect Dis 46(5):668–674

28. Jones RN (2006) Microbiological features of vancomycin in the 21st century: MIC creep, bactericidal/ static activity and applied breakpoints to predict outcomes or detect resistant strains. Clin Infect Dis 42(Suppl 1):13–24

29. Holmes NE, Turnidge JD, Munckholf WJ, Robinson JO, Korman TM, O'Sullivan MVN et al (2011) Antibiotic choice may not explain poorer outcomes in patients with *Staphylococcus aureus* bacteremia and high vancomycin minimum inhibitory concentrations. J Infect Dis 204(3):340–347

30. Zheng X, Qi C, Arrieta M, O'leary A, Wang D, Shulman ST (2010) Lack of increase in vancomycin resistance of pediatric methicillin-resistant *Staphylococcus aureus* isolates from 2000 to 2007. Pediatr Infect Dis J 29(9):882–884

31. Jacoby GA (1997) Extended-spectrum β-lactamases and other enzymes providing resistance to oxyimino-β-lactams. Infect Dis Clin North Am 11(4):875–887

32. Hague R (2011) What is the threat from extended spectrum β-lactamaze-producing organisms in children? Arch Dis Child 96(4):325–327

33. Tamma PD, Jenh AM, Milstone AM (2011) Prolonged β-lactam infusion for Gram-negative infections. Pediatr Infect Dis J 30(4):336–337

34. Overturf GD (2010) Carbapenemnases: a brief review for pediatric infectious disease specialist. Pediatr Infect Dis J 29(1):68–70

35. Bradley JS, Jackson MA, Committee on Infectious Diseases, American Academy of Pediatrics (2011) Clinical report—the use of systemic and topical fluoroquinolones. Pediatrics 128(4):1034–1045

36. Tamma PD, Lee CK (2009) Use of Colistin in children. Pediatr Infect Dis J 28(6):534–535

37. Jajoo M, Kumar V, Jain M, Kumari S, Manchandra V (2011) Intravenous colistin administration in neonates. Pediatr Infect Dis J 30(3):218–221

38. Yahav D, Lador A, Paul M, Leibovici L (2011) Efficacy and safety of tigecycline: a systematic review and meta-analysis. J Antimicrob Chemother 66(9):1963–1971

39. Schwaber MJ, Lev B, Israeli A, Solter E, Smollan G, Rubinowitch B et al (2011) Containment of a country-wide outbreak of carbapenem-resistant *Klebsiella pneumoniae* in Israeli hospitals via a nationally implemented intervention. Clin Infect Dis 52(7):848–855

# An Encephalitis Primer

Dominic Kelly

**10**

**Abstract**

For most clinicians managing a patient with encephalitis will be an infrequent event. The range of differential diagnoses in a child presenting with a clinical syndrome that could be encephalitis is broad. In addition, there are many pathogens that can cause encephalitis. In view of these facts, a systematic approach to clinical assessment, investigation and treatment, based on epidemiology and an understanding the limits of investigations, is crucial to effective management of this condition. This review focuses on encephalitis aetiology and investigation; HSV encephalitis; mycoplasma encephalitis; and flavivirus encephalitis as a globally important cause of disease.

## 10.1 Introduction

Encephalitis presents several challenges to the clinician. Firstly. the presenting signs and symptoms of central nervous system (CNS) disease (e.g. altered consciousness/behaviour, seizures, focal neurology) and infection (e.g. fever, rash) are common to a wide variety of non-infectious and infectious conditions which are hard to distinguish on clinical grounds alone. Secondly, the pathogens able to cause encephalitis are numerous. Finally, at presentation children are usually becoming progressively more unwell and

may have significant neurological and respiratory compromise for which timely intervention and support will prevent secondary damage.

In view of these facts a systematic approach to clinical assessment, investigation and treatment, based on epidemiology and an understanding the limits of investigations, is crucial to the effective management of this condition. This short review will focus on encephalitis aetiology and investigation, HSV encephalitis (due to its prevalence and the existence of treatment), mycoplasma encephalitis (frequently implicated in encephalitis but for which there are significant uncertainties around pathogenesis and treatment), flavivirus encephalitis (as a globally important cause of disease) and 'Louping ill' (out of local interest).

Several comprehensive reviews of childhood encephalitis have been published recently [1, 2]

D. Kelly (✉)
Children's Hospital, Oxford, UK
e-mail: dominic.kelly@paediatrics.ox.ac.uk

N. Curtis et al. (eds.), *Hot Topics in Infection and Immunity in Children IX,*
Advances in Experimental Medicine and Biology 764, DOI 10.1007/978-1-4614-4726-9_10,
© Springer Science+Business Media New York 2013

133

and recommendations on investigation and management produced in the US and UK [3, 4].

## 10.2 Regional Epidemiology Together with a History of Travel, Immunisation Status, Immune Compromise and Animal Exposure Guides Initial Investigation

Although viral causes are most common, encephalitis or encephalitis-like syndromes can also be caused by bacteria, mycobacteria, fungi and parasites.

---

UK and North American studies demonstrate the importance of the following pathogens in childhood encephalitis (treatable causes are underlined) [5–7]:

1. Herpes viruses (HSV, VZV, EBV)

2. Respiratory viruses (e.g. influenza, para-influenza, adenovirus)

3. Enteroviruses

4. *Mycoplasma pneumoniae*

A history of travel, lack of immunisation, immune compromise and animal exposure extends this list to include:

5. Lack of immunisation—Measles, mumps

6. Travel—Arboviruses (e.g. Japanese encephalitis, West Nile virus, tick borne encephalitis; St Louis encephalitis)

7. Immunocompromise—HIV, Listeria, Toxoplasma, CMV, HHV6, JC, Cryptococcus

8. Animal exposure—*Borrelia burgdorferi*, Rabies, *Bartonella henselae* (Cat-scratch disease)

*HSV* herpes simplex virus, *VZV* Varicella Zoster virus, *EBV* Epstein Barr virus, *HIV* human immunodeficiency virus, *CMV* cytomegalovirus, *HHV6* human herpes virus 6

---

Other pathogens may cause clinical syndromes very similar to encephalitis and these, although usually considered separately, are an important consideration in the differential diagnosis of treatable causes. Such pathogens include; bacteria causing meningitis (or a meningo-encephalitis) such as *Streptococcus pneumoniae*, *Neisseria meningitidis* and *Haemophilus influenzae*; *Plasmodium falciparum* causing malaria; *Mycobacterium tuberculosis* causing TB meningitis.

The relative importance of these pathogens varies by region of the world and season [8]. In central, northern and eastern Europe, tick borne encephalitis is the a prevalent pathogen associated with encephalitis. In areas of South-East Asia, such as Vietnam, Japanese encephalitis predominates. Such is the importance of tick borne encephalitis and Japanese encephalitis in these regions that in some countries, vaccines against these pathogens are part of the routine immunisation schedule.

Whilst not exhaustive, the above list, combined with a full history, allows investigation for the pathogen causing encephalitis to be focused on the most likely organisms. However it is important to note that even with extensive investigation no definite pathogen is identified in a significant proportion of cases [5, 9, 10].

Of the many organisms causing encephalitis only a limited number have specific treatment (these are highlighted in the list above). Despite the prevalence of arboviruses in many regions of the world there is currently little to support specific treatment for any of these organisms. There is a single case report of an individual surviving rabies associate with a specific intensive care unit based treatment protocol but this is the exception and has not been replicated [11].

## 10.3 Initial Investigation and Empirical Treatment should be Focussed on Non-infectious Differential Diagnoses Together with Common or Treatable Infectious Causes

The extent of investigation will depend on resources available and region. Confirmation of encephalitis is based on the combination of findings from cerebrospinal fluid (CSF) microscopy, CNS imaging, identification of a specific pathogen (e.g. HSV from CSF PCR) and in some cases EEG. The importance of investigation for non-infectious causes lies in the lack of specificity of the initial presenting clinical syndrome and the wide variety of other treatable causes (Table 10.1). Various guidelines have been published and the list of first-line investigations

**Table 10.1** Non-infectious differential diagnoses for children presenting with possible encephalitis. (Adapted from Long 2010 [2]) (With kind permission from Springer Science + Business Media B.V.)

| Non-infectious, para-infectious and autoimmune | Reye syndrome |
| --- | --- |
| | Acute disseminated encephalomylitis |
| | Acute necrotizing encephalopathy |
| Neoplasia | Primary or metastatic |
| | Paraneoplastic diseases |
| Cerebrovascular | Ischemic stroke |
| | Subdural/epidural hematoma |
| | Vasculitis |
| Systemic | Metabolic conditions |
| | Connective tissue disorders |
| | Drug intoxication |
| Other | Epilepsy |
| | Head injury |
| | Confusion migraine |

used in one UK centre, for the child presenting with encephalopathy which may be encephalitis, is given here (Table 10.2).

Empirical anti-microbial treatment should include cover for likely bacterial pathogens causing CNS infection. In the UK, penicillin would cover *N. meningitidis* and most *S. pneumoniae*, however the possibility of *H. influenzae*, *Staphylococcus aureus*, penicillin resistant *S. pneumoniae* and Gram negative pathogens means that a third generation cephalosporin (eg ceftriaxone) is indicated. For the severely unwell child it is reasonable to cover *Listeria* with amoxicillin until the clinical picture is clearer as, although rare, it is described in immunocompetent children. High dose intravenous acyclovir treatment is mandatory for any child with suspected encephalitis. Other treatments depend on the specific history and risk factors for the individual case.

## 10.4  HSV Encephalitis is Treatable but Still Problematic

HSV is the commonest non-seasonal cause of encephalitis (HSVE) and is treatable. Despite this HSV often provides the clinician with diagnostic

and therapeutic dilemmas. The outcome of HSVE in the pre-acyclovir era was devastating, with a mortality of 70 % and neuro-disability almost the rule amongst survivors. In 1986 a comparison of 10 days of acyclovir demonstrated a significantly improved outcome in comparison to vidarabine [12]. Of 69 individuals recruited to the trial the mortality in the 32 randomised to acyclovir treatment was 28 % versus 54 % in the vidarabine group. Of the survivors, 38 % in the acyclovir group versus 14 % in the vidarabine group were 'functioning normally'. In these trials the duration of treatment was 10 days and there were no relapses. However one of the authors of this study noted that 'many patients were still febrile at the conclusion of treatment suggesting that a longer duration of therapy …may be advisable' [13]. Relapses of patients treated with acyclovir have been reported and in addition neonatal trials indicated the improved outcomes of 21 days of treatment versus 10 days (although the latter group also had a lower dose of acyclovir) [14]. These factors have led to the current recommendations for treatment duration of 14–21 days. The major modifiable factor associated with poor outcome in HSVE is now delay in starting acyclovir and clinicians should have a low threshold for starting acyclovir in any child with a compatible syndrome.

Detection of HSV DNA on PCR testing of CSF is the current 'gold standard' of diagnosis. However a negative CSF PCR is found in a number of children in the early stages of disease (see Table 10.3). Diagnostic work-up should therefore include an LP at least 72 h after the onset of symptoms even if this means a repeat LP for children who presented at <72 h. An important issue is that a number of children with subsequently proven HSV will have a normal CSF white cell count [15]. Whilst scanning is usually abnormal in HSVE, this may not be the case early in the course of the illness particularly with CT [16]. A wide variety of imaging findings have been described from the classic fronto-temporal lesions through to more diffuse lesions. The decision to discontinue acyclovir treatment in a child with an encephalitis-like illness needs to be considered in the light of these diagnostic uncertainties and should be based on clinical recovery

**Table 10.2** The investigation of children with unexplained encephalopathy. **a** Initial investigation and **b** targeted further investigation in relation to risk-factors from detailed history. (Adapted from clinical guidelines at the Children's Hospital, Oxford)

▼

| **Investigation** | |
| --- | --- |
| **Radiology** | : CT (+/- contrast) or MRI if available |
| | <u>**Sedation contraindicated - contact anaesthetist for GA if necessary**</u> |
| **Blood** | : FBC/film, UEs/glucose/Ca/Mg, LFTs/coags |
| | Blood gas, **ammonia (urgent),** lactate, amino acids (biochemistry 'to store') |
| | Blood culture |
| | Serology (>2ml) for EBV, CMV, Mycoplasma, Lyme, HIV (specific consent required) |
| **CSF** | : Opening pressure, gram stain, culture, gluc/protein, lactate, PCR (HSV/VZV/Entero/Adeno) |
| **Urine** | : Culture, organic/amino acids/toxicology (biochemistry 'to store') |
| **Throat swab** | : PCR (Entero/Adeno) |
| **NPA** | : PCR (Influenza/Parainfluenza/Adeno/RSV/Entero) <u>only if compatible symptoms</u> |

a

**Box 1 Additional ID investigations**
Many uncommon infectious causes of encephalopathy have specific treatments(box 3) and are important to consider in terms of history of potential risk factors (e.g. travel, animal exposure, immune compromise, immunisations (MMR)) +/- investigation including;

<u>Immunocompromise</u>
| | |
| --- | --- |
| CMV | - blood/CSF pcr |
| HHV6 | - CSF pcr |
| Measles (unvaccinated) | - serology (serum/saliva) |
| Cryptococcus neoformans | - Cryptococcal antigen (blood/CSF), India ink stain (CSF) |
| Mycobateria tuberculosis | - CSF for AFBs/TB culture |
| Toxoplasma gondii | - serology |

<u>Travel</u> – **country-dependent**
| | |
| --- | --- |
| Malaria | - thick and thin film |
| Rickettsial diseases (Typhus, spotted fevers) | - serology |
| Rabies if animal bite/ bat exposure | - discuss investigation |
| Arboviruses (JE, WNV, TBE) | - serology / CSF pcr |
| Measles (unvaccinated) | - serology (serum/saliva) |
| Mycobaterium tuberculosis (contact history) | - CSF for AFBs/TB culture |

<u>Animal exposure</u>
| | |
| --- | --- |
| Bartonella henselae (cat-scratch) | - serology |
| Rabies if animal bite/ bat exposure | - discuss investigation |

b

together with (i) normal MRI, (ii) negative CSF PCR at >72 h post-onset of symptoms and negative EEG [1]. In practice if all of these investigations are normal and there is complete recovery then discontinuing acyclovir around 10 days may be considered in that the risk of HSVE is relatively low and a course of treatment has been given that was shown to be efficacious in the original trials. If there are significant clinical, CSF, MRI or EEG abnormalities consistent with HSVE, and no alternative diagnosis, then a full 21-day course of treatment is warranted despite a negative CSF PCR [1, 17]. An algorithm to facilitate the decision to discontinue acyclovir treatment has been described previously [17] and this issue is discussed in detail in a recent review of childhood encephalitis [1].

With sensitive CSF PCR testing, cases of possible HSVE may be identified in clinical settings where the illness appears very mild compared to

**Table 10.3** Case series which document the occurrence of initial negative HSV PCR in individuals with subsequent confirmed herpes simplex encephalitis. [15, 39–41]

| Study | Total number with HSE | Initial negative HSV PCR | |
|---|---|---|---|
| Weil et al. [41] Clin Infect Dis | 11 | 3 | Negative day 1–3, positive day 4–7 |
| Stugahl et al. [40] Scan J Infect Dis | 10 | 2 | Positive 4–7 days later |
| Guffond et al. [39] Clin Infect Dis | 22 | 1 | Negative day 3, positive day 8 |
| De Tiege et al. [15] Clin Infect Dis | 38 (children) | 8 | Negative day 0–3 |

typical HSVE or there are possible alternative diagnoses [18]. Whilst it has been speculated that some of these may be reactivation, of what is a persistent neurotropic virus, identification of HSV in CSF remains a clear indication for a full course of acyclovir treatment.

The availability of valaciclovir, a highly bio-available alternative to intravenous acyclovir, has prompted some discussion about its role in treating invasive HSV disease. This is particularly relevant where a prolonged course of intravenous treatment requires secure venous access which in some cases can become more challenging as treatment progresses. There is a case report of oral valaciclovir use in one child with confirmed HSVE during the last week [19] and an uncontrolled series of cases from Vietnam treated with oral valaciclovir [20]. In the latter six individuals with confirmed HSVE were treated with oral valaciclovir and two died (days 2 and 3) whilst the other four completed treatment. These four had negative CSF PCRs by day 10 and CSF aciclovir concentrations were considered to be therapeutic, although levels fell with increasing duration of treatment. At present this agent cannot be recommended at any stage of treatment of HSVE.

Even with acyclovir there are still significant sequelae to HSVE and adjunctive treatments have been considered. Steroids have been associated with clinical recovery in some case reports [21] and improved outcomes in some case series [22]. A randomised trial is currently in progress in Europe [23]. At the present time corticosteroids remain of unproven benefit in HSVE.

## 10.5  *Mycoplasma pneumoniae* is Frequently Identified in the Setting of Encephalitis

*M. pneumoniae* is a common respiratory tract infection in childhood and is associated with a wide variety of neurological complications including encephalitis. In some studies of encephalitis *M. pneumoniae* has accounted for 5–13 % of cases [24, 25]. It's relative rarity in other studies may relate to limitations of current diagnostic methods [26] and also the occurrence of disease in cycles (3–4 years in the UK) longer than most epidemiological studies [27]. The detection of organisms via culture or PCR in CSF suggests that direct invasion of the central nervous system plays an important role in pathogenesis in at least some individuals [24, 26]. However immunological mechanisms are likely to be important in many children including cases of ADEM [28]. In most centres testing is limited to serology whose limited specificity and sensitivity reduces the ability to determine the role of this agent in encephalitis. In one of the most detailed case series, PCR identified *M. pneumoniae* DNA in CSF, or in throat swabs in conjunction with positive serological evidence of infection, in almost 7 % of cases of childhood encephalitis (11 out of 159 children) [24]. Around two-thirds of these children had significant neurological sequelae at follow-up. In a further 25 % of cases of encephalitis there was evidence for a possible role of *M. pneumoniae* although many of these children also had other pathogens detected as well. Not all children had a respiratory prodrome and the lack of such a prodrome was associated with the detection of *M. pneumoniae* by PCR in the CSF (rather than confirmation by serology with a negative CSF PCR). Given the evidence regarding neurological sequelae it is reasonable to consider empirical treatment for

*M. pneumoniae* in settings and seasons where infection is common. A CNS penetrating agent should be used and options include azithromycin [29], fluoroquinolones [30] and doxycycline although there is no definitive evidence for efficacy [3].

## 10.6 Arboviruses Including Japanese Encephalitis are Rarely Seen in the UK but are Globally Important Causes of Viral Encephalitis

Arbovirus (Arthropod Borne virus) is a colloquial term referring to viruses that are sustained in cycles of transmission between host vertebrate species and vector arthropods such as mosquitoes or ticks. A number of these viruses can cause encephalitis in humans, who are usually not the primary host but an evolutionary 'dead-end' for the virus. The major families of 'arboviruses' are Bunyaviridae (La Crosse virus), Togaviridae (e.g. Eastern and Western Equine encephalitis and Chikungunya) and Flaviviridae (e.g. Japanese encephalitis, West Nile virus, St Louis encephalitis, tick borne encephalitis). Globally it is viruses from the genus flavivirus, within the family Flaviviridae, which are responsible for the greatest burden of disease. The combination of vector ecology, host ecology and virus biology result in very specific geographic and seasonal patterns for these infections and the most significant flaviviruses vary by region. The mosquito-transmitted flaviviruses that typically cause encephalitis are Japanese encephalitis, West Nile virus and St Louis encephalitis. Although only a minority of infections result in significant disease, encephalitis is one of the characteristic disease syndromes of these viruses [31].

Japanese encephalitis is estimated to cause up to 67,900 cases of encephalitis annually [32] despite the fact that there is an effective vaccine which has been incorporated into routine immunisation programmes in many areas of the world. This virus is endemic in many areas of Asia and causes sporadic disease in the western Pacific and north Australia. Cases are predominantly in children, with significant neurological morbidity

and a case fatality rate of 20–30 %. Between 30–50 % of survivors suffer neuropsychological sequelae. Disease in returning travellers is rare. Whether this is due to exposure or immunisation of travellers is unclear but the short incubation period (5–15 days) reduces the probability of Japanese encephalitis being a significant consideration except in the first 1–2 weeks after return. The prevalence of this disease in certain regions has driven attempts to find an effective treatment including trials of steroids and interferon-2 alpha but to date none have been shown to have efficacy [33, 34]. Effective immunisation programmes of individuals exposed to infection is the mainstay of prevention for this disease. For many years Japanese encephalitis was known as Japanese B encephalitis although the use of this term has now mostly lapsed. The origin of this term was as a distinction from type A epidemic encephalitis (encephalitis lethargica).

Since the introduction of its mosquito vector into the US in 1999 West Nile virus has become an important diagnostic consideration in the North America [35]. The arrival of this pathogen to the USA demonstrates the dynamic epidemiology of this virus family related to the interdependence of host, vector and virus. Despite suggestive experiments in animal models with immunoglobulin treatment and some human data there remains no proven treatment for infection with this virus [36].

Some other flaviviruses rely on tick vectors. In many areas of central and eastern Europe and across Russia into China one of several strains of tick borne encephalitis make up one of the most common aetiologies of encephalitis [37]. The incidence of disease has been increasing over recent years perhaps in response to climatic variables. Several countries in this region include vaccines against this virus in routine immunisation schedules.

## 10.7 'Louping ill': the UK Flavivirus

Many regions of the world have syndromes of viral encephalitis related to arboviruses sustained in life-cycles adapted to very specific local ecologies and the UK is no exception. The 'Louping ill' virus is the only flavivirus endemic in the UK [38]. It is closely related to tick borne encephalitis virus. The life-cycle involves tick vector (*Ixodes ricinus*) which is found mainly in areas of rough upland grazing or moorland. 'Louping ill' is an infection of the nervous system and occurs mostly in sheep and grouse. Human cases are rare and usually related to occupational exposure (abattoir workers, shepherds, vets). A range of syndromes have been described one of which is an encephalitis. There is no specific treatment and the name relates to the leaping behaviour of the sheep which develop symptomatic disease manifest by 'spring(ing) into the air.'

## 10.8 Conclusion

For the clinician, managing a child with an encephalitis-like syndrome a detailed history (including travel, immunisation status, animal exposure and evidence of immune-compromise) will facilitate prompt and effective investigation and treatment of important causative pathogens. Despite this, in a significant number of cases no aetiology will be identified. Key issues in management are prompt empiric treatment with antibiotics and acyclovir, consideration of other likely treatable causes (infectious and non-infectious) and good supportive care.

## References

1. Thompson C, Kneen R, Riordan A, Kelly D, Pollard AJ (2012) Encephalitis in children. Arch Dis Child 97(2):150–161
2. Long SS (2011) Encephalitis diagnosis and management in the real world. Adv Exp Med Biol 697:153–173
3. Tunkel AR, Glaser CA, Bloch KC, Sejvar JJ, Marra CM, Roos KL et al (2008) The management of encephalitis: clinical practice guidelines by the Infectious Diseases Society of America. Clin Infect Dis 47(3):303–327
4. Kneen R, Michael BD, Menson E, Mehta B, Easton A, Hemingway C et al (2012) National guideline for the management of suspected viral encephalitis in children. J Infect 64(5):449–477
5. Kolski H, Ford-Jones EL, Richardson S, Petric M, Nelson S, Jamieson F et al (1998) Etiology of acute childhood encephalitis at The Hospital for Sick Children, Toronto, 1994–1995. Clin Infect Dis 26(2):398–409
6. Glaser CA, Gilliam S, Schnurr D, Forghani B, Honarmand S, Khetsuriani N et al (2003) In search of encephalitis etiologies: diagnostic challenges in the California Encephalitis Project, 1998–2000. Clin Infect Dis 36(6):731–742
7. Granerod J, Ambrose HE, Davies NW, Clewley JP, Walsh AL, Morgan D et al (2010) Causes of encephalitis and differences in their clinical presentations in England: a multicentre, population-based prospective study. Lancet Infect Dis 10(12):835–844
8. Granerod J, Crowcroft NS (2007) The epidemiology of acute encephalitis. Neuropsychol Rehabil 17(4–5):406–428
9. Granerod J, Tam CC, Crowcroft NS, Davies NW, Borchert M, Thomas SL (2010) Challenge of the unknown. A systematic review of acute encephalitis in non-outbreak situations. Neurology 75(10):924–932
10. Fowler A, Stodberg T, Eriksson M, Wickstrom R (2008) Childhood encephalitis in Sweden: etiology, clinical presentation and outcome. Eur J Paediatr Neurol 12(6):484–490
11. Willoughby RE Jr., Tieves KS, Hoffman GM, Ghanayem NS, Amlie-Lefond CM, Schwabe MJ et al (2005) Survival after treatment of rabies with induction of coma. N Engl J Med 352(24):2508–2514
12. Whitley RJ, Alford CA, Hirsch MS, Schooley RT, Luby JP, Aoki FY et al (1986) Vidarabine versus acyclovir therapy in herpes simplex encephalitis. N Engl J Med 314(3):144–149
13. Whitley RJ, Kimberlin DW (2005) Herpes simplex encephalitis: children and adolescents. Semin Pediatr Infect Dis 16(1):17–23
14. Kimberlin DW, Lin CY, Jacobs RF, Powell DA, Corey L, Gruber WC et al (2001) Safety and efficacy of high-dose intravenous acyclovir in the management of neonatal herpes simplex virus infections. Pediatrics 108(2):230–238
15. De Tiege X, Heron B, Lebon P, Ponsot G, Rozenberg F (2003) Limits of early diagnosis of herpes simplex encephalitis in children: a retrospective study of 38 cases. Clin Infect Dis 36(10):1335–1339
16. Baskin HJ, Hedlund G (2007) Neuroimaging of herpesvirus infections in children. Pediatr Radiol 37(10):949–963
17. Kelly D, Kroll JS (2004) Encephalitis—beyond aciclovir. Adv Exp Med Biol 549:177–183
18. Elbers JM, Bitnun A, Richardson SE, Ford-Jones EL, Tellier R, Wald RM et al (2007) A 12-year prospective study of childhood herpes simplex encephalitis: is there a broader spectrum of disease? Pediatrics 119(2):399–407
19. Chan PK, Chow PC, Peiris JS, Mak AW, Huen KF (2000) Use of oral valaciclovir in a 12-year-old boy

with herpes simplex encephalitis. Hong Kong Med J 6(1):119–121

20. Pouplin T, Pouplin JN, Van Toi P, Lindegardh N, Rogier van Doorn H, Hien TT et al (2011) Valacyclovir for herpes simplex encephalitis. Antimicrob Agents Chemother 55(7):3624–3626

21. Musallam B, Matoth I, Wolf DG, Engelhard D, Averbuch D (2007) Steroids for deteriorating herpes simplex virus encephalitis. Pediatr Neurol 37(3):229–232

22. Kamei S, Sekizawa T, Shiota H, Mizutani T, Itoyama Y, Takasu T et al (2005) Evaluation of combination therapy using aciclovir and corticosteroid in adult patients with herpes simplex virus encephalitis. J Neurol Neurosurg Psychiatry 76(11):1544–1549

23. German Trial of Aciclovir and Corticosteroids in Herpes Simplex Encephalitis (GACHE). http://www.klinikum.uni-heidelberg.de/index.php?id=4223&L=en

24. Bitnun A, Ford-Jones EL, Petric M, MacGregor D, Heurter H, Nelson S et al (2001) Acute childhood encephalitis and Mycoplasma pneumoniae. Clin Infect Dis 32(12):1674–1684

25. Koskiniemi M, Vaheri A (1989) Effect of measles, mumps, rubella vaccination on pattern of encephalitis in children. Lancet 1(8628):31–34

26. Bitnun A, Richardson SE (2010) Mycoplasma pneumoniae: Innocent Bystander or a True Cause of Central Nervous System Disease? Curr Infect Dis Rep 12(4):282–290

27. Tong CY, Menson E, Lin JP, Lim M (2011) Prevalence of mycoplasma encephalitis. Lancet Infect Dis 11(6):425–426

28. Narita M (2009) Pathogenesis of neurologic manifestations of Mycoplasma pneumoniae infection. Pediatr Neurol 41(3):159–166

29. Jaruratanasirikul S, Hortiwakul R, Tantisarasart T, Phuenpathom N, Tussanasunthornwong S (1996) Distribution of azithromycin into brain tissue, cerebrospinal fluid, and aqueous humor of the eye. Antimicrob Agents Chemother 40(3):825–826

30. Esposito S, Tagliabue C, Bosis S, Principi N (2011) Levofloxacin for the treatment of Mycoplasma pneumoniae-associated meningoencephalitis in childhood. Int J Antimicrob Agents 37(5):472–475

31. Solomon T (2004) Flavivirus encephalitis. N Engl J Med 351(4):370–378

32. Campbell GL, Hills SL, Fischer M, Jacobson JA, Hoke CH, Hombach JM et al (2011) Estimated global incidence of Japanese encephalitis: a systematic review. Bull World Health Organ 89(10):766–774, 74A–74E

33. Hoke CH Jr., Vaughn DW, Nisalak A, Intralawan P, Poolsuppasit S, Jongsawas V et al (1992) Effect of high-dose dexamethasone on the outcome of acute encephalitis due to Japanese encephalitis virus. J Infect Dis 165(4):631–637

34. Solomon T, Dung NM, Wills B, Kneen R, Gainsborough M, Diet TV et al (2003) Interferon alfa-2a in Japanese encephalitis: a randomised double-blind placebo-controlled trial. Lancet 361(9360):821–826

35. Murray KO, Mertens E, Despres P (2010) West Nile virus and its emergence in the United States of America. Vet Res 41(6):67

36. Jackson AC (2004) Therapy of West Nile virus infection. Can J Neurol Sci 31(2):131–134

37. Mansfield KL, Johnson N, Phipps LP, Stephenson JR, Fooks AR, Solomon T (2009) Tick-borne encephalitis virus—a review of an emerging zoonosis. J Gen Virol 90(Pt 8):1781–1794

38. HPA (2011) Louping ill. 2011 (cited 15 Jan 2012). http://www.hpa.org.uk/Topics/InfectiousDiseases/InfectionsAZ/LoupingIll/

39. Guffond T, Dewilde A, Lobert PE, Lefebvre DC, Hober D, Wattre P (1994) Significance and Clinical Relevance of the Detection of Herpes-Simplex Virus-DNA by the Polymerase Chain-Reaction in Cerebrospinal-Fluid from Patients with Presumed Encephalitis. Clin Infect Dis 18(5):744–749

40. Studahl M, Bergstrom T, Hagberg L (1998) Acute viral encephalitis in adults—a prospective study. Scand J Infect Dis 30(3):215–220

41. Weil AA, Glaser CA, Amad Z, Forghani B (2002) Patients with suspected herpes simplex encephalitis: rethinking an initial negative polymerase chain reaction result. Clin Infect Dis 34(8):1154–1157

# The Evidence Behind Prophylaxis and Treatment of Wound Infection After Surgery

**11**

## Mona A. Al-Dabbagh and Simon Dobson

**Abstract**

Surgical site infections (SSIs) represent a serious post surgical complication. They are the leading cause of healthcare-related infections in developing countries and the second most common healthcare-related infection in developed countries. Here we discuss the epidemiology of and risk factors for SSIs together with the current evidence supporting the use of antibiotic prophylaxis for the prevention of wound infection after surgery.

## 11.1 Introduction

In the early nineteenth century, Ignez Semmelweis demonstrated that washing hands in the obstetrical clinic resulted in a dramatic reduction in the rate of puerperal sepsis, and then he published a book "Etiology, Concept and Prophylaxis of Childbed Fever" of his findings [1]. Afterwards, Louis Pasteur and Joseph Lister revolutionized the concept of wound infection prevention using antiseptic procedures in surgery. Antibiotic prophylaxis was established in the 1960s, when it was demonstrated that antibiotics cause maximum suppression of infection if given before bacteria gain access to tissue [2]. Since then, many advances have developed to prevent and control wound infection after surgery.

Because of the increasing number of operative procedures, and despite strict infection control policies, SSI remains a leading cause of morbidity and mortality in modern health care settings. It is the second most common type of health care–associated infection (HAI) after urinary tract infection in developed countries, while it is the leading cause of HAI in developing countries [3, 4]. SSI occurs in 2–5 % of patients undergoing inpatient surgery in the United States [3, 5]; and its reported prevalence varies between 1.13 infections per 100 procedures in developed countries to 5.6 per 100 surgical procedures in developing countries, of which 48 % are complex SSI [4–6]. Previously published reports have shown that patients who develop SSIs, in comparison to those with non-infected surgeries, are up to 60 % more likely to spend time in the ICU, five

S. Dobson (✉) · M. A. Al-Dabbagh
Division of Infectious and Immunological Diseases, Department of Pediatrics, BC Children's Hospital, Vancouver, Canada
e-mail: sdobson@cw.bc.ca

M. A. Al-Dabbagh
King Abdulaziz Medical City, Jeddah, Saudi Arabia
e-mail: dabbaghM@ngha.med.sa

N. Curtis et al. (eds.), *Hot Topics in Infection and Immunity in Children IX*,
Advances in Experimental Medicine and Biology 764, DOI 10.1007/978-1-4614-4726-9_11,

times more likely to be readmitted to the hospital, and have a twofold increase risk of death [7]. In addition, 77 % of deaths in patients with SSI are attributed directly to SSI [8].

Infection of surgical wounds is defined as infection following a surgical procedure that occurs within 30 postoperative days, if no implant was placed, and up to 1 year if an implant was placed [8]. SSI is classified according to the degree of microbial contamination into the following [8, 9]:

a. Clean: Closed, uninfected wound with no evidence of acute inflammation. This includes an uninterrupted viscus (respiratory, gastrointestinal, biliary, or urinary tracts) during a clean procedure.
b. Clean contaminated: Elective entry of a viscus (respiratory, gastrointestinal, biliary, or urinary tracts) under controlled conditions with minimal spillage, and no evidence of infection under aseptic conditions.
c. Contaminated: Includes gross spillage from gastrointestinal tract, open accidental wounds, and major breaks in aseptic conditions. Usually associated with acute, nonpurulent inflammation.
d. Dirty: Includes preoperative perforation of viscera with retained tissue or foreign material, or fecal contamination, or presence of old penetrating traumatic wound. Usually associated with the presence of purulent inflammation.

The rate of SSI increases significantly from 7.6 % episodes per 100 surgical procedures in clean wounds to 39.2 episodes per 100 surgical procedures in dirty wounds [4]. In addition, the rate of SSI varies widely by the type of surgical procedure. It is highest with intra-abdominal operations followed by cardiovascular surgeries [10].

## 11.2   Aetiology of SSI

Infection of surgical wounds is mainly caused by the patient's endogenous flora, and it is believed that infection is acquired at the time of surgery by direct inoculation. The most common single pathogen contributing to SSI is *Staphylococcus aureus*, causing 20–37 % of all SSI [5, 11, 12]. Of these, 49–54 % are *methicillin-resistant S. aureus* (MRSA) strains; making MRSA the single most common pathogen isolated from SSI in community hospitals, with an increasing trend over recent years [4, 5, 11]. *Gram negative bacilli (Escherichia coli, Pseudomonas aeruginosa, Enterobacter species, Klebsiella species, Acinetobacter species and Proteus mirabilis)* are also considered major players, causing just under half of all SSIs [4, 13]. Other organisms causing SSI include *Coagulase negative staphylococci, Enterococci, Streptococci, Candida species,* and *anaerobic organisms* [5, 11, 13]. Infection with endogenous gut flora when surgery involves opening a viscus, and the infection is usually polymicrobial. Exogenous contamination of the wounds can rarely be acquired from the operating room personnel or the operation room environment [14, 15].

## 11.3   Risk Factors for SSI

Patient-related risk factors (such as diabetes, obesity, smoking, and known colonisation with resistant organisms) are well known predisposing risk factors for SSI. These, together with the surgical characteristics (introduction of foreign material, amount of tissue damage, duration of surgery, shaving) and pathogen characteristics (degree of contamination, microbial burden and virulence) form a complex relationship posing patients to higher likelihood of SSI [12, 13, 16]. Thus many important factors should be considered perioperatively to minimize the risk of surgical site infection. These include prophylactic antimicrobial administration, use of aseptic surgical techniques and proper surgical site cleaning, proper hair removal using clippers if required and the avoidance of shaving, cessation of smoking 30 days before surgery, ensuring proper oxygenation during surgery, blood sugar control and avoiding hyperglycemia in patients undergoing cardiac surgeries, and avoidance of hypothermia in patients undergoing colorectal surgeries [8, 12, 13, 17–24]. While all are important, the current evidence behind prophylactic antimicrobial administration in surgical patients will be discussed in more details in the following section.

## 11.4 Perioperative Antimicrobial Prophylaxis

The use of perioperative antimicrobial prophylaxis is a proven intervention to reduce the risk of SSI in elective surgical procedures [25–28]. This aims to reduce the burden of possible pathogens at or in close proximity to the surgical incision at a critical time. The main principles that should be followed to maximise the benefit of antimicrobial prophylaxis include [8, 13] (i) using antimicrobial prophylaxis in elective surgeries with a high risk for infection or if SSI would have a high risk of deleterious outcomes, (ii) using a prophylactic agent that is safe, inexpensive, and bactericidal with activity against the most probable infective pathogens in the surgical procedure, (iii) timing the infusion of the antimicrobial agent so that a bactericidal concentration of the drug is present in serum and tissue at the time of incision and, (iv) maintaining a therapeutic serum and tissue level of the antimicrobial agent throughout the operation and until the incision is closed.

### 11.4.1 Indications for Perioperative Antimicrobial Prophylaxis

The National Institute of Health Service and Clinical Excellence (NICE) in the United Kingdom recommends antimicrobial prophylaxis to be given before clean surgeries that involve placement of a prosthesis or implant, clean-contaminated surgeries, and contaminated surgeries [29]. Antimicrobial prophylaxis for surgery should be used if evidence from clinical trials is available, in which the surgery requires entry into a viscus or implantation of a prosthetic

device, and in operations where SSI would pose major consequences [8, 17]. However, a recent meta-analysis demonstrated that prophylactic antimicrobial therapy is significantly associated with reduced risk of post-operative wound infection in 23 different types of surgeries, but the relative risk of wound infection did not vary between the different levels of surgery cleanliness; suggesting that prophylactic antimicrobial use is effective in reducing the risk of wound infection for all types of surgery including the ones with no available evidence from clinical trials [25]. Yet, the use of antimicrobial prophylaxis is associated with cost, potential adverse events, and possible development of antibiotic resistance; so this has to be well considered when using prophylactic antibiotics in clean surgeries.

Cardiothoracic surgeries are generally considered as clean surgeries, but they usually involve placement of implants. In addition, deep infection after cardiac surgeries is usually associated with catastrophic outcomes. Antimicrobial therapy in the settings of dirty wounds is considered part of the treatment because infection is already established [29]. In surgeries involving total hip and knee replacement, 13 patients need to be treated with antimicrobial prophylaxis in order to prevent one case of SSI [27].

For surgical intra-abdominal infections (IAIs), the current Canadian practice guidelines recommend antimicrobial prophylaxis in clean-contaminated abdominal surgeries for colorectal resection; small intestinal surgeries; oesophageal surgeries (obstruction, dilation, or sclerotherapy for varieces); high risk gastro-duodenal surgery (cancer, active bleeding, low gastric acidity, obstruction, obesity); high risk biliary-tract surgeries (acute cholecystitis, cholelithiasis or obstructive jaundice, open billiary-tract surgeries, old age, diabetes or obesity); and perforated, gangrenous or necrotising appendicitis [17]. However, a meta-analysis on antimicrobial prophylaxis for prevention of postoperative infection after appendectomy recommended prophylaxis in all patients undergoing operation for appendicitis with no apparent difference in the nature of the removed appendix [30].

**Table 11.1** Suggested prophylactic antimicrobial regimens according to the type of surgery and infective pathogens

| Type of surgical procedure | Infective pathogen | Prophylactic agent | Adult dose[a] | Timing of the infusion[b] | Duration (h) | Comments |
|---|---|---|---|---|---|---|
| Cardiac surgery [12, 31, 32, 37, 38, 44, 47] | S. aureus and CONS[c] | Cefazolin or Cefuroxime | 1–2 g IV, 1.5 g IV | Within 60 min before incision | 24–48 | Vancomycin as an acceptable alternative if a prosthetic material is used |
| | | | | | | β-lactam or penicillin allergy: vancomycin + aminoglycoside |
| | | | | | | High risk for MRSA infection[d]: vancomycin (1 g IV) + cephalosporin |
| | | | | | | If an aminoglycoside is given, do not repeat the dose |
| Non-cardiac thoracic surgery [8, 12, 31, 32] | S. aureus, CONS, Streptococcus pneumonia, and GNB | Cefazolin or cefuroxime | 1–2 g IV, 1.5 g IV | Within 60 min before incision | ≤24 | High risk for MRSA infection[d]: vancomycin (1 g IV) + aminoglycoside |
| Abdominal surgery: [8, 12, 17, 28, 31, 32, 42, 54] | | | | | | |
| Esophageal, gastroduodenal or biliary tract surgery | GNB, Gram positive cocci | Cefazolin | 1–2 g IV | 30–60 min before incision | ≤24 | β-lactam or penicillin allergy: clindamycin with either an aminoglycoside or a fluoroquinolone, High risk for MRSA infection[d]: vancomycin (1 g IV) + aminoglycoside |
| Colorectal surgery | GNB, Enterococci, and anaerobes | (IV): Cefazolin +, metronidazole or cefoxitin or cefotetan or ampicillin/ sulbactam | 1–2 g IV, 500 mg IV, 2 g IV, 1–2 g IV, 3 g IV | 30–60 min before incision | ≤24 | |
| Appendectomy | GNB, and anaerobes | (PO): Neomycin + erythromycin base | 1–2 g IV, 500 mg IV, 2 g IV, 1–2 g IV, 3 g IV | 30–60 min before incision | ≤24 | |
| | | Cefazolin +, metronidazole or cefoxitin or cefotetan or ampicillin/ sulbactam | | | | |
| Orthopaedic surgery [8, 12, 31, 32, 45, 48, 49, 55–57] | S. aureus, CONS, Streptococci, and GNB | Cefazolin or cefuroxime or teicoplanin | 1–2 g IV, 1.5 g IV, 10 mg/ kg IV | No tourniquet: 30 min before surgery, Tourniquet:10 min before surgery | ≤24 | Vancomycin as an acceptable alternative if a prosthetic material is used, β-lactam or penicillin allergy: vancomycin or clinamycin [58] IV) |
| Urologic surgery [8, 12, 31, 32] | | | | | | |
| Open surgeries or laparoscopy | GNB and Enterococci | Cefazolin | 1–2 g IV | Within 60 min of incision | ≤24 | High risk for MRSA infection[d]: vancomycin (1 g IV) + aminoglycoside |
| Cystoscopy[e] or upper urinary tract instrumentation | GNB and Enterococci | Ciprofloxacin or TMP/SXT | 500 mg PO or 400 mg IV, 1 DS tablet | | | |

**Table 11.1** (continued)

| Type of surgical procedure | Infective pathogen | Prophylactic agent | Adult dose[a] | Timing of the infusion[b] | Duration (h) | Comments |
|---|---|---|---|---|---|---|
| Neurosurgeries [12, 31, 32, 59] | S. aureus, CONS | Cefazolin or cefuroxime | 1–2 g IV, 1.5 g IV | Within 60 min of incision | ≤24 | High risk for MRSA infection[d]: vancomycin (1 g IV) |

*GNB* Gram-negative bacilli, *CONS* coagulase-negative staphylococci, *MRSA* Methicillin-resistant *S. aureus*, *TMP/SXT* Trimethoprim/Sulfamethoxazole

[a]Repeat dose intraoperatively if prolonged surgery more than 4 h (every 3–4 h for cefuroxime and cefazolin, every 8 h for vancomycin and metrronidazole, and every 6 h for clindamycin)

[b]Two hours are allowed for the administration of vancomycin and fluoroquinolones. If vancomycin is used in cardiac surgeries, the dose should be administered within 60–16 min prior to incision

[c]Gram-negative bacilli are rarely reported in SSI after cardiac surgery, contamination occurs during the saphenous vein harvesting [60]

[d]High risk for MRSA infection: known MRSA colonization or coming from facilities with high prevalence of MRSA infection

[e]This includes any cystoscopy with manipulation, or high risk cystoscopy with positive urine cultures, transrectal prostatic biopsy, or preoperative urine catheter

## 11.4.2   Choice of Prophylactic Antimicrobial Regimens

The choice of prophylactic antibiotic regimen depends on the nature of surgery and the infective pathogens likely to cause the infection. In general, the chosen prophylactic agent should be safe, cost-effective, have good tissue penetration and bactericidal against expected pathogens [8, 17].

Table 11.1 summarises the recommended prophylactic antimicrobial regimens according to the type of surgery and the likely infective pathogens. In intra-abdominal surgical procedures, antimicrobial prophylaxis should include coverage for *S. aureus, Gram-negative bacilli* and *anaerobes* from the distal gastrointestinal tract [17, 31]. First generation cephalosporin (cefazolin) is effective in most procedures [31]. Metronidazole should be added in distal surgical procedures to cover anaerobic organisms [17, 28, 31]. Alternative regimens include cefoxitin, cefotetan or ampicillin/sulbactam alone [31, 32]. A recent meta-analysis on antimicrobial prophylaxis in colorectal surgeries demonstrated that the addition of aerobic coverage to anaerobic coverage and vice versa both resulted in statistically significant improvements in SSI rates, which supports the current recommended regimens [28]. Interestingly, the study also demonstrated that SSI was significantly lower when giving combined oral and intravenous antibiotic prophylaxis compared to intravenous alone, or oral alone [28]. The use of mechanical bowel preparation is currently not recommended before colorectal surgeries, as there is lack of evidence to support its use in preventing postoperative infectious complication, and it can be associated with rare but serious complications [33].

A meta-analysis of 28 placebo-controlled trials of cardiothoracic prophylaxis demonstrated that second-generation cephalosporins (cefamandole and cefuroxime) resulted in an approximate one and one-half-fold lower rate of SSI compared to cefazolin [26]. However, subsequent trials failed to discriminate between the different cephalosporins [34–36]. Because of this and the fact that Cefazolin has better activity against Staphylococci, as well as its availability and lower cost, it is the recommended prophylactic antimicrobial agent in cardiac surgeries [37].

In the era of community-associated MRSA (CA-MRSA), patients at high risk for MRSA infection (known MRSA colonisation or coming from facilities with high prevalence of MRSA infection) should receive perioperative vancomycin prophylaxis for prevention of MRSA infection [17, 37, 38].

Decolonisation with nasal mupirocin 5 days before surgery was examined in many presurgical patients. A meta-analysis showed that mupirocin prophylaxis in nasal *S. aureus* carriers was associated with significantly lower rate of nosocomial infections due to *S. aureus* among surgical patients, but secondary analysis restricted to SSIs only showed non significant results. The rate of infections caused by microorganisms other than *S. aureus* was also significantly higher in the treatment group. The authors suggested that in people who are nasal carriers of *S. aureus*, the use of mupirocin ointment results in a statistically significant reduction in *S. aureus* infections [39]. Another meta-analysis supported mupirocin use in non general surgery cases (e.g., cardiothoracic surgery, orthopedic surgery, and neurosurgery), but no benefit was found in general surgical cases [40]. Until rapid screening tests for *S. aureus* colonisation are widely available, mupirocin is currently recommended as a routine prophylactic measure for all patients undergoing cardiac surgical procedures [37]. Mupirocin use for decolonisation is still a controversial topic, as increasing incidence of mupirocin resistance is a potential issue. Moreover, it is not clear how many surgical patients need to be treated with prophylactic mupirocin in order to prevent one case of SSI, and it is not clear if prophylactic mupirocin should be administered to all pre-surgical patients or only to those colonised with *S. aureus*.

### 11.4.3 Timing of Perioperative Prophylactic Antimicrobial Infusion

The timing and dosing of the antibiotic infusion should be adjusted to attain peak serum and tissue concentrations at the critical moment of incision [17]. In 1992, Classen et al. demonstrated in a prospective study of 2,847 subjects that the lowest rates of SSIs occurred in the group of patients who received antimicrobial prophylaxis within 2 h of incision [41]. Afterwards, Weber and colleagues examined in a prospective cohort study the rate of SSI by the timing of surgical prophylaxis after cefuroxime (and metronidazole in colorectal cases) infusion in 3,836 surgical proce-

dures. They found that the most effective time for prophylactic antimicrobial infusion is between 30 and 60 min before surgery, while there was a significantly higher odds of SSI when pre-operative antimicrobial prophylaxis was administered less than 30 min (adjusted OR = 1.95; 95 %CI, 1.4–2.8; $p < 0.001$), or between 60 and 120 min (adjusted OR = 1.74; 95 %CI, 1.0–2.9; $p = 0.035$) before surgery [42]. The association between the prophylaxis timing and the occurrence of SSI was also assessed prospectively in a multicenter study involving 4,472 randomly selected cardiac, hip/knee arthroplasty, and hysterectomy cases. Results showed that the best protection was seen when the antibiotic was given within 30 min of incision [43].

The effect of timing of the prophylactic vancomycin infusion on the incidence of SSI was evaluated in 2,048 patients undergoing cardiac bypass graft or valve replacement surgery. Patients who received vancomycin 16–60 min before the beginning of surgery had a lower rates of postoperative infection than those who received vancomycin 0 and 15 min minutes preoperatively. Reduction in the rate of SSI was also noticed among those who received vancomycin 16–60 min before surgery compared to the ones who received it 61–120 min, 121–180 min, and more than 180 min before surgery, but this reduction was not statistically significant [44].

Administration of perioperative antibiotics 30–60 min prior to incision is the current recommended timing in the Canadian practice guidelines for surgical intra-abdominal infections [17]. The Society of Thoracic Surgeons Practice Guidelines recommend prophylactic antibiotics to be administered in cardiac surgery patients within 60 min of skin incision [37]. This timing is also recommended by the Surgical Infection Prevention Project that is developed by the Centers for Medicare & Medicaid Services in collaboration with the Centers for Disease Control and Prevention [32].

In aseptic orthopaedic surgeries, prophylactic antibiotics should administered within 30 min before incision and at least 10 min before tourniquet inflation [45]. Development of SSI does not differ significantly if the prophylactic antibiotic is given before inflation of the tourniquet or shortly after inflation of the tourniquet [46].

### 11.4.4 Duration and Frequency of Perioperative Antimicrobial Prophylaxis

Therapeutic serum and tissue levels should be maintained throughout surgery and ideally until closure of the incision; thus in cases of prolonged surgical procedures (more than 3–4 h), prophylactic antibiotics may need to be readministered intraoperatively [17, 31]. Additional intraoperative doses are to be given at intervals 1–2 times the half-life of the drug, with the exception of aminoglycosides, when the dose should not be repeated [24, 31, 37]. In the absence of an established infection, or bowel perforation, or a penetrating bowel trauma operated within 12 h, antimicrobial prophylaxis should be limited to 24 h or less [17].

A recent meta-analysis of 12 studies involving 7,893 adult patients undergoing open heart surgery compared short-term (<24 h) with longer-term antibiotic prophylaxis (≥24 h). Longer-term antibiotic prophylaxis reduced the risk of sternal SSI by 38 % (risk ratio 1.38, 95 % confidence interval (CI) 1.13–1.69) and deep sternal SSI by 68 % (risk ratio 1.68, 95 % CI 1.12–2.53), with no significant differences in mortality, infections overall and adverse events. This suggests that perioperative antibiotic prophylaxis of ≥24 h may be more efficacious in preventing sternal SSIs in patients undergoing cardiac surgery compared to shorter regimens [47].

It was demonstrated in a meta-analysis of antimicrobial prophylaxis in colorectal surgery that there is no advantage to longer antibiotic dosing [28]. Sub-group analysis of three studies that specifically compared a single preoperative dose of antibiotic to either a second intraoperative dose, or early postoperative dose, or both, also showed no advantage with extended dosing. This suggests that a single dose of antimicrobial prophylaxis is equivalent to multiple perioperative doses in the prevention of SSI, which questions the rec-

ommendation of giving a second dose in longer operations [28].

The use of a single dose of antimicrobial prophylaxis was shown to be effective in many other trials. A prospective randomised study evaluated the efficacy of single versus multiple doses of teicoplanin as antimicrobial prophylaxis for arthroplasties in 616 patients. Single dose teicoplanin was found to be more effective as prophylaxis for total hip or knee arthroplasty compared with multiple doses [48]. Another study also demonstrated no difference in the rate of postoperative SSI after clean orthopedic surgery when comparing single dose versus multiple doses of prophylactic antibiotics [49]. Furthermore, data from 23 studies that included 8,447 subjects undergoing surgery for closed fracture fixation showed that single dose antibiotic prophylaxis was as effective as multiple doses in reducing the rate of deep SSI [50].

## 11.5 Outcome of Antimicrobial Prophylaxis

The Surgical Infection Prevention and the Surgical Care Improvement Projects aim to decrease the morbidity and mortality associated with postoperative surgical site infections. The project's antimicrobial prophylaxis performance measures suggested that (i) prophylactic antimicrobial should be given within 1 h of surgical incision (or within 60–120 min for fluoroquinolones and vancomycin); (ii) prophylactic antimicrobial choice should be consistent with published guidelines; (iii) prophylactic antimicrobials should be discontinued within 24 h of surgery [32]. Hospitals that improve compliance with the different components of appropriate antimicrobial prophylaxis reported decrease in the rates of SSI [51, 52]. The Surgical Infection Prevention Project performed a large study that included 35,543 surgical cases from 56 hospitals on the impact of improved infection control and antimicrobial prophylaxis process measures. Implementation of these measures resulted in a 27 % reduction in the average rate of SSI in the first 3 months after surgery [51]. In contrast, non adherence to the Surgical

Site Infection Prevention Guidelines in elective general surgical, neurological, and orthopedic procedures with more than two errors in antibiotic prophylaxis measures was significantly associated with increased rate of SSI (odds ratio 4.030; 95 % CI, 1.02–15.96) [53]. Furthermore, implementing a comprehensive infection control program for prevention of SSI after cardiac surgery demonstrated that prophylactic antimicrobial administration was a protective factor against deep sternal SSI [52].

## 11.6   Summary

SSI is a leading cause for healthcare associated infections worldwide. As discussed in this review, SSI can be prevented by implementing the recommended infection control measures. Perioperative antimicrobial prophylaxis is one of the most well studied measures with proven benefits in preventing SSI. The current recommendations are to provide prophylactic antibiotics according to the recommended guidelines. The chosen antibiotic should have activity against the pathogens likely to be encountered in the procedure. The prophylactic antibiotic should be administered within one hour of surgical incision (within 30–60 min in general surgeries), and be discontinued within 24 h of the surgery. An exception is in cardiac surgeries, for which most guidelines recommend discontinuation 24–48 h after the surgery.

## References

1. Wyklicky H, Skopec M (1983) Ignaz Philipp Semmelweis, the prophet of bacteriology. Infect Control 4(5):367–370
2. Burke JF (1961) The effective period of preventive antibiotic action in experimental incisions and dermal lesions. Surgery 50:161–168
3. Wenzel RP (2007) Health care-associated infections: major issues in the early years of the 21st century. Clin Infect Dis 45(Suppl 1):85–88
4. Allegranzi B, Bagheri Nejad S, Combescure C, Graafmans W, Attar H, Donaldson L et al (2011) Burden of endemic health-care-associated infection in developing countries: systematic review and meta-analysis. Lancet 377(9761):228–241
5. Anderson DJ, Sexton DJ, Kanafani ZA, Auten G, Kaye KS (2007) Severe surgical site infection in community hospitals: epidemiology, key procedures, and the changing prevalence of methicillin-resistant Staphylococcus aureus. Infect Control Hosp Epidemiol 28(9):1047–1053
6. Anderson DJ, Chen LF, Sexton DJ, Kaye KS (2008) Complex surgical site infections and the devilish details of risk adjustment: important implications for public reporting. Infect Control Hosp Epidemiol 29(10):941–946
7. Kirkland KB, Briggs JP, Trivette SL, Wilkinson WE, Sexton DJ (1999) The impact of surgical-site infections in the 1990s: attributable mortality, excess length of hospitalization, and extra costs. Infect Control Hosp Epidemiol 20(11):725–730
8. Mangram AJ, Horan TC, Pearson ML, Silver LC, Jarvis WR (1999) Guideline for prevention of surgical site infection, 1999. Hospital Infection Control Practices Advisory Committee. Infect Control Hosp Epidemiol 20(4):250–278; quiz 79–80
9. Culver DH, Horan TC, Gaynes RP, Martone WJ, Jarvis WR, Emori TG et al (1991) Surgical wound infection rates by wound class, operative procedure, and patient risk index. National Nosocomial Infections Surveillance System. Am J Med 91(3B):152–157
10. Stulberg JJ, Delaney CP, Neuhauser DV, Aron DC, Fu P, Koroukian SM (2010) Adherence to surgical care improvement project measures and the association with postoperative infections. JAMA 303(24):2479–2485
11. Hidron AI, Edwards JR, Patel J, Horan TC, Sievert DM, Pollock DA et al (2008) NHSN annual update: antimicrobial-resistant pathogens associated with healthcare-associated infections: annual summary of data reported to the National Healthcare Safety Network at the Centers for Disease Control and Prevention, 2006–2007. Infect Control Hosp Epidemiol 29(11):996–1011
12. Anderson DJ, Kaye KS, Classen D, Arias KM, Podgorny K, Burstin H et al (2008) Strategies to prevent surgical site infections in acute care hospitals. Infect Control Hosp Epidemiol 29(Suppl 1):51–61
13. Anderson DJ (2011) Surgical site infections. Infect Dis Clin North Am 25(1):135–153
14. Lowry PW, Blankenship RJ, Gridley W, Troup NJ, Tompkins LS (1991) A cluster of legionella sternal-wound infections due to postoperative topical exposure to contaminated tap water. N Engl J Med 324(2):109–113
15. Richet HM, Craven PC, Brown JM, Lasker BA, Cox CD, McNeil MM et al (1991) A cluster of Rhodococcus (Gordona) Bronchialis sternal-wound infections after coronary-artery bypass surgery. N Engl J Med 324(2):104–109
16. Haley RW, Culver DH, Morgan WM, White JW, Emori TG, Hooton TM (1985) Identifying patients at high risk of surgical wound infection. A simple mul-

tivariate index of patient susceptibility and wound contamination. Am J Epidemiol 121(2):206–215

17. Chow AW, Evans GA, Nathens AB, Ball CG, Hansen G, Harding GK et al (2010) Canadian practice guidelines for surgical intra-abdominal infections. Can J Infect Dis Med Microbiol 21(1):11–37

18. Al-Niaimi A, Safdar N (2009) Supplemental perioperative oxygen for reducing surgical site infection: a meta-analysis. J Eval Clin Pract 15(2):360–365

19. Zerr KJ, Furnary AP, Grunkemeier GL, Bookin S, Kanhere V, Starr A (1997) Glucose control lowers the risk of wound infection in diabetics after open heart operations. Ann Thorac Surg 63(2):356–361

20. Jones JK, Triplett RG (1992) The relationship of cigarette smoking to impaired intraoral wound healing: a review of evidence and implications for patient care. J Oral Maxillofac Surg 50(3):237–239; discussion 9–40

21. Qadan M, Akca O, Mahid SS, Hornung CA, Polk HC Jr (2009) Perioperative supplemental oxygen therapy and surgical site infection: a meta-analysis of randomized controlled trials. Arch Surg 144(4):359–366; discussion 66–67

22. Tanner J, Norrie P, Melen K (2011) Preoperative hair removal to reduce surgical site infection. Cochrane Database Syst Rev 11:CD004122

23. Bickel A, Gurevits M, Vamos R, Ivry S, Eitan A (2011) Perioperative hyperoxygenation and wound site infection following surgery for acute appendicitis: a randomized, prospective, controlled trial. Arch Surg 146(4):464–470

24. Alexander JW, Solomkin JS, Edwards MJ (2011) Updated recommendations for control of surgical site infections. Ann Surg 253(6):1082–1093

25. Bowater RJ, Stirling SA, Lilford RJ (2009) Is antibiotic prophylaxis in surgery a generally effective intervention? Testing a generic hypothesis over a set of meta-analyses. Ann Surg 249(4):551–556

26. Kreter B, Woods M (1992) Antibiotic prophylaxis for cardiothoracic operations. Meta-analysis of thirty years of clinical trials. J Thorac Cardiovasc Surg 104(3):590–599

27. AlBuhairan B, Hind D, Hutchinson A (2008) Antibiotic prophylaxis for wound infections in total joint arthroplasty: a systematic review. J Bone Joint Surg Br 90(7):915–919

28. Nelson RL, Glenny AM, Song F (2009) Antimicrobial prophylaxis for colorectal surgery. Cochrane Database Syst Rev (1):CD001181

29. National Institute for Health and Clinical Excellence (NICE) developed by National Collaborating Centre for Women's and Children's Health (2008) NICE Clinical Guideline on Surgical site infection; prevention and treatment of surgical site infection. United Kingdom

30. Andersen BR, Kallehave FL, Andersen HK (2005) Antibiotics versus placebo for prevention of postoperative infection after appendicectomy. Cochrane Database Syst Rev 20(3):CD001439

31. Antimicrobial prophylaxis for surgery (2009) Treat Guidel Med Lett 7(82):47–52

32. Bratzler DW, Hunt DR (2006) The surgical infection prevention and surgical care improvement projects: national initiatives to improve outcomes for patients having surgery. Clin Infect Dis 43(3):322–330

33. Eskicioglu C, Forbes SS, Fenech DS, McLeod RS (2010) Preoperative bowel preparation for patients undergoing elective colorectal surgery: a clinical practice guideline endorsed by the Canadian Society of Colon and Rectal Surgeons. Can J Surg 53(6):385–395

34. Curtis JJ, Boley TM, Walls JT, Hamory B, Schmaltz RA (1993) Randomized, prospective comparison of first- and second-generation cephalosporins as infection prophylaxis for cardiac surgery. Am J Surg 166(6):734–737

35. Doebbeling BN, Pfaller MA, Kuhns KR, Massanari RM, Behrendt DM, Wenzel RP (1990) Cardiovascular surgery prophylaxis. A randomized, controlled comparison of cefazolin and cefuroxime. J Thorac Cardiovasc Surg 99(6):981–989

36. Townsend TR, Reitz BA, Bilker WB, Bartlett JG (1993) Clinical trial of cefamandole, cefazolin, and cefuroxime for antibiotic prophylaxis in cardiac operations. J Thorac Cardiovasc Surg 106(4):664–670

37. Engelman R, Shahian D, Shemin R, Guy TS, Bratzler D, Edwards F et al (2007) The Society of Thoracic Surgeons practice guideline series: Antibiotic prophylaxis in cardiac surgery, part II: Antibiotic choice. Ann Thorac Surg 83(4):1569–1576

38. Garey KW, Lai D, Dao-Tran TK, Gentry LO, Hwang LY, Davis BR (2008) Interrupted time series analysis of vancomycin compared to cefuroxime for surgical prophylaxis in patients undergoing cardiac surgery. Antimicrob Agents Chemother 52(2):446–451

39. van Rijen M, Bonten M, Wenzel R, Kluytmans J (2008) Mupirocin ointment for preventing Staphylococcus aureus infections in nasal carriers. Cochrane Database Syst Rev 4:CD006216

40. Kallen AJ, Wilson CT, Larson RJ (2005) Perioperative intranasal mupirocin for the prevention of surgical-site infections: systematic review of the literature and meta-analysis. Infect Control Hosp Epidemiol 26(12):916–922

41. Classen DC, Evans RS, Pestotnik SL, Horn SD, Menlove RL, Burke JP (1992) The timing of prophylactic administration of antibiotics and the risk of surgical-wound infection. N Engl J Med 326(5):281–286

42. Weber WP, Marti WR, Zwahlen M, Misteli H, Rosenthal R, Reck S et al (2008) The timing of surgical antimicrobial prophylaxis. Ann Surg 247(6):918–926

43. Steinberg JP, Braun BI, Hellinger WC, Kusek L, Bozikis MR, Bush AJ et al (2009) Timing of antimicrobial prophylaxis and the risk of surgical site infections: results from the Trial to Reduce Antimicrobial Prophylaxis Errors. Ann Surg 250(1):10–16

44. Garey KW, Dao T, Chen H, Amrutkar P, Kumar N, Reiter M et al (2006) Timing of vancomycin prophylaxis for cardiac surgery patients and the risk of surgical site infections. J Antimicrob Chemother 58(3):645–560

45. Hunfeld KP, Wichelhaus TA, Schafer V, Rittmeister M (2003) (Evidence-based antibiotic prophylaxis in aseptic orthopedic surgery). Orthopade 32(12):1070–1077

46. Akinyoola AL, Adegbehingbe OO, Odunsi A (2011) Timing of antibiotic prophylaxis in tourniquet surgery. J Foot Ankle Surg 50(4):374–376

47. Mertz D, Johnstone J, Loeb M (2011) Does duration of perioperative antibiotic prophylaxis matter in cardiac surgery? A systematic review and meta-analysis. Ann Surg 254(1):48–54

48. Kanellakopoulou K, Papadopoulos A, Varvaroussis D, Varvaroussis A, Giamarellos-Bourboulis EJ, Pagonas A et al (2009) Efficacy of teicoplanin for the prevention of surgical site infections after total hip or knee arthroplasty: a prospective, open-label study. Int J Antimicrob Agents 33(5):437–440

49. Ali M, Raza A (2006) Role of single dose antibiotic prophylaxis in clean orthopedic surgery. J Coll Physicians Surg Pak 16(1):45–48

50. Gillespie WJ, Walenkamp GH (2010) Antibiotic prophylaxis for surgery for proximal femoral and other closed long bone fractures. Cochrane Database Syst Rev (3):CD000244

51. Dellinger EP, Hausmann SM, Bratzler DW, Johnson RM, Daniel DM, Bunt KM et al (2005) Hospitals collaborate to decrease surgical site infections. Am J Surg 190(1):9–15

52. Graf K, Sohr D, Haverich A, Kuhn C, Gastmeier P, Chaberny IF (2009) Decrease of deep sternal surgical site infection rates after cardiac surgery by a comprehensive infection control program. Interact Cardiovasc Thorac Surg 9(2):282–286

53. Young B, Ng TM, Teng C, Ang B, Tai HY, Lye DC (2011) Nonconcordance with surgical site infection prevention guidelines and rates of surgical site infections for general surgical, neurological, and orthopedic procedures. Antimicrob Agents Chemother 55(10):4659–4663

54. Solomkin JS, Mazuski JE, Bradley JS, Rodvold KA, Goldstein EJ, Baron EJ et al (2010) Diagnosis and management of complicated intra-abdominal infection in adults and children: guidelines by the Surgical Infection Society and the Infectious Diseases Society of America. Clin Infect Dis 50(2):133–164

55. Marculescu CE, Osmon DR (2005) Antibiotic prophylaxis in orthopedic prosthetic surgery. Infect Dis Clin North Am 19(4):931–946

56. American Academy of Orthopaedic Surgeons (2004) Recommendations for the use of intravenous antibiotic prophylaxis in primary total joint arthroplasty

57. W-Dahl A, Robertsson O, Stefansdottir A, Gustafson P, Lidgren L (2011) Timing of preoperative antibiotics for knee arthroplasties: Improving the routines in Sweden. Patient Saf Surg 5:22

58. Fujiwara K, Suda S, Ebina T (2000) (Efficacy of antibiotic prophylaxis in clean neurosurgical operations: a comparison of seven-day versus one-day administration). No Shinkei Geka 28(5):423–427

59. Morofuji Y, Ishizaka S, Takeshita T, Toyoda K, Ujifuku K, Hirose M et al (2008) (Efficacy of antimicrobial prophylaxis in neurosurgical operations). No Shinkei Geka 36(9):769–774

60. Farrington M, Webster M, Fenn A, Phillips I (1985) Study of cardiothoracic wound infection at St. Thomas' Hospital. Br J Surg 72(9):759–762

# Infectious Risks Associated with Biologics

David Isaacs

## Abstract

Biologics are generally either custom-designed monoclonal antibodies against specific target cells (e.g. B-cells) or target cytokines (e.g. tumour necrosis factor, TNF) or they are receptor constructs (fusion proteins) based on naturally-occurring cytokine or cell receptors. Biologics are mostly used in adult rheumatology but are increasingly used in paediatrics. There are significant concerns about safety and also about cost. The main safety concerns are about increased risk of infection and malignancy.

The use of TNF antagonists is associated with increased risk of serious infections with intracellular organisms, particularly mycobacteria, but also intracellular bacteria, fungi and Pneumocystis. B-cell antagonists like rituximab can cause progressive multifocal leukoencephalopathy. IL-6 antagonists are associated with increased rates of common bacterial infections and the complement pathway antagonist eculizumab with meningococcal infection.

The risk of some infections associated with biologics can be reduced, by screening patients starting TNF antagonists for latent tuberculosis and giving them cotrimoxazole prophylaxis against Pneumocystis, and by immunising against VZV, hepatitis B, meningococci and pneumococci. However, the risk of the biologics causing serious infection in children is unknown and needs study. Children should not be started on the biologics without careful consideration of the risks and without fully informed consent.

## 12.1 What is a Biologic?

The US FDA has a Center for Biologics Evaluation and Research which regulates a diverse array of complex products they term biologic agents [1]. The term biologics (or biologicals [2]) can be used to include a wide range of medicinal

D. Isaacs (✉)
Department of Infectious Disease and Microbiology, Children's Hospital at Westmead, Westmead, Australia
e-mail: davidi@chw.edu.au

N. Curtis et al. (eds.), *Hot Topics in Infection and Immunity in Children IX*,
Advances in Experimental Medicine and Biology 764, DOI 10.1007/978-1-4614-4726-9_12,
© Springer Science+Business Media New York 2013

products including vaccines, blood and blood components, somatic cells, gene therapy, tissues, and recombinant therapeutic proteins created by biological processes (as distinguished from chemistry). Biologics can be composed of sugars, proteins, or nucleic acids or complex combinations of these substances, or may be living entities such as cells and tissues.

However, in most cases, the term biologics is used more restrictively for a class of genetically engineered medications produced by means of biological processes involving recombinant DNA technology. These medications are usually one of three types:

1. Recombinant human proteins. Examples are erythropoetin, growth hormone and biosynthetic human insulin and its analogues (all hormones).
2. Monoclonal antibodies. Custom-designed antibodies (using hybridoma technology or other methods) designed to counteract or block specific targets of the inflammatory response, either target cells (e.g. B-cells) or target cytokines (e.g. tumour necrosis factor or TNF).
3. Receptor constructs (fusion proteins), usually based on a naturally-occurring receptor for a cytokine (e.g. etanercept targets the TNF-alpha receptor) or for a cell receptor (e.g. belatacept blocks T-cell activation by targeting the CD 28/CTAL-4 receptor).

All of the currently used biologics have to be administered parenterally, either sub-cutaneously or by intravenous infusion, anything from daily (anakinra) to every few weeks, with obvious implications for the paediatric population. Most data come from use in adults, but paediatric use is increasing [2].

The biologics, when used in this sense, have arguably had most impact in adult rheumatology and notably in the treatment of rheumatoid arthritis, giving rise to the term biologic disease-modifying agents for rheumatic diseases or bDMARDs. However, they are being used increasingly in a range of fields, to treat inflammatory conditions (e.g. TNF-alpha inhibitors like infliximab for inflammatory bowel disease), malignancy (e.g. trastuzumab or Herceptin for breast cancer), and to treat transplant rejection

(e.g. monoclonals directed at B-cells like rituximab). Almost every discipline now uses biologics to treat one or more disease with an immunologic contribution to pathogenesis.

However, the advent of biologic therapeutics has also raised significant concerns about safety and cost. The cost of biologic therapies is dramatically higher than the cost of conventional pharmacological medications. Furthermore, biologics are mainly used to treat chronic conditions such as rheumatoid arthritis or inflammatory bowel disease, or for the treatment of refractory cancer for the remainder of life. Safety concerns include increased risk of malignancy, particularly lymphoma, and worrying reports of demyelination, hepatotoxicity and severe allergic reactions to the biologics. This paper, however, will concentrate on the infectious risks of the biologics.

For all the biologics, the exact degree of increased susceptibility to infection is difficult to quantify. Randomised controlled trials and meta-analyses tend to consider "all infections", "serious infections" and "death" as their outcomes [3]. Most patients on biologics are also taking or were taking corticosteroids, methotrexate and other medications with profound effects on the immune system. Other important variables include underlying disease state, age, and prior and future exposure to pathogenic organisms such as mycobacteria, *Pneumocystis* and fungi.

## 12.2 Biologics and Infection

Biologics target the immune system. We know that patients with congenital or acquired immune deficiency are at increased risk of infections. Persons with severe congenital or acquired T-cell immune deficiency are also at increased risk of malignancy, particularly lymphoma. It is far from surprising, therefore, that biologics have been shown to be associated with increased risk of infections and of lymphoma.

What should intrigue infectious disease physicians is that different biologics are associated with different patterns of susceptibility to different infectious agents. Relating the nature of

**Table 12.1** Biologics acting as inhibitors of tumour necrosis factor

| Generic name | Trade name | Indications | Technology |
|---|---|---|---|
| Adalimumab | Humira | Rheumatoid arthritis, ankylosing spondylitis, psoriasis, Crohn's | Humanised monoclonal antibody |
| Etanercept | Etebrel | Rheumatoid arthritis, ankylosing spondylitis, psoriasis | Recombinant human TNF-receptor fusion protein |
| Infliximab | Remicade | Rheumatoid arthritis, ankylosing spondylitis, psoriasis, Crohn's | Monoclonal antibody |
| Golimumab | Simponi | Rheumatoid arthritis, ankylosing spondylitis, psoriatic arthritis | Monoclonal antibody |

the infections caused by a particular biologic to the target of the biologic should help increase our knowledge of the normal human defences to infection.

A corollary of the previous statement is that there are some congenital immune deficiency diseases which affect a relatively specific part of the immune system and that knowledge of the nature infections experienced by persons with these disorders may predict the nature of the infections likely to be caused by use of a specific monoclonal antibody and vice versa.

## 12.3 Tumour Necrosis Factor Antagonists

Tumour necrosis factor (TNF), previously called cachectin and then tumour necrosis factor-alpha, is a systemic inflammatory cytokine able to induce apoptotic cell death, to induce inflammation, and to inhibit tumorigenesis and viral replication. It is produced mainly by macrophages, but also by a variety of other cell types including lymphoid cells, mast cells, endothelial cells, cardiac myocytes, adipose tissue, fibroblasts, and neuronal tissue. Large amounts of TNF are released in response to lipopolysaccharide, other bacterial products, and interleukin-1 (IL-1). TNF can bind to two distinct receptors, TNF-R1 (TNF receptor type 1; CD120a; p55/60) and TNF-R2 (TNF receptor type 2; CD120b; p75/80), TNF-R1 being found in most tissues but TNF-R2 only in cells of the immune system [4, 5]. Binding of TNF to receptors leads to activation of a complex array of inter-acting and often conflicting inflammatory responses including enhancement of activated B-cells, pathways involved in cell proliferation and both anti-apoptotic and pro-apoptotic responses. This necessitates a sophisticated signalling system and diverse factors such as cell type, concurrent stimulation of other cytokines, or the amount of reactive oxygen species can shift the balance in favour of one pathway or another. Such complicated signalling ensures that, whenever TNF is released, various cells with vastly diverse functions and conditions can all respond appropriately to inflammation [4, 5]. TNF appears to be important in augmentation of both the adaptive and innate immune systems.

There is no known human state of TNF deficiency, but a "knock-out" mouse model was first developed in 1997, with the genetically engineered mice lacking the gene to produce TNF [6]. TNF-deficient mice or those lacking TNF receptors have markedly increased susceptibility to mycobacterial infection [7] and also to other intra-cellular pathogens including Listeria [8], Salmonella, some viruses but also to some extra-cellular pathogens like *Streptococcus pneumoniae* [9]. Interestingly, in IRAK-4 deficiency in humans, a defect in Toll receptor signal transduction leads to diminished amounts of TNF and other pro-inflammatory cytokines, and patients are susceptible to infections with *S. pneumoniae* and Salmonella. To interfere with TNF in humans seems risky in the extreme, but the important role played by TNF in causing and prolonging inflammation in rheumatic and other auto-immune or inflammatory disorders led researchers to develop a number of different biologics which interfere with TNF. These are shown in Table 12.1.

**Table 12.2** Biologics, their targets and their infectious risks

| Target | Names of biologics | Technology | Indications | Infectious risks |
|---|---|---|---|---|
| Tumour necrosis factor | Adalimumab | Humanised Mab to TNF | Arthritis | Mycobacteria |
| | Etanercept | Receptor fusion protein | Psoriasis | Fungi |
| | Infliximab | Mab to TNF | Crohn's | Viruses |
| | Golimumab | Mab to TNF | | PCP, listeria |
| B-cell | Rituximab | Mab to CD20 | Transplant rejection, cancer, etc | PML, HepB reactivation |
| | Epratuzumab | Mab to CD22 | Lupus | Unknown |
| T-cell activation | Abatacept | Fusion protein to CTLA4 | Transplant rejection | Unknown |
| | Belatacept | | | PML, EBV-related proliferative disease |
| IL-6 | Tocilizumab | Mab to IL-6 | Arthritis | Slight increase, no specific organisms |
| IL-1 | Anakinra | IL-1 receptor antagonist | Arthritis | Cellulitis, pneumonia |
| Complement | Ecalizumab | Mab to C5 | PNH | Meningococcus |

*EBV* Epstein-Barr virus, *IL* interleukin, *Mab* Monoclonal antibody, *PCP Pneumocystis carinii* (now *jirovecii*) pneumonia, *PML* Progressive multifocal leucoencephalopathy, *PNH* Paroxysmal nocturnal haemoglobinuria, *TNF* tumour necrosis factor

As predicted from the mouse model, use in humans of biologics that target TNF (adalimumab, infliximab) or TNF-receptor (etanercept) has been associated with increased susceptibility to mycobacterial infections, both tuberculosis and atypical mycobacterial infections [3, 10, 11]. Persons with latent TB are at high risk of reactivation, and the Product Information for all the TNF biologics in Table 12.1 states that patients should be screened for latent TB before commencing the biologic. A patient shown to have latent TB (positive tuberculin skin test and/or interferon-gamma release assay, normal physical examination and chest X-ray), should be commenced on chemoprophylaxis with one or more anti-tuberculous drugs. Monotherapy with isoniazid or rifampicin is usual, but it may be advisable to use two or three drugs, depending on risk factors for reactivation, notably the patient's degree of immunosuppression.

For TNF antagonists, as for other biologics, the exact degree of increased susceptibility is hard to quantify. In most of the analysed studies and comparisons, there were no significant differences in safety outcomes between adalimumab and control groups [3, 10]. Serious infections were significantly more frequent in adalimumab patients in only one study [11] with a RR (95 % CI) of 7.64 (1.02–57.18) and a NNH of 30.2. Similarly, based on 4 RCTs (1,231 patients treated with golimumab and 483 with placebo) no significant differences were noted between golimumab and placebo regarding serious adverse events, infections, serious infections, lung infections, tuberculosis, cancer, withdrawals due to adverse events and inefficacy and deaths [12].

On the other hand, there are increasing numbers of reports of serious infections associated with TNF antagonist biologics. These are primarily with intracellular pathogens: tuberculosis and atypical mycobacterial infections, fungal infections and viral infections. The fungal infections have been with Aspergillus, Candida and in the USA with histoplasmosis, coccidiomycosis and blastomycosis, and failure to recognise fungal infections early prompted an FDA warning [13]. Fungal infections are often disseminated rather than localised, have a high mortality (12 of 240 patients died who developed histoplasmosis while taking TNF antagonist biologics and were reported to the FDA) [3] (Table 12.2).

TNF antagonist biologics may be associated with reactivation of chronic hepatitis B infection, and with severe HSV and VZV infections, including sometimes the macrophage activation syndrome [14].

The pattern of infections with TNF antagonist biologics resembles those of a patient with T-cell deficiency, and there are reports of patients taking infliximab and other TNF antagonist biologics developing *Pneumocystis jirovecii* pneumonia (PJP, PCP) a median of 8 weeks after commencing therapy [15]. In addition, there are individual case reports of patients on infliximab developing Listeria infections.

Patients with T-cell defects may also have increased susceptibility to extra-cellular pathogens and this appears to be the case for TNF biologics, in as much as there are reports of pneumonia, pyelonephritis, septic arthritis and septicaemia in association with their use in adult rheumatology patients [10, 11], although the biologics may be only one of multiple factors contributing to these infections.

## 12.4  B-cell Antagonists

There is increasing use of rituximab, a monoclonal antibody which targets CD20, a protein found mainly on B-cells, in the treatment of lymphomas, leukaemias, transplant rejection and some autoimmune disorders.

B-cell antagonists can cause hypogammaglobulinaemia with a risk of sinopulmonary infections. While initial reports of the use of rituximab were optimistic regarding safety, there are emerging concerns about reports of reactivation of hepatitis B [16], sometimes leading to fulminant disease, and of the development of progressive multifocal leukoencephalopathy (PML) due to JC virus [17].

The JC virus (JCV) is a human polyomavirus genetically similar to BK virus and SV40. It was discovered in 1971 and named after the two initials of a patient with progressive multifocal leukoencephalopathy (PML). The virus causes PML and other diseases only in persons with immunodeficiency, as in AIDS or during immunosuppressive treatment (e.g. organ transplant patients). The virus is very common in the general population, infecting 40–60 % of humans [18, 19]. Most people acquire JCV in childhood or adolescence. It is found in high concentrations in urban sewage worldwide, leading some researchers to suspect contaminated water as a typical route of infection. PML is a demyelinating disease, affecting the white matter, destroying oligodendrocytes and producing intranuclear inclusions. PML is similar to another demyelinating disease, multiple sclerosis, but progresses much more quickly. In a recent report of 57 cases of PML associated with rituximab, the mortality was 90 % [17].

Epratuzumab, which targets CD22 on human and malignant B-cells, has been used in lupus and in oncology patients. The preliminary data have not suggested any particular infectious risk, but increasing use may reveal a pattern similar to rituximab.

## 12.5  Interleukin-6 Antagonists

Interleukin-6 (IL-6) is both a pro-inflammatory and anti-inflammatory cytokine. It is secreted by T cells and macrophages to stimulate immune response to trauma, especially burns or other tissue damage leading to inflammation. In terms of host response to a foreign pathogen, IL-6 has been shown, in mice, to be required for resistance against *Streptococcus pneumoniae* [20].

The (IL-6) antagonist tocilizumab has been used mainly in adults with rheumatoid arthritis and currently has a benign safety profile. A Cochrane meta-analysis of eight randomised controlled trials (3,334 participants; 2,233 tocilizumab and 1,101 controls) found that the tocilizumab group was a non-significant 1.18 times more likely to have any infection or infestation [21]. However, a non-Cochrane systematic review of six studies found that the higher but usual dose of 8 mg/kg was associated with a significant increase in risk of infection (Odds Ratio of any infection = 1.30, of serious infection = 1.78) [22]. There was no clear pattern to the infections incurred by the tocilizumab group in either meta-analysis.

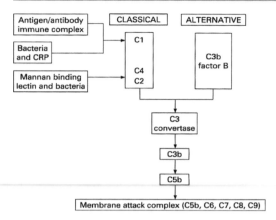

**Fig. 12.1** Diagrammatic representation of the complement cascade

## 12.6    Interleukin-1 Antagonists

Interleukin-1 (IL-1) is produced in response to inflammatory stimuli including infection and mediates various physiological responses, including inflammatory and immunologic reactions, notably lymphocyte proliferation and fever. Anakinra (Kineret) is an IL-1 receptor antagonist, used in rheumatoid and sometimes in juvenile idiopathic arthritis. A Cochrane systematic review found 2,876 patients in trials, but the incidence of serious infections was only reported for 1,900 patients. The incidence of serious infections was "clinically higher", but not statistically different, in the anakinra (25/1,366 patients, 1.8 %) versus the placebo group (3/534 patients, 0.6 %). The pattern was mainly bacterial infections (cellulitis, bone and joint infections and pneumonia) [23].

## 12.7    Complement Pathway Antagonists

The monoclonal antibody eculizumab is being used increasingly to treat adult patients with paroxysmal nocturnal haemoglobinuria, whose red cells are susceptible to complement-mediated lysis. Eculizumab targets the C5 component of the complement cascade. We know that there are familial disorders that can cause deficiency of any of the terminal components or membrane attack complex of the complement cascade (C5 to C9).

The complement cascade occurs on the bacterial cell surface and the membrane attack complex punches a hole in the bacterial cell surface (see Fig. 12.1). Congenital deficiency in C5, C6, C7, C8 or C9 predisposes to meningococcal or gonococcal infections [24]. It is not surprising, therefore, that 3 of 196 adult patients receiving eculizumab developed meningococcal infection [25], and that meningococcal immunisation is strongly recommended for all patients starting eculizumab.

## 12.8    T-cell Activation Antagonists

Two drugs have been developed that block T-cell activation. Abatacept and belatacept are fusion proteins composed of an immunoglobulin linked to the extracellular domain of CTLA-4, which is a molecule crucial for T-cell co-stimulation, selectively blocking the process of T-cell activation. Abatacept and belatacept differ by only two amino acids. They are intended to provide extended graft survival. Preliminary studies have not reported significant infections with abatacept, but there have been concerning reports that belatacept has been associated with cases of PML and of EBV-related lymphoproliferative disease. The FDA has warned that concurrent use of abatacept and TNF antagonist appears to increase the risk of serious infection compared with TNF antagonists alone.

## 12.9    Diagnosis and Treatment

It is obviously critical to consider infections, including opportunist infections, in any patient who develops compatible symptoms when taking a biologic, and to follow the usual infectious disease practice of trying to obtain tissue to make a microbiological diagnosis.

## 12.10    Prevention

Appropriate preventative measures will depend on the nature of the biologic and the pattern of infections anticipated.

1. Screening for tuberculosis with Tuberculin Skin Test and/or interferon-gamma release assay: most relevant for TNF biologics, but also recommended for anakinra and tocilizumab.
2. *Pneumocystis* prophylaxis: consider with TNF biologics
3. Immunisation
   a. Meningococcal conjugate vaccine: all taking eculizumab, consider for B-cell biologics
   b. Pneumococcal conjugate vaccine: all taking biologics
   c. Hepatitis B: TNF biologics, B-cell antagonists (rituximab).

# References

1. FDA Center for Biologics Evaluation and Research (2007–10-29) (2011) What are "biologics" questions and answers. http://www.fda.gov/AboutFDA/CentersOffices/CBER/ucm125684.htm. Accessed 1 Nov 2011 (Last updated 4/30/2009)
2. Beresford MW, Baildam EM (2009) New advances in the management of juvenile idiopathic arthritis—2: the era of biologicals. Arch Dis Child Educ Pract Ed 94:151–156
3. Singh JA, Christensen R, Wells GA et al (2009) Biologics for rheumatoid arthritis: an overview of Cochrane reviews. Cochrane Database Syst Rev 2009(4): CD007848. doi:10.1002/14651858.CD007848.pub2
4. Locksley RM, Killeen N, Lenardo MJ (2001) The TNF and TNF receptor superfamilies: integrating mammalian biology. Cell 104:487–501
5. Bouwmeester T, Bauch A, Ruffner H et al (2004) A physical and functional map of the human TNF-alpha/NF-kappa B signal transduction pathway. Nat Cell Biol 6:97–105 (England)
6. Marino MW, Dunn A, Grail D et al (1997) Characterization of tumor necrosis factor-deficient mice. Proc Natl Acad Sci USA 94:8093–8098
7. Flynn J, Goldstein MM, Chan J et al (1995) Tumor necrosis factor-α is required in the protective immune response against M. tuberculosis in mice. Immunity 2:561–572
8. Rothe J, Lesslauer W, Lötscher H et al (1993) Mice lacking the tumour necrosis factor receptor 1 are resistant to TNF-mediated toxicity but highly susceptible to infection by Listeria monocytogenes. Nature 364:798–802
9. Wellmer A, Gerber J, Ragheb J et al (2001) Effect of deficiency of tumor necrosis factor alpha or both of its receptors on *Streptococcus pneumoniae* central nervous system infection and peritonitis. Infect Immun 69:6881–6886
10. Navarro-Sarabia F, Ariza-Ariza R, Hernandez-Cruz B, Villanueva I (2005) Adalimumab for treating rheumatoid arthritis. Cochrane Database Syst Rev 2005(3):CD005113. doi:10.1002/14651858.CD005113.pub2
11. Keystone EC, Kavanaugh AF, Sharp JT et al (2004) Radiographic, clinical, and functional outcomes of treatment with Adalimumab (a human anti-tumor necrosis factor monoclonal antibody) in patients with active rheumatoid arthritis receiving concomitant methotrexate therapy. A randomized, placebo-controlled, 52-week trial. Arth Rheumat 50:1400–1411
12. Singh JA, Noorbaloochi S, Singh G (2010) Golimumab for rheumatoid arthritis. Cochrane Database Syst Rev 2010(1):CD008341. doi:10.1002/14651858.CD008341
13. Food, Drug Administration (2011) FDA: Manufacturers of TNF-blocker drugs must highlight risk of fungal infections. FDA, 2008. http://www.fda.gov/NewsEvents/Newsroom/PressAnnouncements/2008/ucm116942.htm. Accessed 1 Nov 11
14. Skripak JM, Rodgers GL, Martucci C, Goldsmith DP (2011) Disseminated simplex (HSV) infection precipitating macrophage activation syndrome (MAS) in a child with systemic juvenile idiopathic arthritis (SJIA) undergoing therapy with infliximab. Pediatric Rheumatology Online Journal. #Abstract 58. 2003. Link: http://www.pedrheumonlinejournal.org/June/24.htm. Accessed 1 Jun 2011
15. Komano Y, Harigai M, Koike R et al (2009) *Pneumocystis jiroveci* pneumonia in patients with rheumatoid arthritis treated with infliximab: a retrospective review and case-control study of 21 patients. Arthritis Rheum 61:305–312
16. Evens AM, Jovanovic BD, Su YC et al (2010) Rituximab-associated hepatitis B virus (HBV) reactivation in lymphoproliferative diseases: meta-analysis and examination of FDA safety reports. *Ann Oncol* (29 Nov 2010, Epub ahead of print)
17. Carson KR, Evens AM, Richey EA et al (2009) Progressive multifocal leukoencephalopathy after rituximab therapy in HIV-negative patients: a report of 57 cases from the Research on Adverse Drug Events and Reports project. Blood 113:4834–4840
18. Kean JM, Rao S, Wang M, Garcea RL (2009) Seroepidemiology of human polyomaviruses. PLoS Pathog 5:e1000363. Link: http://www.plospathogens.org/article/info%3Adoi%2F10.1371%2Fjournal.ppat.1000363
19. Egli A, Infanti L, Dumoulin A et al (2009) Prevalence of Polyomavirus BK and JC Infection and Replication in 400 Healthy Blood Donors. J Infect Dis 199:837–846
20. van der Poll T, Keogh CV, Guirao X, Buurman WA, Kopf M, Lowry SF (1997) Interleukin-6 gene-deficient mice show impaired defense against pneumococcal pneumonia. J Infect Dis 176:439–444
21. Singh JA, Beg S, Lopez-Olivo MA (2010) Tocilizumab for rheumatoid arthritis. Cochrane

Database Syst Rev 2010(7):CD008331. doi:10.1002/14651858.CD008331.pub2

22. Campbell L, Chen CV, Bhagat SS, Parker RA, Ostor AJK (2010) Risk of adverse events including serious infections in rheumatoid arthritis patients treated with tocilizumab: a systematic literature review and meta-analysis of randomized controlled trials. Rheumatology. doi:10.1093/rheumatology/keq343

23. Mertens M, Singh JA (2009) Anakinra for rheumatoid arthritis. *Cochrane* Database Syst Rev 2009(1): CD005121. doi:10.1002/14651858.CD005121.pub3

24. Ross SC, Densen P (1984) Complement deficiency states and infection: epidemiology, pathogenesis and consequences of neisserial and other infections in an immune deficiency. Medicine 63:243–273

25. Dmytrijuk A, Robie-Suh K, Cohen MH, Rieves D, Weiss K, Pazdur R (2008) FDA report: eculizumab (Soliris) for the treatment of patients with paroxysmal nocturnal hemoglobinuria. Oncologist 13:993–1000 (Epub 2008 Sep 10)

# Cat Scratch Disease and Other *Bartonella* Infections

# 13

Kenneth M. Zangwill

**Abstract**

First described in 1931, cat scratch disease remains the most commonly identified clinical syndrome associated with *Bartonella* infection. Over the last 20 years, however, the discovery and use of modern diagnostic tests has greatly expanded our understanding of the pathogenesis, clinical spectrum, and treatment options for *Bartonella* infections of all types. Indeed, each varies substantially depending on the infecting species and the immune status of the host.

## 13.1 Introduction

*Bartonella* species cause a wide range of clinical syndromes which vary substantially depending on the infecting species and immune status of the infected. Well known as a cause of trench fever and bartonellosis (a febrile illness endemic in the high valleys of the Andes mountains [1]), their recognition as human pathogens in the United States was virtually forgotten until the early 1990s when new diagnostic tests linked this genus to cat scratch disease. Today, the use of such tests has significantly broadened the clinical spectrum of cat scratch disease and also confirmed the role of *Bartonella* as the cause of bacillary angiomatosis (BA), an entity seen mainly among the immunocompromised.

K. M. Zangwill (✉)
Pediatric Infectious Diseases, David Geffen School of Medicine at UCLA, Harbor-UCLA Medical Center, Los Angeles, USA
e-mail: kzangwill@uclacvr.labiomed.org

## 13.2 Making the Link Between *Bartonella* and Cat Scratch Disease

Cat scratch disease (CSD) was first described in a French boy seen in 1931 [2], but > 50 years elapsed before evidence of its bacterial cause emerged. In 1983, Warthin-Starry staining revealed small bacilli among 34 of 39 lymph node samples from patients with CSD and in subcutaneous nodules from a patient with AIDS with unusual neovascular lesions subsequently recognized as BA [3, 4]. In 1988, it was proposed that CSD and BA may result from infection by the same organism based on the similarity of the morphologic and staining characteristics of these bacilli and that a history of cat scratches was often present in these patients. Importantly, antiserum raised against the "CSD bacillus" reacted with organisms in tissue samples from BA patients [5].

In 1990, molecular probe techniques (polymerase chain reaction (PCR)) suggested that BA

N. Curtis et al. (eds.), *Hot Topics in Infection and Immunity in Children IX*,
Advances in Experimental Medicine and Biology 764, DOI 10.1007/978-1-4614-4726-9_13,
© Springer Science+Business Media New York 2013

was caused by a bacterium similar to *Bartonella quintana* [6]. Other reports described morphologically similar-appearing organisms from patients with a vascular abnormality of the liver known as peliosis hepatitis [7] and febrile bacteremia [8], subsequently noted to be caused by both *B. quintana* and *B. henselae* [9–12]. Soon thereafter, a newly developed immunofluorescent antibody serologic test developed for BA was positive in 36 of 41 (88 %) serum samples from persons with suspected CSD, compared with only 6 % among healthy controls [13]; findings soon corroborated in a field trial among CSD patients in Connecticut, USA [14]. During this time, it was reported that CSD skin test antigen contained *Bartonella* nucleic acid sequences [15]. These observations confirmed the etiologic link between *Bartonella* and CSD and BA. Subsequently, organisms have been visualized microscopically or detected using PCR at the site of skin inoculation, lymph nodes, bone, eye, liver, and spleen in patients with CSD [16–21] and in nearly every organ, including the brain, among those with BA [22]. Regardless, the organism is only infrequently successfully cultured from clinical specimens among patients with either of these conditions.

## 13.3 Pathogenesis

*Bartonella* are facultative, intracellular Gram-negative bacilli with relatively fastidious growth characteristics. To maximize growth, tissue samples ideally should be directly plated onto solid media supplemented with hemin and incubated in a hypercapneic atmosphere. Most standard protocols for detection of such fastidious bacteria will allow for growth of *Bartonella*, but only after at least 7–10 days of incubation [23]. Currently at least 22 species have been described, 12 are known to cause disease in humans and the complete genome is available for 10 separate species.

In general, *Bartonella* establish infection within the reservoir host and eventually establish persistence within the erythrocyte. Several observations suggest that *Bartonella* may persist intracellularly: (1) persistent bacteremia has been noted in humans and a variety of mammals,

(2) isolation is enhanced by lysis centrifugation (which disrupts cellular membranes), (3) BA and trench fever patients may experience clinical relapse, and (4) better clinical response has been noted with antimicrobial agents that penetrate intracellularly. The defining pathogenic characteristics of human infection vary substantially with the infecting *Bartonella* species and also with immune status of the host. Transmission to humans can occur directly via an arthropod vector (e.g. sandflies and bartonellosis) or indirectly via inoculation (perhaps from infected flea feces) through intact skin via a cat scratch or other mucous membrane contact (CSD and BA). Many virulence factors involved in human infection have been identified that contribute to cellular invasion, erythrocyte persistence, pathologic angiogenesis, and evasion from host immune responses. A comprehensive review of *Bartonella* pathogenetic mechanisms has been recently published [24].

These primary clinical histopathologic manifestations range from focal granulomatous changes (CSD), multifocal angioproliferative changes (BA), endovascular multiplication (endocarditis), and an (proposed) exaggerated inflammatory response without evidence of bacterial invasion (meningoencephalitis). In CSD specifically, the lymph node histology reveals patchy, necrotic, granulomatous change with stellate abscesses and variable infiltration of leukocytes. This is in contrast to BA which typically reveals proliferation of small blood vessels with large, "plump" cuboidal endothelial cells with intermixed leukocyte infiltration and focal necrosis without granulomas, similar to that seen with bartonellosis.

## 13.4 Ecology

Various reservoirs for *Bartonella* species have been identified including feline, bovine, canine, cetacean, and rodent species. Most individual species, however, are relatively restricted in their distribution. Epidemiologic studies of patients with CSD and BA have shown that contact with cats, particularly kittens in the case of CSD, is associated with disease [14, 25]. Cats from households of such patients are more likely to have antibody against *B. henselae* than are control cats, seroprevalence of anti-*Bartonella* antibody among community cats can be very high (depending on the region) and overt feline bacteremia among community cats may exceed 50 % persisting for up to a year [14, 26–29]. Younger cats and stray cats are more likely to be infected with *Bartonella* than are older or domesticated cats [30].

It appears that each pathogenic *Bartonella* species may be transmitted by an arthropod vector, with variation by species. These vectors include flies, fleas, keds, lice, sandflies, and ticks, but most infections appear to be accidental in animal and human hosts. *B. quintana* and *B. bacilliformis* are transmitted by the human body louse and the sandfly, respectively. For *B. henselae*, the cat flea (*Ctenocephalides felis*) has been confirmed as a vector for CSD and/or BA through epidemiologic [14, 25] and experimental [14, 27, 31, 32] evidence. These data show that cat fleas can harbor *Bartonella* as detected by culture and PCR, cats bacteremic with *B. henselae* are more likely than nonbacteremic cats to be infested with fleas, and fleas can successfully transmit to *Bartonella* cats in a laboratory setting.

## 13.5 Cat Scratch Disease

Although definitive evidence is not available, it appears that the majority of infections caused by *Bartonella* species result in CSD, usually due to *B. henselae*. Large case series and systematic evaluations have defined the epidemiology of CSD. United States population-based surveillance data indicate that the incidence over all age groups is 3–4 per 100,000 population. Approximately 60 % of cases occur in persons < 18 years of age with children 5–10 years of age at greatest risk (~ 10 per 100,000 population) [14, 33]. Familial and geographic clusters have been reported and incidence peaks in the winter months, with some geographic variability. Hospitalization for this illness occurs most commonly for surgical drainage or excision of affected lymph nodes with an annual incidence of ~0.6/100,000 children < 18 years of age in the United States [34]. A case-

**Table 13.1** Cat scratch disease: classic manifestations. (Data from [14, 38, 46, 66]

| Clinical finding | Percent of patients |
|---|---|
| Lymphadenopathy | >90 |
| Upper extremity | 46–52 |
| Neck/jaw | 26–43 |
| Groin | 6–18 |
| Suppuration | 15–30 |
| Fever | 26–60 |
| Inoculation site detectable | 25–90 |
| Malaise/fatigue/anorexia/headache | 10–45 |
| Myalgia/arthropathy/tendinitis | 3–11 |
| Parinaud oculoglandular syndrome | 5 |
| Skin rash/erythema nodosum | 3–5 |
| Encephalopathy | < 1 |

control evaluation noted that owning a kitten and being scratched, bitten, or licked by a kitten with fleas are the most important risk factors for disease [14]. Nearly 25 % of persons with CSD, however, do not report intimate contact with cats, though person-to-person transmission has not been documented.

Uncomplicated CSD is usually a self-limited illness characterized by lymphadenopathy that occurs after contact with a cat. A few days after a cat-scratch or bite, a papule or wheal frequently develops at the site which is followed by development of regional lymphadenopathy 1–2 weeks later. Concomitant constitutional symptoms occur in up to 60 % of cases including malaise, fever, sore throat, and/or headache (Table 13.1). Children < 15 years of age are more likely to develop lymphadenopathy in the neck, whereas older persons are more likely to have enlarged nodes in the groin or axilla [33]. Up to one-third of patients will report lymphadenopathy at more than one site. Nodes generally are not tender, yet suppuration occurs in 15–30 % of cases with even deep neck space disease reported [35]. Prolonged bacteremia and even widespread dissemination involving multiple organ systems has been reported as well [36, 37]. In most cases of CSD, the clinical illness resolves spontaneously but lymphadenopathy may persist for weeks to months [38].

The widespread use of a serologic test against *Bartonella* species has rapidly and significantly expanded the clinical spectrum of this illness to include nearly every organ system. Delays in diagnosis are commonplace and occur most likely due to general lack of awareness with the potential for infection among patients who have a non-classic presentation. Central nervous system disease most commonly presents as encephalopathy with associated seizures in 40–50 % of patients [21]. Status epilepticus, cranial and peripheral nerve paresis, and intracranial masses also have been reported. Onset is usually 1–2 months after the development of lymphadenopathy, and complete and rapid recovery is the rule, but severe disease with sequelae may occur [21, 39, 40]. Cerebrospinal fluid analysis is usually normal, but lymphocytic pleocytosis is noted in approximately one-third of patients. Ophthalmic disease usually presents as Parinaud oculoglandular syndrome (bulbar conjunctivitis, preauricular lymphadenopathy, and conjunctival granuloma), but other manifestations are well-described and neuroretinitis in the form of a "macular star" is particularly distinctive [41, 42]. Hepatitis and/or splenitis is well-described as well, usually accompanied by prolonged fever, abdominal pain, arthralgias, and/or pleomorphic rashes. Back pain is a common presenting symptom in such children [43–45]. In addition, many patients have been reported with musculoskeletal disease (particularly in women > 20 years of age) or osteomyelitis with vertebral disease and contiguous abscess formation being prominent [46, 47]. Pulmonary disease has also been described in which nearly all patients developed pleural effusion [48]. The majority of patients with non-classic disease recover completely, and no deaths have been directly attributable to CSD. Clinical disease appears to provide lifelong protection insofar as recurrent disease has been described very rarely [43].

## 13.6 Bacillary Angiomatosis

BA is an uncommon manifestation of *B. henselae* or *B. quintana* infection seen primarily in adults with acquired immunodeficiency, usually during the late stages of AIDS [22]. The incidence and seasonal distribution of BA is unknown. Being scratched or bitten by a cat has been established as a risk factor for BA, but overall, one-third of patients did not report such contact in one study [25]. The clinical manifestations of BA are quite varied and may involve nearly every organ system but cutaneous, osseous, hepatic (peliosis hepatis), and splenic disease are the most commonly reported [22]. The range of cutaneous disease is broad, usually including superficial, erythematous, highly vascular, exophytic lesions or subcutaneous nodules. Lesions occur singly, or hundreds of lesions may occur in the same patient. Bone disease usually is painful and localized to tubular bones, with osteolysis noted on roentgenogram. Hepatosplenic disease often presents with prolonged fever, abdominal pain, and substantial weight loss. Only a few instances of BA have been reported among children some of whom were immunocompetent [49–51].

## 13.7 Other Syndromes

In addition to "classic" CSD and BA, *Bartonella* species have been linked to several other syndromes. *B. quintana* and *B. henselae* bacteremia, most commonly seen among the homeless or those infested with scabies or lice, may persist for prolonged periods despite antimicrobial therapy [8, 11, 12, 52]. Endocarditis caused by several *Bartonella* species has been described and now should be included as a potential cause of "culture-negative" HACEK-type endocarditis, i.e., endocarditis caused by fastidious Gram-negative organisms. Such patients usually have one or more underlying conditions, including infection with HIV, chronic alcoholism, or homelessness [53–56]. The homelessness and infestations associated with *Bartonella* infection noted above parallel conditions experienced by soldiers with trench fever in which prolonged bacteremia with *B. quintana* was also reported.

Lastly, *Bartonella* have been implicated in several different central nervous system syndromes which may represent rare forms of CSD or BA, or unique manifestations of *Bartonella* infection altogether. These syndromes, most frequently noted in patients with HIV infection, include aseptic meningitis, transient cranial and/or peripheral nerve dysfunction, aphasia, alteration in mood or affect, psychoses, space-occupying lesions, and an acute psychiatric complex [41, 57, 58].

## 13.8  Diagnosis

The approach to diagnosis of infection with *Bartonella* varies with the clinical presentation. For classic CSD, particularly in immunocompetent hosts, the diagnosis can usually be confidently made with the subacute emergence of regional lymphadenopathy with or without constitutional symptoms in persons with prior traumatic contact with a cat or kitten distal to the affected node. In such instances, serologic testing, molecular diagnostics, or imaging is usually unnecessary.

Other *Bartonella* syndromes, however, require the use of serologic testing or clinical specimens for microbiologic and/or molecular evaluations. Currently, serologic testing is the mainstay of diagnosis and is most commonly done using a commercially available enzyme immunoassay or an indirect immunofluorescence assay for *Bartonella* IgM and IgG antibody levels. The sensitivity of these tests is variable and may be low, but specificity is generally quite high [59, 60]. Among patients with CSD, antibody titers peak at 4–5 months after onset of symptoms and have been shown to persist for up to 3 years [14, 61]. Serologic tests have completely supplanted the use of the CSD skin test, a non-standardized preparation previously available in limited quantities, originally developed in 1946 [2]. Currently available serologic tests are not able, however, to differentiate reliably between species of *Bartonella*. Molecular techniques are also available in many commercial laboratories, but the methods are not standardized.

*Bartonella* species are relatively fastidious and recovery is maximized with the use of lysis centrifugation technique and selective media that contain hemin, followed by growth in a hypercapnic atmosphere for at least 10 days. Tissue samples may reveal *Bartonella* with the use of a range of silver stains and Gram's stain is generally insensitive for this pathogen.

## 13.9  Therapy

The majority of mild-to-moderate cases of CSD resolve in 1–2 months without any antimicrobial therapy. Although *Bartonella* species appear to be susceptible to several antimicrobials when tested *in vitro*, such results do not predict reliably the clinical response to therapy; use of such tests as a clinical tool should therefore be discouraged.

Only one prospective controlled study of therapy for cat scratch disease has been published [62]. This study reported on 29 patients who were randomized to azithromycin (for 5 days) or placebo after diagnosis was confirmed by serologic testing. The outcome measure was change in the (largest) lymph node size, as determined by ultrasound 30 days after initiation of treatment. No other clinical signs or symptoms were reported and drug compliance was not monitored. The authors note that those in the treatment group had significant improvement in nodal size and suggested that azithromycin be considered for therapy of this condition.

This study, however, was flawed in several important areas rendering its conclusions highly suspect [63]. There were only 14–15 patients per group, >50 % of whom in each group received at least one antimicrobial before randomization and the details of type and duration of antimicrobial use are not presented. Very little information was presented with regard to clinical resolution of signs and symptoms (aside from the ultrasound data) and interestingly there was little correlation between physical examination and findings on ultrasonography of lymph nodes. As well, the authors report a statistically significant difference in nodal size at 30 d, but this difference was not present by 60 d. The clinical relevance of the ultrasonographic findings, specifically for deciding about the need for antimicrobial therapy in the absence of demonstrable clinical benefit, is unclear. Anecdotal evidence and retrospective series suggest that some clinical response may be achieved for patients with CSD with trimethoprim-sulfamethoxazole, rifampin, gentamicin, and in adults, ciprofloxacin.

A combination of therapeutic restraint, pain control, and reassurance remains the most prudent approach to the great majority of patients with uncomplicated CSD in whom clinical symptoms and signs, as well as lymphadenopathy, will likely spontaneously resolve over several weeks to months. Antimicrobial therapy may be considered for those patients (1) in whom lymphadenopathy does not resolve over a reasonable period, (2) in whom lymphadenopathy is associated with significant morbidity, such as pain or persistence of debilitating constitutional symptoms, (3) with severe systemic disease; i.e., encephalopathy, osteomyelitis, or neuroretinitis, and/or (4) with an underlying medical disorder complicated by severe CSD. Specific recommendations for these circumstances cannot be definitive as controlled data are unable, but an expert panel has published guidelines [64].

Aspiration of inflamed lymph nodes only should be performed to rule out more serious diagnoses or if a treatable superinfection is suspected. In such cases, either a fine-needle aspiration or excisional biopsy should be done; incision and drainage procedures may promote the development of fistulae. No specific measures regarding antimicrobial prophylaxis of close contacts of the implicated cat itself can be recommended. Also, declawing of the cat or its removal from the household is not necessary.

Patients with BA usually respond to macrolides or tetracyclines. Uncomplicated cutaneous disease in HIV-infected patients requires at least 6–8 weeks of therapy, whereas more severe or invasive disease may require months of therapy or perhaps lifetime suppressive therapy. Others have proposed specific treatment regimens for BA, bacteremia, and endocarditis caused by *Bartonella* that include various combinations of gentamicin, erythromycin, or ceftriaxone [64].

## 13.10 Opportunities

With the wide availability of serologic and molecular diagnostic tools, new information about *Bartonella* infections will likely continue to emerge. Important knowledge gaps remain in nearly all domains of this organism as a human pathogen including the dynamics of its ecology and transmission in nature, human pathogenesis, relationship of the host to the pathogen, and differing responses to antimicrobial therapy. Cats remain the most important reservoir for CSD and BA—and over one-third of all U.S. households include at least one cat [65]—perhaps new information will lead to a preventive approach such as vaccination of cats.

## References

1. Garcia-Caceres U, Barcia FU (1991) Bartonellosis: an immunodepressive disease and the life of Daniel Alcides Carrion. Am J Clin Pathol 95(Supp 1):558–566
2. Carithers H (1970) Cat-scratch disease. Notes on its history. Am J Dis Child 119:200–203
3. Wear DJ, Margileth AM, Hadfield TL et al (1983) Cat-scratch disease: a bacterial infection. Science 225:1403–1405
4. Stoler MH, Bonfiglio TA, Steigbigel RT et al (1983) An atypical subcutaneous infection associated with acquired immune deficiency syndrome. Am J Clin Pathol 80:714–718
5. LeBoit PE, Berger TG, Egbert BM et al (1988) Epithelioid haemangioma-like vascular proliferation in AIDS: manifestation of cat-scratch disease bacillus infection? Lancet 1:960–963
6. Relman DA, Loutit JS, Schmidt TM et al (1990) The agent of bacillary angiomatosis: an approach to the identification of uncultured pathogens. N Engl J Med 323:1573–1580
7. Perkocha, LA, Geaghan SM, Yen TSB et al (1990) Clinical and pathological features of bacillary peliosis hepatis in association with human immunodeficiency virus infection. N Engl J Med 323:1581–1586
8. Slater, LN, Welch DF, Hensel DH et al (1990) A newly recognized fastidious gram-negative pathogen as a cause of fever and bacteremia. N Engl J Med 32:1587–1593
9. Relman DA, Falkow S, Le Boit PE et al (1991) The organism causing bacillary angiomatosis, peliosis hepatis, and fever and bacteremia in immunocompromised patients. N Engl J Med 324:1514
10. Welch DF, Pickett DA, Slater LN et al (1992) *Rochalimaea henselae* sp. nov., a cause of septicemia, bacillary angiomatosis, and parenchymal bacillary peliosis. J Clin Microbiol 30:275–280
11. Lucey D, Dolan M, Moss CW et al (1992) Relapsing illness due to *Rochalimaea henselae* in immunocompetent hosts: implication for therapy and new epidemiological associations. Clin Infect Dis 14:683–688
12. Regnery RL, Anderson BE, Clarridge JE et al (1992) Characterization of a novel *Rochalimaea* species, *R. henselae* sp. nov., isolated from blood of a febrile,

human immunodeficiency virus-positive patient. J Clin Microbiol 30:265–274

13. Regnery RL, Olson JG, Perkins BA et al (1992) Serological response to *"Rochalimaea henselae"* antigen in suspected cat-scratch disease. Lancet 339:1443–1445

14. Zangwill KM, Hamilton DH, Perkins BA et al (1993) Cat-scratch disease in Connecticut: epidemiology, risk factors and evaluation of a new diagnostic test. N Engl J Med 329:8–13

15. Anderson B, Kelly C, Threlkel R et al (1993) Detection of *Rochalimaea henselae* in cat scratch disease skin test antigens. J Infect Dis 168:1034–1036

16. Margileth AM, Wear DJ, Hadfield TL et al (1984) Cat-scratch disease: bacteria at the primary inoculation site. JAMA 252:928–931

17. Anderson B, Sims K, Regnery R et al (1994) Detection of *Rochalimaea henselae* DNA in specimens from cat-scratch disease patients by PCR. J Clin Microbiol 32:942–948

18. Demers DM, Bass JW, Vincent JM et al (1995) Cat-scratch disease in Hawaii: etiology and seroepidemiology. J Pediatr 127:23–36

19. Waldvogel K, Regnery R, Anderson B et al (1994) Disseminated cat-scratch disease: detection of *Rochalimaea henselae* in affected tissue. Eur J Pediatr 153:23–27

20. Le HH, Palay DA, Anderson B, Steinberg JP (1994) Conjunctival swab to diagnose ocular cat-scratch disease. Am J Ophthalmol 118:249–250

21. Carithers HA, Margileth AM (1991) Cat-scratch disease: acute encephalopathy and other neurologic manifestations. Am J Dis Child 145:98–101

22. Koehler JE, Tappero JW (1993) Bacillary angiomatosis and bacillary peliosis in patients infected with human immunodeficiency virus. Clin Infect Dis 17:612–624

23. Welch DF, Hensel RM, Pickett DA et al (1993) Bacteremia due to *Rochilimaea henselae* in a child: practical identification of isolates in the clinical laboratory. J Clin Microbiol 31:2381–2386

24. Harms A, Dehio C (2012) Intruders below the Radar: molecular pathogenesis of *Bartonella* spp. Clin Microbiol Rev 25:42–78

25. Tappero JW, Mohle-Boetani J, Koehler JE et al (1993) The epidemiology of bacillary angiomatosis and bacillary peliosis. JAMA 269:770–775

26. Jameson P, Greene C, Regnery R et al (1995) Prevalence of *Bartonella henselae* antibodies in pet cats through the regions of North America. J Infect Dis 172:1145–1149

27. Koehler JE, Glaser CA, Tappero JW (1994) *Rochalimaea henselae* infection: a new zoonosis with the domestic cat as reservoir. JAMA 271:531–535

28. Breitschwerdt E, Kordick DL (2000) *Bartonella* infection in animals; carriership, reservoir potential, pathogenicity, and zoonotic potential for human infection. Clin Microbiol Rev 13:428–438

29. Kordick DL, Wilson KH, Sexton DJ et al (1995) Prolonged *Bartonella* bacteremia in cats associated with cat-scratch disease. J Clin Microbiol 33:3245–3251

30. Chomel BB, Abbott RC, Chasten RW et al (1995) *Bartonella henselae* prevalence in domestic cats in California: risk factors and association between bacteremia and antibody titers. J Clin Microbiol 33:2445–2450

31. Chomel BB, Kasten RW, Floyd-Hawkins KF et al (1996) Experimental transmission of *Bartonella henselae* by the cat flea. J Clin Microbiol 34:1952–1956

32. Rolain JM, Franc M, Davoust B, Raoult D (2003) Molecular detection of *Bartonella quintana, B. koehlerae, B. henselae, B. clarridgeiae, Rickettsia felis,* and *Wolbachia pipientis* in cat fleas, France. Emerg Infect Dis 9:338–342

33. Hamilton DH, Zangwill KM, Hadler JL et al (1995) Cat-scratch disease—Connecticut,1992–1993. J Infect Dis 172:570–573

34. Reynolds MG, Homan RC, Curns AT et al (2005) Epidemiology of cat-scratch disease hospitalizations among children in the United States. Pediatr Infect Dis J 24:700–704

35. Yeh SH, Zangwill KM, Hall B, McPhaul L, Keller M (2000) Parapharyngeal abscess due to cat-scratch disease. Clin Infect Dis 30:599–601

36. Arvand M, Schad SG (2006) Isolation of *Bartonella henselae* DNA from the peripheral blood of a patient with cat scratch disease up to 4 months after the cat scratch injury. J Clin Microbiol 44:2288–2290

37. Waldvogel K, Regnery R, Anderson BE et al (1994) Disseminated cat-scratch disease: detection *Rochalimaea henselae* in affected tissue. Eur J Pediatr 153:23–27

38. Carithers HA (1985) Cat-scratch disease: an overview based on a study of 1200 patients. Am J Dis Child 193:1124–1133

39. Selby G, Walker GL (1979) Cerebral arteritis in cat-scratch disease. Neurology 29:1413–1418

40. Wong M, Isaacs D, Dorney S (1995) Fever, abdominal pain and an intracranial mass. Pediatr Infect Dis J 14:725–728

41. Wong MT, Doan MJ, Lattuada CP et al (1995) Neuroretinitis, aseptic meningitis, and lymphadenitis associated with *Bartonella (Rochalimaea) henselae* infection in immunocompetent patients and patients infected with human immunodeficiency virus type 1. Clin Infect Dis 21:352–360

42. Roe RH, Michael Jumper J, Fu AD et al (2008) Ocular *Bartonella* infections. Intl Ophthal Clin 48:93–105

43. Margileth AM, Wear DJ, English CK (1987) Systemic cat-scratch disease: report of 23 patients with prolonged or recurrent severe bacterial infection. J Infect Dis 155:390–402

44. Rappaport DC, Cumming WA, Ros PR (1991) Disseminated hepatic and splenic lesions in cat-

scratch disease: imaging features. Am J Roentgenol 156:1227–1228

45. Malatack JJ, Jaffe R (1993) Granulomatous hepatitis in three children due to cat-scratch disease without peripheral adenopathy. Am J Dis Child 147:949–953

46. Maman E, Bickels J, Ephros M et al (2007) Musculoskeletal manifestations of cat scratch disease. Clin Infect Dis 45:1535–1540

47. Hajjaji N, Hocqueloux L, Kerdraon R, Bret L (2007) Bone infection in cat-scratch disease: a r eview of the literature. J Infect 54:417–421

48. Margileth AM, Baehren DF (1998) Chest-wall abscess due to cat-scratch disease (CSD) in an adult with antibodies to *Bartonella* clarridgeiae: case report and review of the thoracopulmonary manifestations of CSD. Clin Infect Dis 27:353–357

49. Myers SA, Prose NS, Garcia JA et al (1992) Bacillary angiomatosis in a child undergoing chemotherapy. J Pediatr 121:574–578

50. Malane MS, Laude TA, Chen CK et al (1995) An HIV-1-positive child with fever and a scalp nodule. Lancet 346:1466

51. Zarraga M, Rosen L, Herschthal D (2011) Bacillary angiomatosis in an immunocompetent child: a case report and review of the literature. Am J Dermatopath 33:513–515

52. Ohl ME, Spach DH (2000) *Bartonella* quintana and urban trench fever. Clin Infect Dis 31:131–135

53. Drancourt M, Birtles R, Chaumentin G et al (1996) New serotype of *Bartonella henselae* in endocarditis and cat-scratch disease. Lancet 347:552–554

54. Spach DH, Kanter AS, Daniels NA et al (1995) *Bartonella (Rochalimaea)* species as a cause of apparent "culture-negative" endocarditis. Clin Infect Dis 20:1044–1047

55. Drancourt M, Mainardi JL, Brouqui P et al (1995) *Bartonella (Rochalimaea) quintana* endocarditis in three homeless men. N Engl J Med 332:419–423

56. Daly JS, Worthington MG, Brenner DJ et al (1993) *Rochalimaea elizabethae* sp. nov. isolated

from a patient with endocarditis. J Clin Microbiol 31:872–881

57. Schwartzman WA, Patnaik M, Barka NE et al (1994) *Rochalimaea* antibodies in HIV- associated neurologic disease. Neurology 44:1312–1316

58. Baker J, Ruiz-Rodriquez R, Whitfield M et al (1995) Bacillary angiomatosis: a treatable cause of acute psychiatric symptoms in human immunodeficiency virus infection. J Clin Psychiatry 56:161–166

59. Hoey JG, Valois-Cruz F, Goldenberg H et al (2009) Development of an immunoglobulin M capture-based enzyme-linked immunosorbent assay for diagnosis of acute infections with *Bartonella henselae*. Clin Vaccine Immunol 16:282–284

60. Vermeulen MJ, Verbakel H, Notermans DW, Reimerink JH, Peeters MF (2010) Evaluation of sensitivity, specificity and cross-reactivity in *Bartonella henselae* serology. J Med Microbiol 59:743–745

61. Metzkor-Colter E, Kletter Y, Avidor B et al (2003) Long-term serological analysis and clinical follow-up of patients with cat scratch disease. Clin Infect Dis 37:1149–1154

62. Bass JW, Freitas BC, Freitas AD et al (1998) Prospective randomized double blind placebo—controlled evaluation of azithromycin for treatment of cat-scratch disease. Pediatr Infect Dis J 17:447–452

63. Zangwill KM (1998) Therapeutic options for cat-scratch disease. Pediatr Infect Dis J 17:1059–1061

64. Rolain JM, Brouqui P, Koehler JE et al (2004) Recommendations for treatment of human infections caused by *Bartonella* species. Antimicrob Agents Chemother 48:1921–1933

65. http://www.humanesociety.org/issues/pet_overpopulation/facts/pet_ownership_statistics.html. Accessed 5 Feb 2012

66. Margileth AM (1968) Cat scratch disease: nonbacterial regional lymphadenitis. The study of 145 patients and a review of the literature. Pediatrics 42(5):803–818 (http://www.ncbi.nlm.nih.gov/pubmed/4972192)

# When to Think of Immunodeficiency?

# 14

Andrew Cant and Alexandra Battersby

**Abstract**

Primary Immunodeficiencies (PIDs), although rare, are serious and heightened clinical suspicion leads to earlier diagnosis and improved outcome. Recognition of PIDs may be difficult as infections are common in young children in particular. Clues to the diagnosis of PID may be found in history, examination and initial basic investigations such as lymphocyte count. Age at presentation, type of infective organism and family history help focus on likely PIDs. Type of infective organism may indicate a specific PID, for example Aspergillus and Chronic Granulomatous Disease and Pneumocystis Jiroveci and SCID amongst others. Diagnostic aids such as 'The 10 Warning Signs of Primary Immunodeficiency' can be useful with failure to thrive, need for IV antibiotics, and family history of severe or unusual infections being the most discriminating. Systemic examination including the recognition of dysmorphic features may also support a particular diagnosis.

## 14.1 Introduction

Infections are common during childhood and young children in particular are often exposed to new pathogens. Responding to recurrent infection is also part of establishing a mature and functional immune system. The paediatrician must decide when the pattern and type of infection is not normal but indicates that a child could have an underlying immune defect so as to strike a balance between over-investigating every child who sneezes whilst not missing a potential immunodeficiency.

Primary Immunodeficiency (PID) is not much less common than conditions such as coeliac disease and cystic fibrosis yet presents with nonspecific signs. Pattern recognition and awareness of the important warning features are key to early investigation and/or referral [1].

The importance of diagnosis cannot be understated, as earlier treatment can prevent end organ

A. Cant (✉) · A. Battersby
Institute of Cellular Medicine, Newcastle University,
Newcastle upon Tyne, UK
e-mail: andrew.cant@nuth.nhs.uk

A. Battersby
e-mail: a.battersby@ncl.ac.uk

N. Curtis et al. (eds.), *Hot Topics in Infection and Immunity in Children IX,*
Advances in Experimental Medicine and Biology 764, DOI 10.1007/978-1-4614-4726-9_14,
© Springer Science+Business Media New York 2013

damage and ensure that treatment, be it curative, such as Stem Cell Transplantation, or supportive, such as immunoglobulin, will enable the patient to lead a healthy normal life. Diagnosis also has implications for the wider family as it should facilitate genetic counselling, so identifying other family members at risk of the same condition.

## 14.2    What is Compromised?

Defects can be classified according to the aspect of the immune system which is affected. Slightly modifying The International Union of Immunological Societies classification, Primary Immunodeficiency can be divided into six categories [2]:
- Combined T and B cell immunodeficiencies
- Predominantly Antibody Deficiencies
- Defects in Innate Immunity, including phagocytic cell and complement defects
- Other well-defined Immunodeficiency Syndromes
- Diseases of Immune Dysregulation
- Autoinflammatory Disorders

The presentation of these conditions differs depending upon which component of the immue system is compromised; however, there are overlapping features. The first step is not to diagnose the specific defect but to be aware of what may be compromised, how it may present and what investigations should be performed. As an approach to PID, we suggest considering the following:
- Age
- Organisms
- Systems involved
- Family history
- Haematology

These components shall now be considered.

## 14.3    Age

The immune system develops and matures as the child grows older, during which time, the child is exposed to many microbes transmitted by parents, siblings, other children at daycare and the wider environment. It is therefore, unsurprising that infections are frequent in young children, but the more serious the defect the more quickly a child with PID becomes ill, so age can itself be a clue to the underlying diagnosis and its severity.

### 14.3.1    Less than 6 months

As the inuterine is sterile, PID is not embryologically fatal. However, once born, infants are exposed to a vast array of potential organisms. They are immunologically naive and are reliant on the innate and primary adaptive responses, as well as protection from transplacentally-acquired maternal antibody.

The more severe the condition, the earlier the presentation. A child with absent T and B cell function will suffer serious consequences from even a usually trivial infection such as Rotavirus enteritis. Presentation so early in life is highly suggestive of T cell deficiency or defect, and these infants often present with failure to thrive and/or severe infection. They are likely to have a severe underlying diagnosis such as severe combined immunodeficiency (SCID), T lymphocyte disorders, DiGeorge, Wiskott Aldrich syndrome and chronic mucocutaneous candidiasis.

Delayed separation of the umbilical cord beyond 14 days is very suggestive of Leukocyte Adhesion Deficiency (LAD), whilst a pronounced milial rash on the face may indicate Hyper IgE syndrome. Severe congenital neutropenia also presents at this time with severe bacterial infection.

Non-immunological manifestations of syndromes such as the cardiac features associated with DiGeorge may also present and prompt the clinician to look for a PID.

### 14.3.2    6 months to 5 years

By the age of 6 months maternal antibodies have waned and the nadir of their protection reached. A lack of antibody production or function due to conditions such as X linked agammaglobulinaemia, CD40 Ligand deficiency or one of the rare autosomally recessively inherited causes of

**Table 14.1** Organism and disease

| Organism | Disease to consider |
|---|---|
| Burkholderia cepacia | CGD |
| Pneumocystis jiroveci | SCID, CD40 L, ICF |
| Aspergillus | CGD, Hyper IgE |
| Candida | Chronic mucocutaneous candidiasis |
| Haemophilius | Specific antibody failure |
| Meningococcus | Complement deficiency |
| Pneumococcus | Asplenia |
| Staphylococcus aureus | CGD, Hyper IgE |

**Table 14.2** Causes of *Pneumocystis jiroveci* pneumonia

| Malignancy and immunosuppression |
|---|
| HIV |
| SCID |
| CD40 ligand |
| MHC II |
| ICF |
| NEMO |
| Hyper IgE |

antibody deficiency becomes apparent after 6 months of age. Forty two percent of patients with an antibody deficiency have presented within the first year of life, whilst 88 % have presented by the age of 3 years [3]. 'Phagocyte disorders' including chronic granulomatous disease as well as the various forms of Haemophagocytic Lymphohistiocytosis such as Chediak Hegashi, also present at this age.

It is also possible there may be late presentations of SCID and this diagnosis should not be excluded on age alone.

### 14.3.3   5 years and Older

By 5 years old, the immune system has developed and matured considerably. There has been a decline in reliance on the innate immune system, whilst adaptive immunity has developed more fully. Children over the age of 5 years are most likely to have diagnoses such as Common Variable Immunodeficiency, specific antibody deficiency and complement disorders.

## 14.4   Microbial Clues

Whilst no organism is truly pathogneumonic of a PID, there are certain organisms which when present should immediately trigger thoughts of immune dysfunction. Table 14.1 highlights some of the organisms which are particularly associated with specific conditions, some of which will now be discussed in more detail.

### 14.4.1   Fungal Clues

Any invasive fungal infection is unusual in a child with a well functioning immune system and must always be considered as a 'red flag'.

*Pneumocystis jiroveci* (*Pneumocystis carinii*) is one such 'red flag' organism. Pneumocystis is virtually unheard of in the immunocompetent host. It is the most likely opportunistic infection in SCID but its presence should also alert the clinician to one of the underlying diagnoses shown in Table 14.2. It occurs most commonly in the 3–6 month age group and it may result in the need for significant respiratory support including artificial ventilation.

Patients with Chronic Granulomatous Disease (CGD) have defective oxygen-independent killing mechanisms and are particularly susceptible to fungal infection. Their infections are indolent and recurrent in nature. They are particularly susceptible to members of the *Aspergillus* genus and the presence of this organism necessitates excluding a diagnosis of CGD. *Aspergillus nidulans* seems to have a particular predilection for CGD and it almost never causes disease in other immunocompromised patients such as those with febrile neutropenia due to chemotherapy [4].

Candida is part of the normal flora of healthy individuals and infants may be colonised from early life. In individuals with well functioning immune systems it rarely causes significant disease. However, as a manifestation of PID it is significant and in particular, it may point to a diagnosis within the spectrum of conditions of chronic mucocutaneous candidiasis [5] or as part of SCID. Invasive candidial infections or persistent candida should be considered part of the spectrum of PID and warrant investigation.

## 14.4.2 Bacterial Clues

The pneumococcus is a challenging organism for even the immunocompetent child. The polysaccharide capsule ensures that the organism is difficult to opsonise and optimal antibody function is imperative to successfully defeat the organism. These features make it particularly relevant in patients with asplenia.

Meningococcal is known to thrive when there is a deficiency in the complement pathway. The incidence of complement deficiency varies greatly from one population to another; in the Israeli population the incidence is three times that of other countries included in the ESID registry with 16.3 % of PID in Israel due to complement deficiency [6]. However, in England and Wales, where there are approximately 1,000 cases of meningococcal disease occurring per year [7], a review of 297 cases of meningococcal septicaemia and meningitis found only one case of complement deficiency and of note this case had a previous serious infection [8]. Investigation for PID in those who have had meningococcal disease should be reserved for:

• Recurrent meningococcal disease
• Serotype W135 or Y
• Previous bacterial infection
• Positive family history (of PID or severe or recurrent meningococcal disease).

The experience with meningococcal disease highlights that the presence of invasive bacterial infection whilst important is not diagnostic of PID.

*Staphylococcus aureus* is a relatively common organism. However, it is well recognised as being a common presenting organism in CGD. Suppurative adenitis was present in 40 % of cases of CGD in a UK registry, and was the commonest presenting feature. When an organism was isolated it was always *S. aureus* [9]. This bacterium was also the commonest organism in liver abscesses in CGD patients [9].

Failure to properly handle mycobacteria may present further information about immune function. The United Kingdom is now one of several countries who have adopted a targeted neonatal BCG programme [10] rather than universal immunisation. This may be crucial in the presentation of PID. A localised reaction at the site of the BCG and the involvement of a single lymph node may be normal. Disseminated BCG infection however, is not seen in the immunocompetent individual and should be considered a manifestation of PID until proven otherwise [11]. Increased susceptibility to mycobacteria is seen in disorders of phagocytic function. Individuals with SCID, CGD and those with disorders of the IFN-γ/IL-12 pathway are particularly susceptible to severe disease [12].

## 14.4.3 Viral Clues

Viruses associated with a particular system will be discussed later in this chapter. At this point it is worth considering how failure to deal appropriately with viruses may be a significant presentation of PID.

The human herpes viruses are an example of a group which have a profound effect on the immunodeficient child. A considerable amount is known about the normal course of infection with various members of the herpes family.

Varicella zoster virus (VZV) infection remains common in childhood in countries where universal vaccination has not been implemented. In the majority of children, it is a mild illness requiring symptomatic treatment only, serious complications are relatively rare with bacterial sepsis being the most common. [13] However, in the child with an underlying PID, particularly a T cell defect even this typical childhood illness can be life threatening. In particular children with Cartilage Hair Hypoplasia are at risk of Varicella Pneumonitis. A family member who has died from VZV should also be considered an unusual consequence of this childhood illness.

Epstein Barr virus (EBV) infects nearly all children and patients are either asymptomatic or suffer an infectious-mononucleosis illness which is self limiting. Overwhelming EBV infections causing Haemophagocytic Lymphohistiocytosis is a common presentation of X-linked proliferative disease; hypogammaglobulinaemia, lymphoma, aplastic anaemia and vasculitis being other ones. Other PIDs such as ITK deficiency may present as atypical lymphoma, such as extra-nodal Hodgkin's disease involving the lungs, induced by EBV.

## 14.5   Systems

### 14.5.1   Respiratory Clues

Respiratory infection is almost universal in childhood. Furthermore the lungs are vulnerable to infection as they have to compromise the immune defence capability to fulfil their gas exchange function. The healthy child has numerous respiratory tract infections a year with a significant number occurring in winter. Additional risk factors such as passive smoking, day-care attendance or atopy put specific cohorts of children at further risk. It is therefore, all too easy to dismiss the child with an immune deficiency who presents with respiratory distress. There are, however, red flags within the respiratory illnesses to alert us to the children who are not run of the mill.

Common conditions, taking uncommon courses should prompt the clinician to consider a diagnosis of PID. For example, respiratory syncytial virus (RSV) has a very predictable course, with the majority of children improving over 7–10 days and with a low mortality rate [14]. Those children who have chronic RSV or persistent bronchiolitis need to be considered to have an immune defect and of those who do, SCID is top of the differential.

Pneumatocoeles (as shown in Fig. 14.1) are unusual in childhood infection and their occurrence should alert the clinician to an unusual diagnosis at the very least. They are typically associated with *S. aureus* and subsequently with Hyper IgE Syndrome. Their presence should be taken to assume a diagnosis of Hyper IgE syndrome until proven otherwise.

Recurrent infection at the same site should always alert the physician that all is not well. It may be that there is an anatomical anomaly which accounts for the recurrent infection, for example an undiagnosed tracho-oesophageal fistula. How-

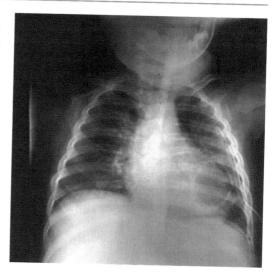

**Fig. 14.1** Pneumatocoeles on chest X-ray. (Source: The Paediatric Immunology Unit, Great North Children's Hospital, Newcastle upon Tyne, NE1 4LP)

ever, it may be that it is the immune system itself, leaving that particular site susceptible to infection. In the case of recurrent pneumonia, several immune deficiencies may present in this manner, namely, XLA, common variable immune deficiency (CVID) and antibody deficiencies.

The commonest cause of bronchiectasis in the UK currently, is Cystic Fibrosis (CF). However, through the introduction of newborn screening, it is hoped that many cases will now be picked up shortly after birth [15]. It is therefore, even more imperative that when bronchiectasis is present, an underlying immune deficiency is considered. The most likely cause is an antibody deficiency. Ideally, this would have been diagnosed prior to the development of bronchiectasis, but even once bronchiectasis is evident the earlier this diagnosis is made the better as antibody deficient patients with established bronchiectasis at the time of diagnosis have a poorer prognosis than those who do not.

### 14.5.2   Gastroenterological Clues

The gastrointestinal tract requires a huge permeable surface area if it is to effectively digest and absorb nutrients. This renders it vulnerable to exploitation by microbes and so a multitude of defence mechanisms have evolved to protect against infection. These comprise both non-specific innate immune responses as well as specific adaptive responses. It is also colonised by bacteria constituting the normal flora which in the immunocompetent host serve to stimulate protective non inflammatory responses rather than pathogenic ones.

Gastrointestinal (GI) symptoms, especially diarrhoea and vomiting are common in young children and repeated exposure to viral pathogens in particular induce protective responses that in time prevent infection. Gut disease as a manifestation of PID is often misdiagnosed as a primary GI problem. Whilst most cases of diarrhoea and vomiting are caused by uncomplicated viral infection in an immunocompetant child, this may be the first presentation of SCID, HIV or a T cell activation defect. Failure to clear an enteric virus such as adenovirus, rotavirus, norovirus or enterovirus should trigger investigation for a PID, especially when associated with chronic failure to thrive, which in infancy may be the first presentation of a severe PID and in particular SCID or HIV.

Chronic diarrhoea with or without the passing of blood or mucous is seen in 15 % of CGD patients. These important symptoms are indistinguishable from those seen in Crohn's Disease. However, there are important differences in management. If CGD is not diagnosed and immunosuppression given without antibiotic and antifungal prophylaxis, severe disseminated infection with organisms such as *Burkholderia cepacia* can ensue.

Hepatic manifestations of PID have been regularly described [16, 17] and these range from abnormal liver enzymes to sclerosing cholangitis (SC) [18]. The most common conditions associated with hepatic involvement are CD-40 ligand deficiency and X-linked Hyper-IgM syndrome. There is a high association of infection with *Cryptosporidum parvum* and the development of liver disease in patients with PID [18] and the co-existence of these two conditions should raise the possibility of PID as a unifying diagnosis. Sclerosing cholangitis is uncommon in the paediatric population, but it is not uncommon in Hyper-IgM syndrome [19] where liver involvement, including SC, occurs in as many as 75 % of patients [17], or CD40-ligand deficiency where SC can be present in around one third of patients [20].

Liver abscesses are an uncommon presentation in the paediatric setting in the developed world. Recurrent, persistent and multiple hepatic abscesses are the hallmark of CGD with *S. aureus* being the most common causative organism [21]. Diagnosis of a hepatic abscess in a child should be considered the starting point of investigation and not the end point. An underlying PID should be sought and in particular CGD must be considered. CGD is X-linked in 70 % of cases [22] but both X-linked and autosomal disease may present in this manner and therefore, gender does not discriminate in this diagnosis.

### 14.5.3 Skin

The skin presents the first part of the immune system; a mechanical barrier to pathogen entry, vital in the defence of the body from infection. It is, therefore, unsurprising, that children with defective immune systems may present with dermatological manifestations.

Eczema is common throughout childhood and it has several associations with PID. It may be a presenting feature in Wiskott Aldrich syndrome (WAS), Hyper IgE Syndrome (Job's), DOCK8 mutation and SCID.

In WAS, eczema is one of the defining features. However, it may be mild and there are often, small petichial haemorrhages and associated bleeding or bruising. This is in contrast to the findings in SCID. In patients with Omenn syndrome, the skin is thick and leathery with associated rubbery, large axillary and inguinal lymph nodes. Figure. 14.2 shows the classic appearance of Omenn's skin. In patients with SCID with MFE the rash is similar to that in Omenn but may well be milder. DOCK8 features extensive severe eczema with food allergy but in a significant proportion of cases this is accompanied by extremely extensive Molluscum contagiosum (as shown in Fig. 14.3) and/or human papilloma viral warts which often become malignant.

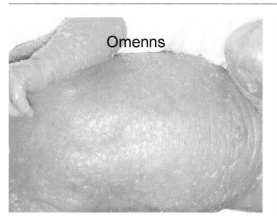

Fig. 14.2 Omenn's skin. (Source: The Paediatric Immunology Unit, Great North Children's Hospital, Newcastle upon Tyne, NE1 4LP)

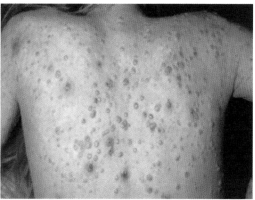

Fig. 14.3 Extensive Molluscum contagiosum. (Source: The Paediatric Immunology Unit, Great North Children's Hospital, Newcastle upon Tyne, NE1 4LP)

Persistent or recurrent skin or mucosal infections, in particular with candida, may be the presentation of SCID, chronic mucocutaneous candidiasis, Autoimmune Polyendocrinopathy-Candidiasis-Ectodermal Dystrophy (APECED) or Hyper-IgE Syndrome. In SCID, there is often superficial candida, in particular in the mouth and napkin area, often with the presence of satellite lesions. Invasive candidiasis, however, is not a feature. In APECED, the presentation of candida is very variable. However, the candida when present may be difficult to treat, with topical treatments offering no relief. Systemic and repeated courses of fluconazole may be required. Associated nail dystrophy, dental enamel hypoplasia and autoimmune disease are pivotal in the diagnosis of this syndrome.

### 14.5.4 Other Features

Paediatricians are trained to recognise patterns of syndromic features. Whilst some facial appearances are striking and proclaim the diagnosis immediately, others are more subtle and must be sought alongside examination of other family members. These fundamental observations are as much a part of PID as any other speciality in paediatrics.

The classic DiGeorge facies are not always immediately recognised but once realised to be present should prompt examination of T cell number and function and look for 22q11 deletion. The classic DiGeorge facies include a button nose, underdeveloped philtrum and small ears [23].

The asymmetrical face with a prominent forehead and deep set eyes, hypertelorism, and cathedral palate [24], alongside the recurrent staphylococcal infection and red, itchy skin should prompt the diagnosis of Hyper IgE syndrome. Even when dysmorphic features are noted it may not always be possible to link them to a diagnosis, but nevertheless, their presence or absence should be sought and noted.

A thorough examination and history of dentition is required. Both a lack of teeth or a persistence of baby teeth may be relevant and if not specifically considered may be easily forgotten. The persistence of baby teeth may appear innocuous but may be part of Hyper IgE Syndrome (Fig. 14.4). The lack of teeth or the presence of conical shaped incisors (Fig. 14.5) should raise suspicion of a defect in the NEMO pathway.

Examination of the mouth by either paediatrician or dentist can yield crucial information. The presence of severe gingivitis may indicate CGD, CVID, Chediak-Higashi, Wiskott-Aldrich or Kostmann Syndrome. Recurrent aphthous ulcers, may often be thought of in connection with GI diseases such as Crohn's but may also be the hallmark of CGD, cyclical neutropenia and Hyper IgM or IgE disease. Oral candidasis beyond infancy can be the result of chronic

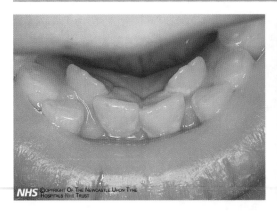

**Fig. 14.4** Typical IgE teeth. (Source: The Paediatric Immunology Unit, Great North Children's Hospital, Newcastle upon Tyne, NE1 4LP)

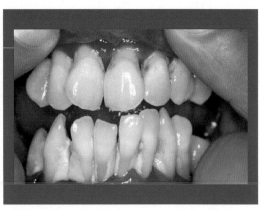

**Fig. 14.5** The conical incisors seen in NEMO defects. (Source: The Paediatric Immunology Unit, Great North Children's Hospital, Newcastle upon Tyne, NE1 4LP)

mucocutaneous candidasis and hyper IgE syndrome. Periodontitis is characteristic of many PIDs including LAD, Wiskott-Aldrich, Chediak-Higashi and selective IgA deficiency. The eyes may be the window to your soul but the oral cavity may offer an insight into the functionality of your immune system. [25]

Partial oculocutaneous albinism is an unusual presentation worldwide and the detection of it should prompt consideration of Chediak-Higashi as this feature is well described [26, 27].

## 14.6 Family History

A family history of PID is a strong clue to a potential diagnosis of PID and is considered one of the three most important warning signs for a potential diagnosis of PID, alongside the need for IV antibiotics and failure to thrive [28]. However, the family history may not be known to the wider family or its importance not appreciated or understood. Therefore, more specific questions need to be asked to elicit relevant family history, and it is often worth asking about family history of disease on more than one occasion as more than one prompt may be needed for a relevant history to be revealed. Consanguinity needs to be specifically asked about in all ethnic groups

as the prevalence of autosomal recessive PID is significantly higher in this group.

Questions should include,
- unexplained death in infancy within the family
- a history of recurrent or severe infections in a family member
- a history of abscesses particularly requiring surgical intervention or recurrent in nature
- the presence of malignancy in family members, e.g. lymphoma in male relatives may indicate XLP or Wiskott-Aldrich syndrome
- autoimmunity
- a history of eczema and bleeding.

Death from infection in infancy must raise the possibility of a diagnosis of SCID. This is particularly pertinent in the male relatives due to the prevalence of X-linked conditions. Autoimmunity is seen in female carriers of certain conditions, for example, in CGD the female carriers have a high incidence of SLE [29]. Autoimmunity should not be dismissed as unconnected. As our understanding develops we are aware of more associations that may be relevant.

## 14.7 Haematological Clues

Contrary to popular perception diagnosing PID does not start with complicated investigations even though they may be part of the diagnostic process. A simple full blood count (FBC) can yield crucial information and guide the physician to further appropriate investigations. Each component of the FBC should be scrutinised.

**White cell count:** Whilst the total white cell count is usually normal even in SCID, a closer look at the differential will often reveal a lymphocyte count $<2 \times 10^9$/litre. It is often forgotten that whilst a total lymphocyte count above $1.0 \times 10^9$/litre is normal in older children and adults, in infancy the lower limit of normal is between 2.5 and $2.7 \times 10^9$/litre. Thus in any infant with an infection, a lymphocyte count below this on two occasions should lead to consideration of a diagnosis of SCID [30]. A neutrophil count of $<1.5 \times 10^9$/litre should cause alarm bells to ring and the possibility of cyclical neutropenia and CD40 Ligand deficiency considered.

**Platelets:** Thrombocytopenia is seen in WAS, and whilst counts of $70–80 \times 10^9$/litre are often seen in young infants with WAS, they fall with age. The platelets themselves are classically small with the mean platelet volume typically being half that of healthy controls (3.8–5.0 fl) [31]. It is imperative, however, that the Coulter counter is calibrated prior to testing in order that accurate results are given. Thrombocytopenia may also be part of a pancytopenic picture.

**Red cells:** The presence of erythrophagocytosis seen on a blood film is unusual and may be due to XLP. Aplastic anaemia may also be associated with XLP and this would be demonstrated on the blood film.

## 14.8 Overall Impression

Throughout this review, individual factors which that could point to a potential PID have been highlighted. They are outlined to prompt thought about an underlying diagnosis but as with all clinical medicine should not be seen in isolation but rather should prompt the physician to re-visit relevant aspects of the history and physical examination

with a new insight. Overall impression, or 'gut instinct' may also alert the paediatrician and this of course should not be ignored.

## 14.9 Who to Investigate?

Throughout this review, there has been a focus on common presentations of PID in childhood. The National Institute for Health drew attention to ten warning signs which indicate PID [32] to further educate physicians about immune deficiency and they are:

- Eight or more new ear infections within 1 year
- Two or more serious infections within 1 year
- Two or more months on antibiotics with little effect
- Two or more episodes of pneumonia within 1 year
- Failure of an infant to gain weight or grow normally (failure to thrive)
- Recurrent deep skin or organ abscesses
- Need for intravenous antibiotics to clear infections
- Two or more deep seated infections e.g. sepsis, meningitis
- A family history of PID
- Persistent thrush in mouth or fungal infection on skin.

However, when evaluated further, the three most critical of these are:

- Failure of an infant to gain weight or grow normally (failure to thrive)
- Need for intravenous antibiotics to clear infection
- Family history of PID [28].

## 14.10 How to Investigate?

It is not possible in this review to describe the full range of possible investigations for PID and so it shall focus on some core and first line investigations.

The importance of the total lymphocyte count in the diagnosis of SCID has already been highlighted. To further investigate, it is necessary to look at the numbers of T and B lymphocytes and

NK cells. Infants with significant T cell defects will have reduced T cell numbers in all subsets, but may have raised, low or normal B cell and NK cells. Lack of naïve T lymphocytes characterises unusual forms of SCID, and the expression of common gamma chain is defect in this form of X linked SCID, so these and other markers can be helpful in delineating the exact type of SCID. Further investigation into T cell function may be required. This can be determined by measurement of lymphocyte proliferation *in vitro* to mitogenic stimulation. In infants with severe T cell defects there will usually be very low (< 10 % of normal) or no response to stimulation. Genetic studies pinpointing the exact gene defect are increasingly useful.

Measurement of serum immunoglobulins is an important part of the diagnostic process for PID. IgG, IgM and IgA should all be measured along with IgG subclassses. Age-specific normal values must be referred to when appraising these results. In young infants, the IgG present may represent transplacentally-acquired antibody. Immunoglobulin measurement is an important part of assessment but the results may be misleading with values not infrequently borderline in infants with well functioning immune systems and immunoglobulin present in low levels in those with SCID. Patients with B cell defects have low levels of immunoglobulins and some have low or absent numbers of B cells. In some antibody deficiencies B cells are present, but the lack of class-switched memory B cells makes a significant antibody deficiency likely.

Further information can be gleaned by the measurement of response to vaccinations. In the child that is already at least partially immunised, measurement of antibody response to tetanus and *Haemophilus influenzae* type b is critical. When measuring response to pneumococcal vaccination serotype specific response should be sought. When vaccinations have not been performed then they can be given and serological responses measured four weeks after.

The diagnostic test for CGD is flow cytometric assay of oxidative metabolism using dihydrorhodamine. The Nitroblue Tetrazolium (NBT) reduction test may also be used in the diagnosis,

but the results are more subjective. It is worth noting that it is a *negative* NBT test that confirms the diagnosis of CGD!

When investigating for defects in the complement pathway it is critical that the investigations are not limited to C3 and C4 which may be considered laboratory standard. A more comprehensive assessment for complement deficiencies involves measuring both classical and alternative pathways (CH100 and AP100).

Protein expression in leucocytes may form an important part of the diagnostic pathway in certain conditions. Flow cytometric analysis of WAS protein expression in leucocytes aids the diagnosis of WAS [33] and, in the case of WAS and XLT, there is correlation between the amount of protein expression and clinical prognosis. Degree of abnormal protein expression does not always correlate with prognosis, however, and this is particular true in XLP.

It is now frequently possible to confirm a genetic diagnosis which may not alter management but may enable a more accurate prognosis to be given and also further information about recurrence to be given to parents and wider family.

The theme throughout this review has been the infective clues and these should never be understated; they may well be the first indication of an underlying problem and provide significant information as to what that problem is.

## References

1. Sewell WAC, Khan S, Dore PC (2006) Early indicators of immunodeficiency in adults and children:protocols for screening for primary immunological defects. Clin Exp Immunol 145:201–203
2. Geha RS, Notarangelo LD et al (2007) The International Union of Immunological Societies (IUIS) Primary Immunodeficiency Diseases (PID) Classification Committee. J Allergy Clin Immunol 120(4):776–794
3. Dow M, Arkwright PD (2004) Primary antibody immunodeficiency in children. J Allergy Clin Immunol 113(2)
4. Bylund J, Goldblatt D, Speert D (2005) Chronic granulomatous disease: from genetic defect to clinical presentation. In: Pollard AJ, Finn A (eds)

Hot topics in infection and immunity in children. Springer, New York

5.  Wan-Fai Ng, Carmichael AJ, Arkwright PD, Abinun M, Cant AJ, Jollies S, Lilic D (2010) Impaired TH17 responses in patients with chronic mucocutaneous candidiasis with and without autoimmune polyendocrinopathy—candidiasis—ectodermal dystrophy. J Allergy Clin Immunol 126(5):1006–1015

6.  Golan H, Dalal I, Garty B-Z, Schlesinger M, Levy J, Handzel Z, Wolach B, Rottem M, Goldberg A, Tamir R, Koren A, Levy Y, Katz Y, Passwell J, Etzioni A (2002) The incidence of primary immunodeficiency syndromes in Israel. Isr Med Assoc J 4(Suppl 11):868–871

7.  Health Protection Agency (2011) Epidemiological Data of Meningococcal 2011. http://www.hpa.org.uk/Topics/InfectiousDiseases/InfectionsAZ/MeningococcalDisease/EpidemiologicalData/. (cited 2011 10/10/2011)

8.  Hoare S, El-Shazali O, Clark JE, Cant AJ (2002) Investigation for complement deficiency following meningococcal disease. Arch Dis Child 86:215–217

9.  Jones LBKR, McGrogan P, Flood TJ, Gennery AR, Morton L, Thrasher A, Goldblatt D, Parker L, Cant AJ (2008) Special Article: Chronic granulomatous disease in the United Kingdom and Ireland: a comprehensive national patient-based registry. Clin Exp Immunol 152:211–218

10. Chief Medical Officer (2005) Changes to the BCG vaccination programme. Department of Health

11. Reichenbach J, Rosenzweig S, Doffinger R, Dupuis S, Holland SM, Casanova JL (2001) Mycobacterial diseases in primary immunodeficiencies. Curr Opin Allergy Clin Immunol 1:503–511

12. Galkina E, Kondratenko I, Bologov A (2007) Mycobacterial infections in primary immunodeficiency patients. In: Shurin MR, Smolkin YS (eds) Immune-mediated diseases. Springer, New York, p 75–81

13. Niklaus H, Mueller DHG, Randall J, Mahalingam R, Nagel MA (2008) Varicella Zoster Virus infection: clinical features, molecular pathogenesis of disease, and latency. Neurol Clin 26:3

14. Ogra PL (2004) Respiratory syncytial virus: the virus, the disease and the immune response. Paediatr Respir Rev 5(Suppl A):S119-S126

15. Newborn blood spot screening in the UK (2005) Department of Health

16. Fiore M, Ammendola R, Gaetaniello L, De Felice C, Iorio R, Vegnente A et al (1998) Chronic unexplained liver disease in children with primary immunodeficiency syndromes. J Clin Gastroenterol 26:187–192

17. Levy J, Espanol-Boren T, Thomas C, Fischer A, Tovo P, Bordigoni P et al (1997) Clinical Spectrum of X-linked hyper-IgM syndrome. J Pediatrics 131:47–54

18. Rodrigues F, Davies EG, Harrison P et al (2004) Liver disease in children with primary immunodeficiencies. J Pediatr 145:333–339

19. Winkelstein JA, Marino MC, Ochs H, Fuleihan R, Scgool PR et al (2003) The X-linked Hyper IgM syndrome: clinical and immunological features of 79 patients. Medicine 82(6):373–384

20. Hadzic N (2000) Liver disease in primary immundeficiencies. J Hepatol 32(Supp 2):9–10

21. Lublin M, Bartlett DL, Danforth DN et al (2002) Hepatic abscess in patients with chronic granulomatous disease. Ann Surg 235:383–391

22. Bylund J, Goldblatt D, Speert D (2005) Chronic granulomatous disease: from genetic defect to clinical presentation. In: Pollard AJ, Finn A (eds) Hot topics in infection and immunity in children. Springer, New York

23. Ryan AK, Wilson DI, Philip N, Levy A, Seidel H et al (1997) Spectrum of clinical features associated with interstitial chromosome 22q11 deletions: a European collaborative study. J Med Genet 34(10):798–804

24. Woellner C, Michael Gertz E, Schäffer AA, Macarena L (2010) Mutations in the signal transducer and activator of transcription 3 (STAT3) and diagnostic guidelines for the Hyper-IgE Syndrome. J Allergy Clin Immunol 125 (2):424–432

25. Szczawinska-Poplonyk A, Gerreth DDS, Breborowicz K, Borysewicz-Lewicka M, Poznam P (2009) Oral manifestations of primary immune deficiencies in children. Oral Surg Oral Med Oral Pathol Oral Radiol Endod 108(3):9–20

26. Spickett G (2008) Immune deficiency disorders involving neutrophils. J Clin Pathol 61:1001–1005

27. Kersseboom R, Brooks A, Weemaes C (2011) Syndromic forms of primary immunodeficiency. Eur J Pediatr 170:295–308

28. Subbarayan ACG, Hughes SM, Gennery AR, Slatter M, Cant AJ, Arkwright PD (2011) Clinical features that identify children with primary immunodeficiency diseases. Pediatrics 127:810–816

29. Cale CM, Morton L, Goldblatt D (2007) Cutaneous and other lupus-like symptoms in carriers of X-linked chronic granulomatous disease: incidence and autoimmune serology. Clin Exp Immunol 148:79–84

30. Hague RA, Rassam S, Morgan G, Cant AJ (1994) Early diagnosis of severe combined immunodeficiency syndrome. Arch Dis Child 70:260–263

31. Ochs HD, Thrasher AJ (2006) The Wiskott-Aldrich syndrome. J Allergy Clin Immunol 117:725–38

32. NIH (2011) http://www.nichd.nih.gov/publications/pubs/primary_immuno.cfm#SignsandSymptoms. (cited 2011 17/10/2011)

33. Griffith LM et al (2009) Improving cellular therapy for primary immune deficiency diseases: recognition, diagnosis, and management. J Allergy Clin Immunol 124(6):1152–1160.e12

# Mycetoma Caused by *Madurella mycetomatis:* A Completely Neglected Medico-social Dilemma

# 15

Alex van Belkum, Ahmed Fahal
and Wendy W. J. van de Sande

## Abstract

Mycetoma is a debilitating disease with a highly particular geographical distribution. The mycetoma belt circles the entire world just above the equator and defines the region with the highest prevalence and incidence. Although the disease is seen in Central America, India and all across Africa, Sudan seems to be the homeland of mycetoma. Mycetoma is an infectious disease caused either by bacteria (actinomycetoma) or true fungi (eumycetoma).

In Sudan most cases are caused by the fungal species *Madurella mycetomatis*. The precise natural habitat of this fungus is still an enigma, but its DNA can easily be found in soil and plant samples in endemic areas. Although the entire human population in these areas are in regular contact with the fungus, most individuals are unaffected. Thus mycetoma is an ideal clinical and experimental model system for the study of host-pathogen interactions. Also, given its relative importance locally, improvements in clinical and laboratory diagnostics and knowledge of the epidemiology of the disease are badly needed. This chapter describes the current state of affairs in the field of eumycetoma caused by *M. mycetomatis*. The value of laboratory research on this disease and future perspective for control and prevention of the infection are discussed.

A. van Belkum (✉)
bioMérieux, Microbiology Unit, La Balme-Les-Grottes,
France
e-mail: alex.vanbelkum@biomerieux.com

A. Fahal
University of Khartoum, Mycetoma Research Centre,
Khartoum, Sudan
e-mail: ahfahal@hotmail.com

W. W. J. van de Sande
Medical Microbiology and Infectious Diseases, Erasmus
MC, Rotterdam, The Netherlands
e-mail: w.vandesande@erasmusmc.nl

## 15.1 Introduction

Mycetoma, regionally also known as "Madura foot", is a chronic granulomatous inflammatory disease of the extremities in particular [1, 2]. However, no single bodily part is fully exempt. First described in a patient from India [3], the infection can be caused either by bacteria (actinomycetoma) or fungi (eumycetoma) [4]. It is characterized by the presence of a chronic sub-

N. Curtis et al. (eds.), *Hot Topics in Infection and Immunity in Children IX*,
Advances in Experimental Medicine and Biology 764, DOI 10.1007/978-1-4614-4726-9_15,
© Springer Science+Business Media New York 2013

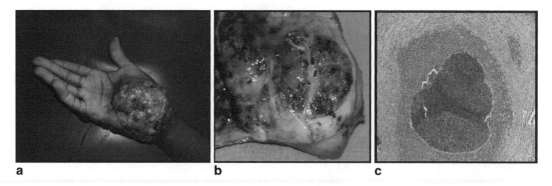

**Fig. 15.1** Mycetoma from lesion to microscopic pathology. **a** Picture of the hand of a patient suffering from myce-toma infection with *Madurella mycetomatis*. Note the *yellowish*, purulent exudate that is characteristic for this disease. Magnification will show the presence of the *black* grains as well. **b** Surgically removed mycetoma lesion in which the macroscopic *black* grains are more clearly visible. **c** Histological analysis of a tissue-embedded *black* grain. The central cement like structure can be seen to be completely surrounded by infiltrated neutrophils

**Fig. 15.2** Cranial mycetoma in a young child. **a** Exterior part of the lesion. **b** MRI picture showing the internal protrusion of the lesion

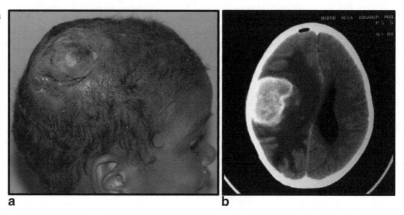

cutaneous mass in which sinuses will develop (Figs. 15.1 and 15.2). Through these sinuses, purulent material and grains of fungal or bacte-rial origin are emitted. Grains can have varying colors, depending on the causative agent. Over-all, as a preliminary diagnostic indicator, bacte-rial grains are white, yellow or red while fungal grains are mostly black or white. Although the presence of grains is characteristic of mycetoma, there is still limited understanding of the physi-cal nature and physiological or pathogenic role of these grains [1, 5].

Most of the causative agents are believed to be soil-dwelling micro-organisms which are acci-dently introduced into the host, possibly follow-ing minor trauma such as a thorn prick [6]. Once inside the body, the causative agent starts to pro-

liferate very slowly. The first symptoms may take several months, even years to become palpable. The mycetoma infection itself is painless, which is why most patients don't seek early medical help, doing so only when the lesion becomes large and the bones become affected. Due to open sinuses and frequent poor hygiene of the open wound, most patients are primarily troubled by other bacterial super-infections, which do cause additional discomfort [7].

Overall, mycetoma is a devastating disease causing obvious physical scars as well as socio-economic and psychological ones. Several fea-tures of mycetoma are presented to help highlight this significantly neglected and very serious trop-ical disease.

## 15.2 The Population at Risk and Approaches to Management

Males are more susceptible than females and children may be affected as well. Late in the disease, dissemination to other bodily locations and severe local tissue destruction (including that of bone) is frequently observed [5, 8–10]. The endemic areas are diverse but mainly tropical. People in large parts of Central Africa, Central America and India are subject to increased risk of infection [1]. Therapy for mycetoma is usually based on local experience and relatively unsophisticated practice as no therapeutic consensus has yet been reached [8, 11, 12]. Treatment is largely guided by relatively subjective clinical observations as laboratory-based diagnosis to guide therapy is lacking. Bacterial actinomycetoma can be treated with antimicrobial agents alone with a high success rate. However, fungal mycetoma is not effectively treated with antifungal agents alone. Surgical intervention (debridement, wide local excision or amputation) is required in combination with post-surgical chemotherapeutic treatment. Success rates are higher when treatment is started early in the infection. Long courses of high doses of antifungals are needed (ketoconazole or itraconazole at 400–800 mg per day) for durations which vary from case to case [13–15]. Treatment success can only be defined on the basis of regular long term follow up to check for recurrences which are rather frequent. This long and unpredictable course makes therapeutic trials difficult to perform. Despite the fact that large numbers of people living in endemic areas risk being infected on a daily basis, essentially nothing is known on disease prevention. More detailed knowledge of the pathophysiology of this disease is badly needed.

## 15.3 Causative Agents

Mycetoma can be caused by a large variety of micro-organisms, both bacterial and fungal in nature. The causative bacterial species mostly belong to the genera *Actinomadura* spp, *Acti-*

*nomyces* spp, *Nocardia* spp and *Streptomyces* spp [1, 4]. There is even more taxonomic diversity among the fungal species that are capable of causing mycetoma. These include species of *Acremonium, Aspergillus, Curvularia, Fusarium, Leptosphaeria, Pyrenochaeta* and several others [11, 16–18]. New and unexpected causative agents are still being discovered on a regular basis. One of the most prominent genera is that of *Madurella* spp, represented in particular by its best known species, *Madurella mycetomatis.* This species is most prevalent in several areas where mycetoma is a clearly endemic disease. For instance, in Central Sudan most mycetoma cases are due to *M. mycetomatis* [19, 20]. This chapter focuses on *M. mycetomatis* mycetoma as it is the best studied epidemiological, clinical, patho-physiological and host response model system to date.

## 15.4 Mycetoma in Children

Mycetoma including those cases caused by *M. mycetomatis* is thought to be relatively rare in children although precise prevalence data from endemic regions are lacking and should be undertaken. Fortunately, recently a large case series has been presented [21] allowing several important conclusions to be drawn but also raising additional questions.

As in adults, boys stand a larger chance of acquiring mycetoma than girls. Of the patients treated, approximately 73 % were boys and in the Central Sudan region where this study originates 80 % of all patients were infected with *M. mycetomatis.* Interestingly, 15 % of the junior patients (age 4–17 years, average 13.7 years) had a family history of mycetoma. Foot, knee, hand, head and neck were the most frequently affected areas and the lesions showed the characteristic sinuses, local sweating, lymphadenopathy and varicose veins. Bone involvement was as frequently observed as in adults and after apparently successful chemical and surgical treatment 18 % of patients still suffered from a relapse. Morbidity was high and in 2 % of the patients a (partial) amputation of a limb had to be performed.

There are several practical problems relating to management of mycetoma in children. Presentation is characteristically late, not only because of fear of the often mutilating surgical treatment but also because of children's complete dependence on adult support and frequent financial constraints. Approximately 45 % of the children had undergone some sort of prior treatment by local physicians or less qualified (para-)medical personnel including people specialized in traditional medicine. In all cases, such interventions had been unsuccessful and in many cases had apparently worsened rather than improved the disease or its prognosis. After treatment, both clinical follow-up and compliance with chemotherapy were problematic and certainly not systematically achieved. Frequent travelling to and from the clinic is a major obstacle for many patients, many of them living in remote areas. For the same reason, chemotherapy was often irregular and incomplete. All these problems are aggravated by the lack of financial resources and medical insurance. In addition, the children suffered from obvious and serious secondary socio-economic consequences: with frequent abandonment of schooling and psychological issues including depression, despair and anxiety.

Other factors may contribute to the clinical impact of mycetoma. Two very important issues are hygiene and protection from infection, which are associated with the nearly complete lack of health education among many affected individuals. There is limited knowledge about disease prevention, recognition and optimal treatment. There is clearly scope for improved education and the necessary tools are not especially costly and may already be available. Health education campaigns can also reach remote areas although this can present challenges. In addition, more advanced research into the pathophysiology of the disease is urgently required as discussed in the remainder of this chapter.

## 15.5 Diagnostic Features of *Madurella mycetomatis*

### 15.5.1 Direct Identification in Tissue Samples

For diagnosis of *M. mycetomatis* mycetoma fine needle aspirates can be taken directly from the lesion. Based on histological analysis of surgical biopsies, within the lesion there are slowly maturing grains in which, for instance, Grocott staining can help visualize fungal material [8, 22]. The mycelium is melanised and the developing grain is usually surrounded by a layer of neutrophils. However, in most laboratories fungal culture of the grains that are expelled from the lesion through the sinuses is more common practice than histopathology which is more difficult. The fungal grains are easy to obtain and can be conveniently inoculated on yeast- and fungus-specific artificial growth media. However, cultures may need to be incubated for weeks in order to be confidently designated as negative. Furthermore, mis-identifications do occur on a regular basis. Reliable serological tests do not exist at the moment although some immunodiffusion tests have been described as has a newly developed but not yet clinically employed ELISA system targeting the human antibody response to the translationally-controlled tumor protein (TCTP) of *M. mycetomatis* [23–27]. This latter test employs recombinant TCTP or *in vitro* synthesized TCTP-specific peptides, derived from the first *M. mycetomatis*-specific antigen so far identified. Further novel *M. mycetomatis*-specific antigens are currently being identified and serological testing may be significantly improved in the near future [28].

### 15.5.2 Growth-based Testing

*M. mycetomatis* is a fungus that grows slowly on artificial growth media. It produces dark, brown-to-black colonies consisting of dense, melanised and, interestingly, always sterile mycelium. Phase variation with respect to melanisation is frequently seen: in these cases only a sector of the colony is melanised with the remaining taking on a brown to beige color. Colonies may acquire spectacular three dimensional structures and the color may vary regionally within a single colony. There is extensive phenotype variation and it is impossible to speciate *M. mycetomatis* reliably on the basis of morphological features of the colony. In contrast to the variability in the colony, microscopic morphology is rather uniform. No distinguishing sporulating structures are observed. *M. mycetomatis* is a sterile fungus, which make determination based on microscopy almost impossible [1, 29]. It appears that many fungi were wrongly identified as *M. mycetomatis* in the past [18] and culture based diagnostics needs to be complemented with alternative information including for instance molecular data.

### 15.5.3   Molecular Testing

Since microbiological, growth-based diagnostics is not unequivocal, new technologies have been explored for their usefulness. One example is molecular testing by PCR. Amplification of the ribosomal internally transcribed spacer (ITS) regions has led to the development of a specific PCR restriction fragment length polymorphism (RFLP) test [17, 19]. This test proved to be important in speciation of the cultured fungus, but also turned out to be very useful in the detection of (non-cultivable) *M. mycetomatis* antigens in soil samples [6]. The test revealed that as a species, *M. mycetomatis* behaves homogeneously in the diagnostic PCR tests and that in case of soil samples from endemic regions as many as 25 % of samples can be scored positive for *M. mycetomatis* DNA at least [6]. The same was documented for acacia thorns indicating the *M. mycetomatis* is environmentally widely-spread. Soil samples from non-endemic areas were negative in these tests.

Another form of molecular testing that can be applied to DNA extracted from *M. mycetomatis* biomass is amplified fragment length polymorphism (AFLP) analysis [30]. This is a technique that helps to PCR-amplify DNA restriction frag-

ment subsets selectively [31]. Variability in the length of such fragments due to sequence variation in the restriction enzyme recognition sites (for example between two isolates of the same species) will define whether or not a species is clonal or genetically diverse. In turn, this may help to define patterns of epidemiological spread of microbes in general and in our case *M. mycetomatis* in particular. Although this technique has not been applied very often in fungal mycetoma research, one study showed that *M. mycetomatis* is genetically quite stable but that on the basis of minor genetic polymorphisms it could be concluded that regional epidemic spread of certain genotypes of *M. mycetomatis* was occurring [30]. Whether such regional clones differ in disease causing potential is currently not known.

### 15.5.4   Antifungal Susceptibility Testing and Response to Treatment

*In vitro* testing for anti-fungal resistance can help tailor adequate chemotherapy for patients. Using newly developed tests it has been demonstrated that *M. mycetomatis* remains highly susceptible to the azole-group of antifungal agents [15, 32]. Low MICs were obtained for ketoconazole, itraconazole, voriconazole and posaconazole. Also low MICs were obtained for terbinafin [32, 33]. Moderate susceptibility was documented for amphotericin B and fluconazole whereas all isolates were completely resistant to 5-flucytosine and the echinocandins studied (caspofungin, anidulafungin and mycafungin) [34]. MICs became elevated for the azoles ketoconazole and itraconazole when melanin was added to the culture medium, suggesting that melanin may have a biological function in shielding the fungi from chemical attack. Interestingly this was not noted for voriconazole. However, although the fungus may appear fully susceptible *in vitro*, this may not predict the *in vivo* infectious setting, especially since the fungus organizes itself into grains *in vivo,* while *in vitro* only hyphal fragments can be used in susceptibility testing. Indeed, clinical trials with ketoconazole, itraconazole, posaconazole and terbinafin, show that variable and even conflicting results are obtained with these antifungals. As an example in a recently published clinical trial in 13 *M. mycetomatis* infected patients treated with 400 mg/day itraconazole, a complete response was observed in only one patient. A partial response was noted in nine and in three the disease remained unchanged. One patient had a recurrence after 18 months [13]. Similar results were obtained with ketoconazole [35], posaconazole [36] and terbinafin [37]. Studies with additional compounds such as artemisinin and tee trea oil have been performed and several of these studies identify new therapeutic opportunities [37].

## 15.6 *Madurella mycetomatis* Virulence Factors

Virulence can be defined as the ability of an infectious agent to cause disease in an affected human individual [38]. This usually is a multi-step process, which ranges from attachment of microbial cells to colonization of the host and, later on, to host invasion and tissue damage. Very little is known on the mechanisms leading to human infection by *M. mycetomatis*. However, some information is available on a limited number of molecules that could play a role in mycetoma pathogenesis.

### 15.6.1 Melanin

Melanin has been shown to be a very important content component of the mycetoma grain. Furthermore as early in 1978 it was shown that as the secreted pigment of mycetoma, melanin was able to transform collagen into a hard matrix, resembling the material of which the fungal grain is composed [39]. Melanin is a complex series of different molecules which can be synthesized via different pathways. The biochemistry is complex but essentially there are three pathways known that lead to the synthesis of the different forms of melanin: pyo-, dopa- and DHN-melanin [40].

Using pathway inhibitors, it has been shown that *M. mycetomatis* produces primarily pyo- and DHN-melanin [41]. Melanin is known to protect these micro-organisms against UV-radiation, enzymatic lysis of the fungal cells, oxidant activities and killing by alveolar macrophages [42]. In *M. mycetomatis* complementary *in vitro* testing demonstrated that melanin has protective effects against the activity of antifungal agents as well. For instance, melanin binds effectively to amphotericin B, thereby potentially blocking its antifungal activity. However, there is a need for further studies into the influence of melanin on the susceptibility to antifungals and, hence, therapeutic success.

### 15.6.2 TCTP

As indicated above, TCTP was the first *M. mycetomatis* molecule to which a specific human response was defined [23]. TCTP is found in many eukaryotes and it has been shown to bind the anti-malaria drug artemisinin. It is thought to be involved in cell cycle control, various stress responses, histamine release and interleukin responses. Furthermore a TCTP variant of the fungus *Cladosporium herbarum*, can induce release of histamine from human basophils, indicating that TCTP may have immunomodulating effects [43]. Although the precise function of TCTP in *M. mycetomatis* has not yet been defined, several findings suggest it may play a role in mycetoma development. For one, the protein is only expressed in the very young grain and not in aged grains, unlike the other two antigens known for *M. mycetomatis*: fructose bisphosphate aldolase (FBA) and pyruvate kinase (PK) [28]. Such stage-specific expression may be associated with invasiveness. Serum anti-TCTP antibody testing also shows some interesting features. It was shown that there were significant differences between the serum response in patients versus controls, as was also noted for FBA. This suggests that immune responses are only or are primarily induced after infection. Additional research into the pathogenic relevance of this molecule is certainly warranted.

## 15.7 Host Susceptibility to *Madurella mycetomatis* Infection

Why do some people develop mycetoma and others not? PCR testing has revealed that in endemic areas M. *mycetomatis* is quite ubiquitous in soil [6]. Furthermore, most indigenous people appear to have antibodies against M. *mycetomatis* [44]. This suggests that regional inhabitants may all be in regular contact with the fungus. Still only a fraction of individuals will become infected in the end. Also the observed differences between frequency of infection in males and females (despite presumably similar exposure) [1], suggests that there must be differences in host susceptibility to infection. There are some conflicting reports about the role of the immune status of the susceptible population. As early as in 1977, local researchers showed that there was a partial impairment of the cell-mediated immune (CMI) response in patients with severe mycetoma [45]. This impaired CMI hypothesis was supported by the fact that mycetoma could more easily and successfully be induced in athymic mice than in immune-competent mice [46]. Furthermore mycetoma was more rapidly induced in immune-suppressed goats than in healthy goats [47]. However, both Bendl *et al.,* and Ahmed *et al.* did not confirm this finding: they showed that mycetoma could be induced in immune-competent animals [48, 49]. In the Mycetoma Research Center in Khartoum, Sudan, it was noted that mycetoma is generally seen in immune-competent individuals with no apparent immunodeficiency. Nevertheless, it was noted, that HIV-infected individuals show a tendency towards accelerated mycetoma progression [50].

Although there were no obvious immune defects noted in mycetoma patients, that does not mean that there are no subtle immunological differences between the people who will develop mycetoma infection and who will not. Most of these differences will have a genetic basis and a number of recent studies have reported several genetic markers associated with the development of fungal mycetoma [44, 51].

## 15.7.1 Neutrophil Function

Mycetoma is a clear example of an inflammatory disease. The combination of histochemistry and histopathology has already shown the obvious infiltration of a large number of neutrophils around developing and fully-developed black M. *mycetomatis* grains [1, 22]. This suggests that neutrophil function may be an important feature of host resistance or susceptibility to mycetoma. However, neutrophil function is complex and the anti-infectious response consists of a large number of activating and inhibitory steps implying that many different genes may contribute to such differences in susceptibility [52]. Genetic comparison of eight genes in these various maturation cascades has helped to identify different marker gene polymorphisms [44]. Among these were complement receptor 1 (CR1), the chemokine interleukin 8 (CXCL8), the interleukin 2 receptor, the chemokine receptor CXCR2, nitric oxide synthase NOS2 and TSP4. Although genetic studies may be biased, for instance based on the selection of patients or the inclusion of inadequate control individuals, functional testing can provide additional support. For instance for interleukin 8 (CXCL8), the genotypes documented in association with mycetoma were correlated with higher CXCL8 production which was also demonstrated phenotypically in the sera of these patients. For the nitric oxide synthase, it was shown immunohistochemically that the enzyme was actively produced at the site of infection. In addition, it was shown that the serum nitrite/nitrate levels were significantly lower in the serum of patients than in the serum of controls. Taken together these data support the concept that variations in neutrophil function are associated with susceptibility to mycetoma.

## 15.7.2 Cathechol-O-methyltranferase and Cytochrome p450 Subfamily 19 Genes

Since the higher incidence eumycetoma caused by M. *mycetomatis* in males cannot easily be explained by behavioral differences between the local men and women, other factors, including sex hormones, could be associated with susceptibility. To monitor for possible deficiencies in hormone synthesis among mycetoma patients, we investigated the types and allele frequencies of the genes encoding for catechol-O-methyltransferase (COMT), cytochrome p450 subfamily 1 (CYP1B1), cytochrome p450 subfamily 17 (CYP17), cytochrome p450 subfamily 19 (CYP19) and hydroxysteroid dehydrogenase 3B (HSD3B) [51]. Significant differences in allele distribution were demonstrated for CYP19 ($p=0.004$) and COMT ($p=0.005$), as well as gender dimorphism for both CYP19 and COMT polymorphisms. The COMT polymorphism was associated with mycetoma lesion size as well. The genotypes obtained for COMT and CYP19 were connected with higher 17β-estradiol production, which was confirmed by significantly elevated serum levels of 17β-estradiol in male patients. In contrast, lower levels of dehydroepiandrosteron (DHEA) were found in mycetoma patients. The *in vitro* growth of *M. mycetomatis* was not influenced by 17β-estradiol, progesterone, DHEA and testosterone excluding the possibility that the catabolism of host hormones plays a role in susceptibility to mycetoma. Apparently, the differences in hormone levels we noted between mycetoma patients and healthy controls were unlikely directly to affect the fungus itself. Indirect effects on the patients' hormone regulated immune states hypothetically form the more likely explanations for their association with mycetoma susceptibility.

Given the above observations and the relative ease with which genetic markers for mycetoma predisposition were obtained, a good strategy for future work may be Genome Wide Association studies [53, 54]. These would entail the comparison of millions of genome polymorphisms that may be consistent between patients and control generating a catalogue of factors that are potentially relevant to mycetoma development and severity. This approach may be helpful in developing innovative prophylactic, diagnostic and therapeutic regimens.

## 15.8   Concluding Remarks and Future Prospects

Our insight in microbial biology and infectious diseases has been enormously boosted by recent technological advances in laboratory methodologies. The first and foremost of these are the next generation sequencing (NGS) technologies [55, 56]. In all fields of infectious disease research, the potential impact of such technologies is becoming obvious. Bacterial whole genome sequences are numerous and for all medically relevant microbes several strain-specific genome sequences are publicly available [57]. This has led to the identification of core versus variable genomes and the characterization of new virulence genes and mechanisms. For *M. mycetomatis* such sequences are not yet available but may be defined soon. This may enhance our insight into fungal biology and help develop novel tools and strategies for prevention and treatment of disease. These fungal genome data need to be complemented with additional host genomic markers that correlate with disease susceptibility or severity. The same genome sequences used for bacterial whole genome assembly will be useful in this field as well. In the end a combination of fungal and host genome polymorphisms may allow for the detailed study of genome-genome interaction in mycetoma, which will further refine our strategies for prevention and cure.

However, it has to be noted that mycetoma is a disease of the poor and although it is a severe disease its overall impact does not compete with those, for example, of HIV or malaria. The number of mycetoma researchers worldwide is extremely small. It is certain that the field will progress significantly over the years to come but the main question still is whether the additional knowledge will attract the funds needed to facilitate real practical and clinical interventions. Until then, we may need instead to consider the implementation of simpler methods including health education to increase medical awareness among the populations at the highest risk of acquiring infection. The provision of footwear in endemic areas may be another simple intervention worth contemplating.

**Conflict of Interest**   The authors declare that no conflicts of interest exist in relation to the current text.

**Acknowledgements**   Mycetoma research in Wendy van de Sande's laboratory was facilitated by a grant in the VENI program provided by the Netherlands Organisation for Scientific Research (NWO).

# References

1. Ahmed AO, van Leeuwen W, Fahal A, van de Sande W, Verbrugh H, van Belkum A (2004) Mycetoma caused by *Madurella mycetomatis*: a neglected infectious burden. Lancet Infect Dis 4(9):566–574
2. McGinnis MR (1996) Mycetoma. Dermatol Clin 14(1):97–104
3. Padhi S, Uppin SG, Uppin MS, Umabala P, Challa S, Laxmi V, Prasad VB (2010) Mycetoma in South India: retrospective analysis of 13 cases and description of two cases caused by unusual pathogens: *Neoscytalidium dimidiatum* and *Aspergillus flavus*. Int J Dermatol 49(11):1289–1296
4. Afroz N, Khan N, Siddiqui FA, Rizvi M (2010) Eumycetoma versus actinomycetoma: diagnosis on cytology. J Cytol 27(4):133–135
5. Wethered DB, Markey MA, Hay RJ, Mahgoub ES, Gumaa SA (1987) Ultrastructural and immunogenic changes in the formation of mycetoma grains. J Med Vet Mycol 25(1):39–46
6. Ahmed A, Adelmann D, Fahal A, Verbrugh H, van Belkum A, de Hoog S (2002) Environmental occurrence of *Madurella mycetomatis*, the major agent of human eumycetoma in Sudan. J Clin Microbiol 40(3):1031–1036
7. Ahmed AO, Abugroun E, Fahal AH, Zijlstra EE, van Belkum A, Verbrugh HA (1998) Unexpected high prevalence of secondary bacterial infection in patients with mycetoma. J Clin Microbiol 36:850–851
8. Ameen M, Arenas R (2009) Developments in the management of mycetomas. Clin Exp Dermatol 34(1):1–7
9. Ahmed AO, Desplaces N, Leonard P, Goldstein F, De Hoog S, Verbrugh H, van Belkum A (2003) Molecular detection and identification of agents of eumycetoma: detailed report of two cases. J Clin Microbiol 41(12):5813–5816
10. Fahal AH, el Toum EA, el Hassan AM, Mahgoub ES, Gumaa SA (1995) The host tissue reaction to *Madurella mycetomatis*: new classification. J Med Vet Mycol 33(1):15–17
11. Ameen M, Arenas R (2008) Emerging therapeutic regimes for the management of mycetomas. Expert Opin Pharmacother 9(12):2077–2085
12. Mahgoub ES, Gumaa SA (1984) Ketoconazole in the treatment of eumycetoma due to *Madurella mycetomii*. Trans R Soc Trop Med Hyg 78:376–379
13. Fahal AH, Rahman IA, El-Hassan AM, Rahman ME, Zijlstra EE (2011) The safety and efficacy of itraconazole for the treatment of patients with eumycetoma due to *Madurella mycetomatis*. Trans R Soc Trop Med Hyg 105(3):127–132
14. Ahmed AO, van de Sande WW, van Vianen W, van Belkum A, Fahal AH, Verbrugh HA, Bakker-Woudenberg IA (2004) In vitro susceptibilities of *Madurella mycetomatis* to itraconazole and amphotericin B assessed by a modified NCCLS method and a viability-base 2,3-Bis(2-methoxy-4-nitro-5-sulfophenyl)-5-((phenylamino)carbonyl)-2H-tetrazolium hydroxide (XTT) assay. Antimicrob Agents Chemother 48(7):2742–2746
15. van de Sande WW, Luijendijk A, Ahmed AO, Bakker-Woudenberg IA, van Belkum A (2005) Testing of the in vitro susceptibilities of *Madurella mycetomatis* to six antifungal agents by using the Sensititre system in comparison with a viability-based 2,3-bis(2-methoxy-4-nitro-5-sulfophenyl)-5- ((phenylamino)carbonyl)-2H-tetrazolium hydroxide (XTT) assay and a modified NCCLS method. Antimicrob Agents Chemother 49(4):1364–1368
16. de Hoog GS, Adelmann D, Ahmed AO, van Belkum A (2004) Phylogeny and typification of *Madurella mycetomatis*, with a comparison of other agents of eumycetoma. Mycoses 47(3–4):121–130
17. Desnos-Ollivier M, Bretagne S, Dromer F, Lortholary O, Dannaoui E (2006) Molecular identification of black-grain mycetoma agents. J Clin Microbiol 44(10):3517–3523
18. Padhye AA, Salkin IF (2011) *Madurella pseudomycetomatis*: an invalidly published name. J Clin Microbiol 49(4):1703
19. Ahmed AO, Mukhtar MM, Kools-Sijmons M, Fahal AH, de Hoog S, van den Ende BG, Zijlstra EE, Verbrugh H, Abugroun ES, Elhassan AM, van Belkum A (1999) Development of a species-specific PCR-restriction fragment length polymorphism analysis procedure for identification of *Madurella mycetomatis*. J Clin Microbiol 37(10):3175–3178
20. Ahmed A, van de Sande W, Verbrugh H, Fahal A, van Belkum A (2003) *Madurella mycetomatis* strains from mycetoma lesions in Sudanese patients are clonal. J Clin Microbiol 41(10):4537–4541
21. Fahal AH, Abu S (2010) Mycetoma in children. Trans R Soc Trop Med Hyg 104:107–112
22. Yousif MA, Hay RJ (1987) Leucocyte chemotaxis to mycetoma agents—the effect of the antifungal drugs griseofulvin and ketoconazole. Trans R Soc Trop Med Hyg 81(2):319–321

23. van de Sande WW, Janse DJ, Hira V, Goedhart H, van der Zee R, Ahmed AO, Ott A, Verbrugh H, van Belkum A (2006) Translationally controlled tumor protein from *Madurella mycetomatis*, a marker for tumorous mycetoma progression. J Immunol 177(3):1997–2005

24. Zaini F, Moore MK, Hathi D, Hay RJ, Noble WC (1991) The antigenic composition and protein profiles of eumycetoma agents. Mycoses 34(1–2):19–28

25. Romero H, Mackenzie DW (1989) Studies on antigens from agents causing black grain eumycetoma. J Med Vet Mycol 27(5):303–311

26. Jiang RS, Hsu CY (2004) Serum immunoglobulins and IgG subclass levels in sinus mycetoma. Otolaryngol Head Neck Surg 130(5):563–566

27. Araujo MJ, Castañeda E (1997) *Madurella mycetomatis* antigen for the serodiagnosis of mycetoma. Rev Iberoam Micol 14(1):31–35

28. de Klerk N, de Vogel C, Fahal A, van Belkum A, van de Sande WW (2011) Fructose-bisphosphate aldolase and pyruvate kinase, two novel immunogens in *Madurella mycetomatis*. Med Mycol (Epub ahead of print)

29. Rajendran C, Baby A, Kumari S, Verghese T (1991) An evaluation of straw-extract agar media for the growth and sporulation of *Madurella mycetomatis*. Mycopathologia 115(1):9–12

30. van de Sande WW, Gorkink R, Simons G, Ott A, Ahmed AO, Verbrugh H, van Belkum A (2005) Genotyping of *Madurella mycetomatis* by selective amplification of restriction fragments (amplified fragment length polymorphism) and subtype correlation with geographical origin and lesion size. J Clin Microbiol 43(9):4349–4356

31. Vos P, Hogers R, Bleeker M, Reijans M, van de Lee T, Hornes M, Frijters A, Pot J, Peleman J, Kuiper M et al (1995) AFLP: a new technique for DNA fingerprinting. Nucleic Acids Res 23(21):4407–4414

32. van Belkum A, Fahal AH, van de Sande WW (2011) *In vitro* susceptibility of *Madurella mycetomatis* to posaconazole and terbinafine. Antimicrob Agents Chemother 55(4):1771–1773

33. N'diaye B, Dieng MT, Perez A, Stockmeyer M, Bakshi R (2006) Clinical efficacy and safety of oral terbinafine in fungal mycetoma. Int J Dermatol 45(2):154–157

34. van de Sande WW, Fahal AH, Bakker-Woudenberg IA, van Belkum A (2010) *Madurella mycetomatis* is not susceptible to the echinocandin class of antifungal agents. Antimicrob Agents Chemother 54(6):2738–2740

35. Mahgoub ES, Gumaa SA (1984) Ketoconazole in the treatment of eumycetoma due to *Madurella mycetomii*. Trans R Soc Trop Med Hyg 78(3):376–379

36. Negroni R, Tobón A, Bustamante B, Shikanai-Yasuda MA, Patino H, Restrepo A (2005) Posaconazole treatment of refractory eumycetoma and chromoblastomycosis. Rev Inst Med Trop Sao Paulo 47(6):339–346

37. van de Sande WW, Fahal AH, Riley TV, Verbrugh H, van Belkum A (2007) In vitro susceptibility of *Madurella mycetomatis*, prime agent of Madura foot, to tea tree oil and artemisinin. J Antimicrob Chemother 59(3):553–555

38. Wassenaar TM, Gaastra W (2001) Bacterial virulence: can we draw the line? FEMS Microbiol Lett 201(1):1–7

39. Findlay GH, Vismer HF (1974) Black grain mycetoma. A study of the chemistry, formation and significance of the tissue grain in *Madurella mycetomi* infection. Br J Dermatol 91(3):297–303

40. Revankar SG, Sutton DA (2010) Melanized fungi in human disease. Clin Microbiol Rev 23(4):884–928

41. van de Sande WW, de Kat J, Coppens J, Ahmed AO, Fahal A, Verbrugh H, van Belkum A (2007) Melanin biosynthesis in *Madurella mycetomatis* and its effect on susceptibility to itraconazole and ketoconazole. Microbes Infect 9(9):1114–1123

42. Martinez LR, Ntiamoah P, Gácser A, Casadevall A, Nosanchuk JD (2007) Voriconazole inhibits melanization in *Cryptococcus neoformans*. Antimicrob Agents Chemother 51(12):4396–4400

43. Rid R, Simon-Nobbe B, Langdon J, Holler C, Wally V, Pöll V, Ebner C, Hemmer W, Hawranek T, Lang R, Richter K, MacDonald S, Rinnerthaler M, Laun P, Mari A, Breitenbach M (2008) *Cladosporium herbarum* translationally controlled tumor protein (TCTP) is an IgE-binding antigen and is associated with disease severity. Mol Immunol 45(2):406–418

44. van de Sande WW, Fahal A, Verbrugh H, van Belkum A (2007) Polymorphisms in genes involved in innate immunity predispose toward mycetoma susceptibility. J Immunol 179(5):3065–3074

45. Mahgoub ES, Gumaa SA, El Hassan AM (1977) Immunological status of mycetoma patients. Bull Soc Pathol Exot Filiales 70(1):48–54

46. Mahgoub ES (1978) Experimental infection of athymic nude New Zealand mice, nu nu strain with mycetoma agents. Sabouraudia 16(3):211–216

47. Gumaa SA, Abu-Samra MT (1981) Experimental mycetoma infection in the goat. J Comp Pathol 91(3):341–346

48. Nishimura K, Miyaji M (1985) Pathogenicity of *Exophiala jeanselmei* for ddY mice. Mycopathologia 91(1):29–33

49. Ahmed AO, van Vianen W, ten Kate MT, van de Sande WW, van Belkum A, Fahal AH, Verbrugh HA, Bakker-Woudenberg IA (2003) A murine model of *Madurella mycetomatis* eumycetoma. FEMS Immunol Med Microbiol 37(1):29–36

50. Greenberg AK, Knapp J, Rom WN, Addrizzo-Harris DJ (2002) Clinical presentation of pulmonary mycetoma in HIV-infected patients. Chest 122(3):886–892

51. van de Sande WW, Fahal A, Tavakol M, van Belkum A (2010) Polymorphisms in catechol-O-methyltransferase and cytochrome p450 subfamily 19 genes predispose towards *Madurella mycetoma-*

*tis*-induced mycetoma susceptibility. Med Mycol 48(7):959–968

52. Kuijpers TW (2002) Clinical symptoms and neutropenia: the balance of neutrophil development, functional activity, and cell death. Eur J Pediatr 161(Suppl 1):S75–82

53. Newport MJ, Finan C (2011) Genome-wide association studies and susceptibility to infectious diseases. Brief Funct Genomics 10(2):98–107

54. Shugart YY, Wang Y, Jia WH, Zeng YX (2011) GWAS signals across the HLA regions: revealing a clue for common etiology underlying infectious tumors and other immunity diseases. Chin J Cancer 30(4):226–230

55. Zhang J, Chiodini R, Badr A, Zhang G (2011) The impact of next-generation sequencing on genomics. J Genet Genomics 38(3):95–109

56. Ostergaard P, Simpson MA, Jeffery S (2011) Massively parallel sequencing and identification of genes for primary lymphoedema: a perfect fit. Clin Genet 80(2):110–116

57. Morelli G, Song Y, Mazzoni CJ, Eppinger M, Roumagnac P, Wagner DM, Feldkamp M, Kusecek B, Vogler AJ, Li Y, Cui Y, Thomson NR, Jombart T, Leblois R, Lichtner P, Rahalison L, Petersen JM, Balloux F, Keim P, Wirth T, Ravel J, Yang R, Carniel E, Achtman M (2010) *Yersinia pestis* genome sequencing identifies patterns of global phylogenetic diversity. Nat Genet 42(12):1140–1143

# Elucidation and Clinical Role of Emerging Viral Respiratory Tract Infections in Children

# 16

Inge Ahout, Gerben Ferwerda and Ronald de Groot

**Abstract**

Acute respiratory viral infections (ARVI's) are the most common infectious disease in humans. With the appearance of molecular techniques the recovery of viruses has dramatically increased. Nowadays virologists can quickly discriminate virological families and related viruses from emerging viruses and consequently identify novel viruses. Many new respiratory viruses have been identified in children in the past 15 years. In this review we shortly discuss novel respiratory viruses and their pathogenic role in pediatric respiratory disease. Advantages and drawbacks of the technique and our current knowledge will be discussed. We will conclude this review with a general discussion on the future role of molecular diagnostic virology in the clinic.

## 16.1 Introduction

### 16.1.1 General (Respiratory Viruses in Pediatrics)

Acute respiratory viral infections (ARVI's) are the most common infectious disease in humans. They occur more frequently in children than in adults (6.1 episodes per year under the age of 1, 3–6 episodes per year between the age of 1–5 and 2.4 episodes per year between the age of 15–19). Disease severity depends on age, underlying condition and type of virus. ARVIs account for huge numbers of doctor's visits and days lost from work and school. They are a leading cause of global mortality and morbidity in children. Moreover, respiratory viral infections are an important driver of unnecessary usage of antibiotics. Unfortunately prevention and treatment of the majority of respiratory virus infections is not possible with the exception of influenza [1].

Although much research has been done on the epidemiology and burden of viral respiratory tract infections the size of the problem is underestimated. Due to the lack of routine testing for (multiple) viruses and the limitation that a majority of infected patients will not visit a doctor.

R. de Groot (✉) · G. Ferwerda · I. Ahout
Department of Pediatrics, Radboud University Medical Centre, Nijmegen, The Netherlands
e-mail: R.deGroot@cukz.umcn.nl

G. Ferwerda
e-mail: J.Ferwerda@cukz.umcn.nl

I. Ahout
e-mail: I.Ahout@cukz.umcn.nl

N. Curtis et al. (eds.), *Hot Topics in Infection and Immunity in Children IX,*
Advances in Experimental Medicine and Biology 764, DOI 10.1007/978-1-4614-4726-9_16,
© Springer Science+Business Media New York 2013

### 16.1.2 General Introduction Molecular Techniques (Discovery Novel Viruses)

With the appearance of molecular techniques the recovery of viruses has dramatically increased. Before their use approximately 50–80 % of the viral tests remained negative [2]. Due to the introduction of PCR and the discovery of novel viruses this proportion decreased to 3–15 %. However, such recovery rates are largely dependent on the selection of the patient group [3–5]. The combination of high sensitivity, multiplex options and quantification was essential for some of the new insights in viral epidemiology. This could not have been achieved with conventional viral diagnostics such as culture and immuneflu-orescence assays (FDA).

The identification of respiratory viruses in a clinical context can also guide diagnostic and treatment strategies. Bonner et al. [6] revealed that a known viral aetiology of disease results in decreased use of additional tests such as X-rays or blood examination, shorter hospital admission and less frequent use of antibiotics. However, molecular diagnostics have also created new dilemmas. For example, the identification of respiratory viruses in asymptomatic children, the occurrence of many viral co-infections, concerns about the pathogenic capacity of certain viruses and the value of quantitative measurements.

### 16.1.3 General Introduction on Emerging Viral Diseases

Some of the advantages of the new genetic (e.g. sequencing) and molecular techniques became clear during outbreaks of novel emerging viruses. Emerging viruses can be classified as (1) previously unknown viruses or (2) previously known viruses that have significantly increased in incidence [7]. Nowadays virologists can quickly discriminate virological families and related viruses from emerging viruses and consequently identify novel viruses.

The introduction of molecular diagnostics in medical virology has led to the identification of

**Table 16.1** Emerging viruses from the last 2 decades

| Virus | Family | Year of discovery |
| --- | --- | --- |
| Hendra-/NipahV | Paramyxovirus | 1995 |
| AIV's | Influenza virus | 1997 |
| hMPV | Paramyxovirus | 2001 |
| SARS-CoV | Coronavirus | 2003 |
| HCoV-NL63 | Coronavirus | 2004 |
| HCoV-HKU1 | Coronavirus | 2005 |
| HBoV | Parvovirus | 2005 |
| HPeV4 | Parechovirus | 2006 |
| HPeV5 | Parechovirus | 2006 |
| HPeV6 | Parechovirus | 2007 |
| KIV/WUV | Polyomavirus | 2007 |
| H1N1V | Influenza virus | 2009 |

many new respiratory viruses in children in the past 15 years (Table 16.1). However, the pathogenicity of these viruses is not always clear and the clinical relevance is often poorly understood. Fredricks and Relman proposed seven rules which are necessary to demonstrate the causative relationship between a virus and disease. These rules are based on Koch's postulates and were adapted for nucleic acid based detection methods, location of the pathogen and quantification (Table 16.2) [8, 9]. These rules can help to interpret research on the role of novel respiratory viruses in disease and guide future research. It should also be stated that the clinical relevance is in some cases apparent, without extensive research to fulfil all requirements.

In this review we briefly discuss novel respiratory viruses and their pathogenic role in pediatric respiratory disease. We will conclude this review with general discussion on the future role of molecular diagnostic virology in the clinic.

## 16.2 Henipavirus (1994–1998)

### 16.2.1 Hendra Virus

The Hendra virus was first detected in a disease outbreak in 1994. It initially presented with a new respiratory disease in horses that was transmitted to two persons one of them died [10]. The virus belongs to the genus of *Henipavirus* within

**Table 16.2** Koch's postulates adapted for viral infections [8, 9]

| A nucleic acid sequence belonging to a putative pathogen should be present in most cases of an infectious disease. Microbial nucleic acids should be found preferentially in those organs or gross anatomic sites known to be diseased, and not in those organs that lack pathology |
| --- |
| Fewer, or no, copy numbers of pathogen-associated nucleic acid sequences should occur in hosts or tissues without disease |
| With resolution of disease, the copy number of pathogen-associated nucleic acid sequences should decrease or become undetectable. With clinical relapse, the opposite should occur |
| When sequence detection predates disease, or sequence copy number correlates with severity of disease or pathology, the sequence-disease association is more likely to be a causal relationship |
| The nature of the microorganism inferred from the available sequence should be consistent with the known biological characteristics of that group of organisms |
| Tissue-sequence correlates should be sought at the cellular level: efforts should be made to demonstrate specific in situ hybridization of microbial sequence to areas of tissue pathology and to visible microorganisms or to areas where microorganisms are presumed to be located |
| These sequence-based forms of evidence for microbial causation should be reproducible |

the family of the *Paramyxoviridae* family. It was initially named morbillivirus and later re-named Hendra virus (HeV) after the suburb where the outbreak occurred [11, 12]. The virus itself is not very contagious. It spreads through direct contact between horses or during intensive contact between humans and severely ill horses. The animal reservoir appears to be the Flying-fox, in this population the Hendra infection is largely subclinical. The breading season of Flying-foxes is a risk period for spread and the human risk group is defined as people with close and intensive contact with horses. Up till now fourteen outbreaks have been reported [13]. In five of these outbreaks humans were involved, resulting in five deaths. So far, only two persons survived an infection. The case fatality rate (CFR) is over 50 % [13]. HeV in humans causes, after an incubation period of 5–21 days, a severe influenza like disease (fever, myalgia and headache) which can progress to

pneumonia, respiratory failure and death [10, 14]. An infection can also result in encephalitis with headache, fever and drowsiness. The encephalitis can occur after initial recovery from the illness. The Hendra virus genome is readily detected in several materials, e.g. blood, urine, nasal- and oropharyngeal swabs by RT-PCR. Next to this standard detection method the virus can be cultured in several cell lines, where it forms syncytia upon infection. ELISA serological tests are used for screening, however their diagnostic sensitivity is not yet established. Immunofluorescense assays and serum neutralization methods can also be used [13, 15].

### 16.2.2 Nipah Virus

In 1998 and 1999 two large outbreaks of respiratory disease in pigs and humans occurred. In Malaysia and Singapore 106 people died [16]. The causative agent had large similarities to the HeV and is the second member of the genus of *Henipavirus* within the *Paramyxoviridae* family. It was named Nipah virus (NiV) after the location of the first human case [17]. This virus had already caused respiratory disease in pigs until late 1996. Like HeV it shares the bat as a natural reservoir. NiV virus is very contagious among pigs and spreads through the respiratory route or directly by the transport of infected pigs. Initially humans became infected via direct contact with pigs, although food borne transmissions were also reported. Initially the case fatality rate was 38.5 % [14, 18]. Since the initial outbreak almost yearly new outbreaks emerged in Bangladesh and India causing fatal encephalitis in humans. Notably, the CFR's of these outbreaks increased to 92 %. Also the transmission changed: starting from pigs, soon cows could transmit the virus. Later human to human and nosocomial transmission was demonstrated [19–21]. It has been hypothesized that there were multiple introductions of viruses in the human population, explaining the unique genetic signature of isolated viruses nowadays [18]. These genetic differences may be the reason for the increase in CFR and differences in clinical manifestations and transmission. The clinical manifestation of a Nipah infection differed per outbreak. Incubation periods differ from an average of 2 weeks in Malaysia to 1 week in Bangladesh. The infection can be asymptomatic, but often starts with influenza-like symptoms of fever, headache, myalgia, vomiting and sore throat. Patients can recover or develop signs of encephalitis or sometimes atypical pneumonia or acute respiratory distress. In severe cases the encephalitis includes the brain stem or progresses to a coma within 24–48 h [18, 19, 22]. Around 20 % of the cases are left with residual neurological symptoms, including personality changes. In comparison with outbreaks in Malaysia and Singapore the Bangladesh and Indian patients experienced more profound respiratory symptoms with case rates of 14, 27, 70 and 51 %, respectively [19]. A Nipah infection can be diagnosed in serum urine and cerebrospinal fluid (CSF) by RT-PCR [23]. Also culture in cell-lines, ELISA for anti-HeV IgG and IgM in serum and CSF, serum neutralization assays or immunofluorescence assays are used [18, 19, 24, 25].

Patients with both Hendra and Nipah virus infections are treated supportive, antiviral therapy is not effective [22]. Prevention is based on careful hygiene, quarantine and safe disposal of animal carcasses [13, 14]. Currently, there are no vaccines available. However, several therapeutic agents seem effective *in vitro* and in some animal models [18].

### 16.2.3  Avian Influenza Virus (1997)

The first cases of avian influenza virus (AIV) infection were reported in 1997 in Honk-Kong [26]. This influenza A (H5H1) originated completely from strains circulating in wild birds and poultry [27]. The avian influenza virus undergoes rapid genetic and antigenic evolution reflected by the occurrence of different clades with distinct phenotypes [28, 29]. The majority of human cases had direct contact with poultry or could be related to outbreaks in wild birds [20]. There is limited transmission from human to human, although some epidemiological studies suggest it is possible [30]. The median age of patients is around 18 years and the mortality rate is extremely high between the age of 10 and 19 years (61 %). Yearly H5N1 outbreaks in humans have been reported in Asia, Africa and Eurasia [31]. These epidemics are all related to outbreaks of avian influenza in wild birds or poultry during the colder seasons [32, 33]. The incubation period of H5N1 is estimated to be 2–7 days [34]. The disease typically manifests as a severe pneumonia which often progresses to respiratory failure and death within 10 days (case fatality rate up to 90 % in children). It appears that in children cases may occur without pneumonia. Detection of viral RNA by (RT-) PCR is the best method for the diagnosis of H5N1, preferably using throat swabs [35]. Because of genetic variably of the virus, primers need to up-dated frequently. The available immune-assays for detection of H5N1 are not sensitive enough for clinical purposes and cannot differentiate between human and avian subtypes of influenza A. Seroconversion after 2–3 weeks can be used to confirm H5N1 infection and can be used for epidemiological studies [35]. Early treatment with oseltamivir is recommended based on some evidence that it increase survival rates [36]. There are differences in susceptibility to oseltamivir between the different clades of H5N1 circulating in different parts of the world. Combination of oseltamivir with amantidine can be given if the circulating H5N1 is susceptible to both agents. Currently, it is possible to produce vaccines that inactivate H5 influenza A strains. However due to the circulation of different clades and the rapidly changing antigenicity of H5N1 the need for the development of a new vaccine remains [34, 37].

### 16.2.4  Metapneumovirus (2001)

The human metapneumovirus (hMPV) was first discovered in the Netherlands in 2001 from a databank of samples from children with respiratory tract infections [38]. hMPV belongs to the genus *metapneumovirus* within the family of *Paramyxoviridae*. It is related to respiratory

**Table 16.3** Symptoms and diagnosis of hMPV mono infections in literature

| Symptoms/diagnosis | Spread in literature[a] |
|---|---|
| Fever[b] | 36–80 |
| Cough | 67–99 |
| Rhinitis | 72–90 |
| Wheezing | 10–73 |
| Respiratory failure | 8 |
| Oxygen 90 % | 32–85 |
| Pharyngitis | 24–66 |
| Bronchitis | 1–68 |
| Bronchiolitis | 11–51 |
| Pneumonia | 3–65 |
| Otitis media | 16 |

[a] Heikkinen et al. [121]; Aberle et al. [122]; Mullins et al. [123]; Chen et al. [124]; Manoha et al. [125]; Williams et al. [126]; Bosis et al. [127]
[b] Different definitions varying from >37.5 to >39 °C

syncytial virus, both belonging to the *pneumoviridae* sub-family. In both retrospective and prospective studies it has been shown that hMPV can be detected in 3.9–14.8 % of respiratory samples from children with respiratory disease [39]. This wide range reflects differences in the tested populations and the level of care. Co-infections with other viruses occur in 15–30 % [40–42]. hMPV is detected in up to 4 % of nasopharyngeal aspirates from healthy children, although percentages of less than 1 % are also frequently published [43]. Serological studies showed that all children by the age of 5 years had been in contact with the virus and that it has been circulating in the human population for over 50 years [38]. It has a seasonal occurrence with a peak incidence just after the influenza and RSV season [43]. Spread is thought to be via direct or close contact with respiratory secretions from an infected person with an incubation period of 3–5 days. Re-infections occur frequently in children, although symptoms are less severe [44]. Symptoms associated with hMPV infections are comparable with RSV (see Table 16.3). hMPV infections are, after RSV, the most frequent cause of bronchiolitis in young children and account for 5–15 % of all hospital admissions [45]. Hospitalization rates are highest among 6–12

month old children, remarkably older than for RSV [46]. There is an association between severe hMPV infection (bronchiolitis) and the development of wheezing in childhood [47].

hMPV can only be cultured in specific cell lines under specific conditions and is time consuming; therefore it has no role in a clinical setting. Real time PCR is the most sensitive test for hMPV detection in NPA and swabs [48] and is therefore the common method in clinical and research settings. RT-PCR also provides semi-quantitative information of the viral load (Ct value), which can be used to monitor treatment in a research setting [49]. Immunofluorescence assays are available for rapid detection of the virus in respiratory specimens; however, these tests are less sensitive than RT-PCR. Serology for hMPV has little additive value in the clinic because most children are seropositive in early childhood. Currently no vaccines against hMPV are available, though several candidates are being pursued [45]. Ribavirin, antiviral therapy, is effective *in vitro* against hMPV, though clinical data are sparce. Currently new therapies such as fusion inhibitors and siRNA's are being tested in murine models [45].

## 16.2.5 Coronavirus (2003–2005)

Human corona viruses related to respiratory disease, 229E and OC43, have been known since the 1930s. They were recognized as the second most common cause of the common cold in humans [50–52]. A new strain of human coronavirus was identified in 2004 from a respiratory sample of a 7 month old infant with bronchiolitis and named NL63 (HCoV-NL63) [53]. The HCoV-NL63 belongs to the genus Coronavirus within the family of Coronavidae. In retrospective cohort studies HCoV-NL63 have been identified in 1.7–9.3 % of respiratory samples from children with respiratory symptoms and occurs worldwide [39]. The virus is often found in combination with other respiratory viruses (57 %) [54, 55]. Peak incidence is found in the winter months and the incubation period is estimated

2–5 days [56, 57]. HCoV-NL63 is associated with mild upper respiratory tract symptoms and less frequent with severe lower respiratory tract symptoms such as bronchiolitis [58]. Some studies have reported an association with croup [55, 59]. HCoV-NL can be detected in respiratory specimens by RT-PCR which is the first choice for diagnosis. Immunoassays are available for rapid detection and distinction of different HCoV strains [60]. Different cell-lines are permissive for viral culture and used in a research setting. Currently no anti-viral treatment against HCoViNL63 is available, although several inhibiting compounds have been identified [61].

A second novel human coronavirus was identified in 2005 in a 71-year-old man with pneumonia in China and named HKU1 after the Hong Kong University where it was found [62]. In a retrospective cohort studies the HCoV-HKU1 was identified in 1–3.1 % of respiratory samples in which no other virus was detected, from children with upper and lower respiratory symptoms [39] with a higher incidence in children younger than 6 months. The peak incidence of HCoV-HKU1 is in spring, early summer and winter with an incubation period of 2 days [63]. HCoV-HKU1 is mainly associated with upper respiratory tract symptoms in children and occasionally with pneumonia and bronchiolitis [64]. The first choice of assay for detection in respiratory specimens is RT-PCR. Coronaviruses exhibit substantial genetic variability hampering the development of pancorona primers and therefore specific primers for each strain have to be used [65]. There is no specific anti-viral therapy available against HCoV-HKU1.

### 16.2.6 Human Bocavirus (2005)

Human bocavirus (HBoV) belongs to the genus *Bocavirus* within the family *Parvoviridae* (and is closely related to the bovine parvovirus and canine minute virus). This virus was identified in 2005 by nucleic acid amplification (PCR) in respiratory tract specimens from Swedish children with lower respiratory tract infections [66]. In this study HBoV was detected in 3.1 % of hos-

pitalized children below the age of three. Other studies detected HBoV in 3–19 % of children with respiratory symptoms depending on the sample type used (NPA and BAL *higher*, nasal swab *lower*) [67] and the age of the patient (*higher* in younger children) [68]. However, HBoV is frequently found in asymptomatic children (up to 40 %) or in combination with other viruses (up to 80 %) in symptomatic children [39]. Based on these findings it is still unclear whether HBoV has a pathogenic role in respiratory disease. One study performed in a PICU suggests that the viral load (high titres) of HBoV may indicate a pathogenic role in (severe) respiratory disease [69]. HBoV has been associated with wheezing in asthmatic children [70]. In general HBoV infection is marked by relatively mild symptoms of the upper respiratory tract such as cough, rhinorrhea and fever. In rare cases it has been associated with lower respiratory tract infection and even respiratory insufficiency [71]. Detection of HBoV is by RT-PCR and the virus can be detected in respiratory as well as gastrointestinal specimens [72]. Diagnostic seroresponses can be used to establish the specific immune response against HBoV during infection, although the clinical relevance is unclear [73]. HBoV can only be cultured on ciliated primary human epithelial cell-lines, and therefore viral isolation is only used in experimental settings [74]. Treatment of HBoV infections is mainly supportive and no specific anti-viral treatment against HBoV is available. Currently there is not enough epidemiological evidence to drive vaccine development against HBoV.

### 16.2.7 Parechovirus (2006–2007)

Human parechoviruses (HPeVs) belong to the genus *Parechovirus* of the family *Picornaviridae*. The first HPeVs, serotype 1 and 2, were identified 50 years ago during a summer diarrhoea outbreak in American children [75]. With the introduction of molecular techniques many new serotypes of HPeVs have been identified in the past 15 years in the stool or NPA of children with gastrointestinal and respiratory disease, and in the cerebrospinal

**Table 16.4** HPeV, discovery and clinical associations

| HPeV type | Known since | Clinical associations |
|---|---|---|
| HPeV1 | 1956 | Mild gastrointestinal and respiratory symptoms, bronchiolitis, pneumonitis, otitis media. Encephalitis, paralysis, myocarditis |
| HPeV2 | 1956 | Mild gastrointestinal and respiratory symptoms, (rare) neonatal sepsis, meningitis, encephalitis |
| HpeV3 | 2004 | Neonatal sepsis, meningitis, encephalitis (transient paralysis) |
| HPeV4 | 2006 | Fever, mild gastrointestinal and respiratory symptoms |
| HPeV5 | 2006 | Fever, mild gastrointestinal and respiratory symptoms (sepsis, Reye's syndrome) |
| HPeV6 | 2007 | Fever, mild gastrointestinal and respiratory symptoms (paralysis, Reye's syndrome) |
| HPeV8 | 2009 | Enteritis |
| HPeV10 | 2010 | Gastro-enteritis |

fluid of children with meningitis and sepsis-like illness (see Table 16.4) [76–79]. Every HPeV serotype has its specific epidemiological and clinical features. All HPeVs infections are very common in children under the age of 1 year and most data are available on HPeV1 and HPeV3 [80]. The median age of infection with HPeV1 is 6.6 months, whereas HPeV3 infections occur at a younger age of 1.3 months. There is also seasonal variability in occurrence, HPeV1 in late summer and early winter season, and HBeV3 mostly in summer. HBeV serotype 5 and 6 have also been associated with respiratory tract symptoms [81–84].

Most HPeVs have are common causes of asymptomatic infection in early childhood and are often found in combination with other viruses, so that the relation with respiratory disease is hard to establish [80]. While the association of HPeV3 with encephalitis, meningitis and neonatal sepsis is widely accepted [85], for most other serotypes the relationship with disease and specific symptoms is less clear (see Table 16.4) [86].

A viral neutralisation assay or culture are time-consuming and not suitable for severe disease such as sepsis and meningitis. Detection by

RT-PCR is only available for HPeV1-3 [87]. Currently amplification and nucleotide sequencing is used to identify specific genotypes in a research setting. The specific antibody response can be used to demonstrate involvement of HPeV in disease if the virus itself cannot be detected. No antiviral treatment against parechoviruses is currently available and only supportive care is given.

### 16.2.8   Polyomavirus (2007)

In 2007 two new members of the *Polyomaviridae* family were discovered in samples of patients with respiratory disease. The first of these new polyomaviruses was identified during a large scale molecular virus screening project in respiratory samples from children and named after the Karolisnka institute where it was discovered (KIV) [88]. The second was identified in a nasopharyngeal aspirate of a 3-year-old child with pneumonia and named Washington University virus (WUV) [89]. Seroprevalence studies show KI in 66 % and WU in 79 % of paediatric sera [46, 90]. The virus has been detected in 1–5 % respiratory samples worldwide in respiratory samples of young symptomatic children [91–93]. However, in 70–80 % of the cases there was a co-infection with other respiratory viruses, and KIV and WUV have been described in asymptomatic HSCT patients [94]. Based on these results it is difficult to assign symptoms and pathogenicity to both of them and more epidemiological evidence is needed. In most studies the viruses have been associated with both upper and lower respiratory tract infections in children. Detection of WUV and KIV in respiratory samples can be undertaken by RT-PCR. Thus far there is no indication for treatment of either of these viruses nor vaccine development.

### 16.2.9   Influenza A H1N1 Virus (2009)

In late march 2009 a novel influenza A (H1N1) virus was identified in America. This virus was subsequently recognised as the cause of an outbreak of respiratory illness in Mexico [95]. The

**Table 16.5** Symptoms of H1N1

| Presentation | Literature[a] (%) |
|---|---|
| Fever (>38 °C) | 81–94 |
| Cough | 69–82 |
| Gastro-intestinal symptoms | 8–32 |
| Rhinorrhea | 31–62 |
| Diarrhea | 8–23 |
| Wheezing | 12–25 |

[a] Libster et al. [97], Hackett et al. [128], Jain et al. [129]

novel flu virus showed reassortment of swine, avian and human strains, and appeared to be very infectious between humans [96]. After the initial detection several other countries reported H1N1 infections. In June 2009 the WHO declared a pandemic with spread over at least two continents. At the start of the pandemic the virus appeared to be very virulent with a high mortality rate, especially in young adults and children [97, 98]. However, in the Northern Hemisphere the virus behaved more like a seasonal influenza virus. H1N1 disease had the highest attack rate in young children causing relatively mild disease [99]. The pH1N1 was able to outcompete the seasonal flu so that, in the influenza season 2009–2010, over 99 % of the influenza positive isolates in Europe and America were pandemic H1N1 influenza A [101].

In general the symptoms resembled those of other winter viruses: fever, cough, sore throat, myalgia and headache. Symptoms at presentation for hospitalised patients are shown in Table 16.5. Spread occurs up to 8 days after the start of symptoms although this may be prolonged in immunocompromised patients and children [102].

H1N1 infection can be diagnosed by RT-PCR on respiratory samples and this appears to be the most sensitive method. In case of high suspicion of H1N1 infection with a negative PCR result, the virus can be cultured or infection proven by documenting seroconversion [103–105].

During the pandemic of H1N1 were treated with oseltamivir (Tamiflu®) and zanamivir. This treatment reduced the duration of symptoms, the occurrence of otitis media and progression into

severe disease, especially when administered early in the course of disease [106]. Also the prophylactic use of anti viral agents is effective in reducing the occurrence of H1N1 infections in exposed individuals. However, oseltamivir and multi drug resistant viruses are emerging [107]. In several countries children have been vaccinated [108, 109]. H1N1 vaccination induced an effective and long lasting humoral immune response [108, 109]. The vaccine seemed to reduce the risk of infection and decreased severity of disease in children, however because of the rapid spread of the H1N1 pandemic most people were vaccinated during the pandemic making efficacy studies complex [111, 112].

## 16.3 Discussion on Molecular Diagnostics of Respiratory Viruses and Their Clinical Use

In this review we have discussed newly identified and emerging viruses from the past 2 decades. These viruses could be subdivided in three categories, based on the evidence for their pathogenicity in respiratory disease in children. First, emerging viruses causing epidemics with high mortality, such as AIV, Hendra and Nipah virus, were clearly associated with a pathogenic role in disease. These epidemic-causing viruses are often of zoonotic origin (transmission from animals to humans). The second group comprises viruses that fulfil the modified Koch's postulates [8, 9]. Most novel respiratory viruses are not completely characterised according to the postulates due to the extensive and costly research needed to achieve this. In this perspective hMPV is unique among the recently discovered respiratory viruses, because all criteria have been fulfilled [111]. Third are viruses that were found during screening for new respiratory viruses in respiratory samples with molecular techniques, such as human bocavirus, the novel polyomaviruses, parechoviruses and some coronaviruses. For most of these viruses their pathogenic role as an important respiratory pathogen is less clear. Although these viruses are present in respiratory samples of children with respiratory disease however, they are also often present in asymptomatic children or found in combination with other viruses. Many studies were performed retrospectively, or without the proper control cohorts of asymptomatic children. In epidemiological studies based on seroconversion it is apparent that a first encounter with these viruses occurs early in childhood without (severe) respiratory tract infections. Especially in this last category of viruses, in which the association with respiratory disease is less clear, large prospective epidemiological studies are needed to further specify the pathogenicity and health burden of these viruses in children.

The highly sensitive molecular techniques for identification and detection of novel viruses are a powerful tool for epidemiological studies, especially when used in multiplex platforms. Their ability to quantify the viral burden in infection may be used as additional information in determining the role of a virus in respiratory disease. For some viruses a positive correlation between viral load and disease severity is described [69, 115]. However whether viral load correlates with disease severity in general remains a point of debate. Viral load appears to be lower in viral-viral co-infection compared to viral-mono-infections, the mechanism behind this and the clinical relevance requries further investigation [116, 117]. Studies show that viral load decreases during the course of disease, and this can be used as marker for the therapeutic effect of anti-viral compounds. A drawback of the high sensitivity of molecular diagnostics is that PCR signals remain positive after recovery from an illness, sometimes even for several weeks. Because young children have frequent viral infections of the upper respiratory tract, the value of a positive PCR test can be limited.

Interaction of viruses with bacteria present in the nasopharynx can result in enhanced disease severity. This is well known for influenza and *Streptococcus pneumoniae*, and other respiratory bacteria [118]. How other (novel) respiratory viruses interact with bacteria and how this leads to enhanced disease is less well known. In study-ing the pathogenicity of viruses these interactions should be taken into account and implemented in new epidemiological studies. In this context 'old' viruses, like rhinovirus, can be seen in a new perspective and the causality with severe respiratory disease should be re-evaluated [118–120].

The introduction of molecular detection of viruses has led to the discovery of many new human respiratory viruses and improvement in diagnostics. Novel molecular techniques, like sequencing of the complete virome, will offer new insight in viral infections but also new challenges in proving causality in human disease.

# References

1. Ruuskanen O, Lahti E, Jennings LC, Murdoch DR (2011) Viral pneumonia. Lancet 377:1264–1275
2. Glezen WP, Loda FA, Clyde Jr, Senior RJ, Sheaffer CI, Conley WG et al (1971) Epidemiologic patterns of acute lower respiratory disease of children in a pediatric group practice. J Pediatr 78:397–406
3. Zalm van der MMM, Uiterwaal CPM, Wilbrink BP, de Jong BMM, Verheij TJM et al (2009) Respiratory Pathogens in Respiratory Tract Illnesses During the First Year of Life: A Birth Cohort Study. Pediatr Infect Dis J 28:472–476
4. Henrickson KJ (1998) Viral pneumonia in children. Semin Pediatr Infect Dis 9:217–233
5. Fabbiani M, Terrosi C, Martorelli B, Valentini M, Bernini L, Cellesi C et al (2009) Epidemiological and clinical study of viral respiratory tract infections in children from Italy. J Med Virol 81:750–756
6. Bonner AB, Monroe KW, Talley LI, Klasner AE, Kimberlin DW (2003) Impact of the rapid diagnosis of influenza on physician decision-making and patient management in the pediatric emergency department: results of a randomized, prospective, controlled trial. Pediatrics 112:363–367
7. Kumar D (2010) Emerging viruses in transplantation. Curr Opin Infect Dis 23:374–378
8. Fredericks DN, Relman DA (1996) Sequence-based identification of microbial pathogens: a reconsideration of Koch's postulates. Clin Microbiol Rev 9:18–33
9. Lipkin WI (2009) Microbe hunting in the 21st century. Proc Natl Acad Sci U S A 106:6–7
10. Selvey LA, Wells RM, McCormack JG, Ansford AJ, Murray K, Rogers RJ et al (1995) Infection of humans and horses by a newly described morbillivirus. Med J Aust 162:642–645
11. Murray K, Rogers R, Selvey L, Selleck P, Hyatt A, Gould A et al (1995) A novel morbillivirus pneu-

monia of horses and its transmission to humans. Emerg Infect Dis 1:31–33

12. Murray K, Selleck P, Hooper P, Hyatt A, Gould A, Gleeson L et al (1995) A morbillivirus that caused fatal disease in horses and humans. Science 268:94–97

13. Hess IMR, Massey PD, Walker B, Middleton DJ, Wright TM ( 2011) Hendra virus: what do we know? N S W Public Health Bull 22:118–122

14. Field H, Young P, Yob JM, Mills J, Hall L, Mackenzie J (2001) The natural history of Hendra and Nipah viruses. Microbes Infect 3:307–314

15. Chiang CF, Lo MK, Rota PA, Spiropoulou CF, Rollin PE (2010) Use of monoclonal antibodies against Hendra and Nipah viruses in an antigen capture ELISA. Virol J 7:115

16. CDC ( 1999) Outbreak of Hendra-like virus–Malaysia and Singapore, 1998–1999. MMWR Morb Mortal Wkly Rep 48:265–269

17. CDC (1999) Update: outbreak of Nipah virus–Malaysia and Singapore, 1999. MMWR Morb Mortal Wkly Rep 48:335–337

18. Lo MK, Rota PA (2008) The emergence of Nipah virus, a highly pathogenic paramyxovirus. J Clin Virol 43:396–400

19. WHO (2010) Nipah virus fact sheet (revised in July 2009). Wkly Epidemiol Rec 85:64–67

20. Mounts AW, Kwong H, Izurieta HS, Ho Y, Au T, Lee M et al (1999) Case-control study of risk factors for avian influenza A (H5N1) disease, Hong Kong, 1997. J Infect Dis 180:505–508

21. Tan CT, Tan KS (2001) Nosocomial transmissibility of Nipah virus. J Infect Dis 184:1367

22. Goh KJ, Tan CT, Chew NK, Tan PS, Kamarulzaman A, Sarji SA et al (2000) Clinical features of Nipah virus encephalitis among pig farmers in Malaysia. N Engl J Med 342:1229–1235

23. Guillaume V, Contamin H, Loth P, Georges-Courbot MC, Lefeuvre A, Marianneau P et al (2004) Nipah virus: vaccination and passive protection studies in a hamster model. J Virol 78:834–840

24. Daniels P, Ksiazek T, Eaton BT (2001) Laboratory diagnosis of Nipah and Hendra virus infections. Microbes Infect 3:289–295

25. Bossart KN, McEachern JA, Hickey AC, Choudhry V, Dimitrov DS, Eaton BT et al (2007) Neutralization assays for differential henipavirus serology using Bio-Plex protein array systems. J Virol Methods 142:29–40

26. (1997) Isolation of avian influenza A(H5N1) viruses from humans–Hong Kong, May-December 1997. MMWR Morb Mortal Wkly Rep 46:1204–1207

27. Subbarao K, Klimov A, Katz J, Regnery H, Lim W, Hall H et al (1998) Characterization of an avian influenza A (H5N1) virus isolated from a child with a fatal respiratory illness. Science 279:393–396

28. Chen H, Smith GJ, Li KS, Wang J, Fan XH, Rayner JM et al (2006) Establishment of multiple sublineages of H5N1 influenza virus in Asia: implications for pandemic control. Proc Natl Acad Sci U S A 103:2845–2850

29. Webster RG, Govorkova EA (2006) H5N1 influenza–continuing evolution and spread. N Engl J Med 355:2174–2177

30. Tran TH, Nguyen TL, Nguyen TD, Luong TS, Pham PM, Nguyen VC et al (2004) Avian influenza A (H5N1) in 10 patients in Vietnam. N Engl J Med 350:1179–1188

31. Beigel JH, Farrar J, Han AM, Hayden FG, Hyer R, de Jong MD et al (2005) Avian influenza A (H5N1) infection in humans. N Engl J Med 353:1374–1385

32. Kilpatrick AM, Chmura AA, Gibbons DW, Fleischer RC, Marra PP, Daszak P (2006) Predicting the global spread of H5N1 avian influenza. Proc Natl Acad Sci U S A 103:19368–19373

33. Ducatez MF, Olinger CM, Owoade AA, De Landtsheer S, Ammerlaan W, Niesters HG et al (2006) Avian flu: multiple introductions of H5N1 in Nigeria. Nature 442:37

34. Abdel-Ghafar AN, Chotpitayasunondh T, Gao Z, Hayden FG, Nguyen DH, de Jong MD et al (2008) Update on avian influenza A (H5N1) virus infection in humans. N Engl J Med 358:261–273

35. de Jong MD, Simmons CP, Thanh TT, Hien VM, Smith GJ, Chau TN et al (2006) Fatal outcome of human influenza A (H5N1) is associated with high viral load and hypercytokinemia. Nat Med 12:1203–1207

36. Schunemann HJ, Hill SR, Kakad M, Bellamy R, Uyeki TM, Hayden FG et al (2007) WHO Rapid Advice Guidelines for pharmacological management of sporadic human infection with avian influenza A (H5N1) virus. Lancet Infect Dis 7:21–31

37. Girard MP, Osterhaus A, Pervikov Y, Palkonyay L, Kieny MP (2008) Report of the third meeting on "influenza vaccines that induce broad spectrum and long-lasting immune responses", World Health Organization, Geneva, Switzerland, 3–4 December 2007. Vaccine 26:2443–2450

38. van den Hoogen BG, van Doornum GJ, Fockens JC, Cornelissen JJ, Beyer WE, de Groot R et al (2003) Prevalence and Clinical Symptoms of Human Metapneumovirus Infection in Hospitalized Patients. The J Infect Dis 188:1571–1577

39. Brodzinski H, Ruddy RM (2009) Review of new and newly discovered respiratory tract viruses in children. Pediatr Emerg Care 25:352–360; quiz 61–3

40. Viazov S, Ratjen F, Scheidhauer R, Fiedler M, Roggendorf M (2003) High prevalence of human metapneumovirus infection in young children and genetic heterogeneity of the viral isolates. J Clin Microbiol 41:3043–3045

41. Dollner H, Risnes K, Radtke A, Nordbo SA (2004) Outbreak of human metapneumovirus infection in norwegian children. Pediatr Infect Dis J 23:436–440

42. Maggi F, Pifferi M, Vatteroni M, Fornai C, Tempestini E, Anzilotti S et al (2003) Human metapneumovirus associated with respiratory tract infections in a 3-year study of nasal swabs from infants in Italy. J Clin Microbiol 41:2987–2991

43. Williams JV, Harris PA, Tollefson SJ, Halburnt-Rush LL, Pingsterhaus JM, Edwards KM et al (2004) Human Metapneumovirus and Lower Respiratory Tract Disease in Otherwise Healthy Infants and Children. N Engl J Med 350:443–450

44. Pelletier G, Dery P, Abed Y, Boivin G (2002) Respiratory tract reinfections by the new human Metapneumovirus in an immunocompromised child. Emerg Infect Dis 8:976–978

45. Kahn JS (2006) Epidemiology of Human Metapneumovirus. Clin Microbiol Rev 19:546–557

46. Papenburg J, Boivin G (2010) The distinguishing features of human metapneumovirus and respiratory syncytial virus. Rev Med Virol 20:245–260

47. Garcia DF, Hiatt PW, Jewell A, Schoonover SL, Cron SG, Riggs M et al (2007) Human metapneumovirus and respiratory syncytial virus infections in older children with cystic fibrosis. Pediatr Pulmonol 42:66–74

48. Landry ML, Cohen S, Ferguson D (2008) Prospective study of human metapneumovirus detection in clinical samples by use of light diagnostics direct immunofluorescence reagent and real-time PCR. J Clin Microbiol 46:1098–1100

49. Maertzdorf J, Wang CK, Brown JB, Quinto JD, Chu M, de Graaf M et al (2004) Real-time reverse transcriptase PCR assay for detection of human metapneumoviruses from all known genetic lineages. J Clin Microbiol 42:981–986

50. Tyrell DJ, Struve FA, Schwartz ML (1965) A methodological consideration in the performance of process and reactive schizophrenics on a test for organic brain pathology. J Clin Psychol 21:254–256

51. Hamre D, Procknow JJ (1966) A new virus isolated from the human respiratory tract. Proc Soc Exp Biol Med 121:190–193

52. McIntosh K, Dees JH, Becker WB, Kapikian AZ, Chanock RM (1967) Recovery in tracheal organ cultures of novel viruses from patients with respiratory disease. Proc Natl Acad Sci U S A 57:933–940

53. van der Hoek L, Pyrc K, Jebbink MF, Vermeulen-Oost W, Berkhout RJ, Wolthers KC et al (2004) Identification of a new human coronavirus. Nat Med 10:368–373

54. Boivin G, Coulombe Z, Wat C (2003) Quantification of the influenza virus load by real-time polymerase chain reaction in nasopharyngeal swabs of patients treated with oseltamivir. J InfectDis 188:578–580

55. van der Hoek L, Sure K, Ihorst G, Stang A, Pyrc K, Jebbink MF et al (2005) Croup is associated with the novel coronavirus NL63. PLoS Med 2:e240

56. Bastien N, Robinson JL, Tse A, Lee BE, Hart L, Li Y (2005) Human coronavirus NL-63 infections in children: a 1-year study. J Clin Microbiol 43:4567–4573

57. Bastien N, Anderson K, Hart L, Van Caeseele P, Brandt K, Milley D et al (2005) Human coronavirus NL63 infection in Canada. J Infect Dis 191:503–506

58. Arden KE, Nissen MD, Sloots TP, Mackay IM (2005) New human coronavirus, HCoV-NL63, associated with severe lower respiratory tract disease in Australia. J Med Virol 75:455–462

59. Han TH, Chung JY, Kim SW, Hwang ES (2007) Human Coronavirus-NL63 infections in Korean children, 2004–2006. J Clin Virol 38:27–31

60. Gerna G, Vitulo P, Rovida F, Lilleri D, Pellegrini C, Oggionni T et al (2006) Impact of human metapneumovirus and human cytomegalovirus versus other respiratory viruses on the lower respiratory tract infections of lung transplant recipients. J Med Virol 78:408–416

61. Pfefferle S, Schopf J, Kogl M, Friedel CC, Muller MA, Carbajo-Lozoya J et al (2011) The SARS-Coronavirus-Host Interactome: Identification of Cyclophilins as Target for Pan-Coronavirus Inhibitors. PLoS Pathog 7:e1002331

62. Woo PC, Lau SK, Chu CM, Chan KH, Tsoi HW, Huang Y et al (2005) Characterization and complete genome sequence of a novel coronavirus, coronavirus HKU1, from patients with pneumonia. J Virol 79:884–895

63. Ren L, Gonzalez R, Xu J, Xiao Y, Li Y, Zhou H et al (2011) Prevalence of human coronaviruses in adults with acute respiratory tract infections in Beijing, China. J Med Virol 83:291–297

64. Esper F, Weibel C, Ferguson D, Landry ML, Kahn JS (2006) Coronavirus HKU1 infection in the United States. Emerg Infect Dis 12:775–779

65. Vijgen L, Moes E, Keyaerts E, Li S, Van Ranst M (2008) A pancoronavirus RT-PCR assay for detection of all known coronaviruses. Methods Mol Biol 454:3–12

66. Allander T, Tammi MT, Eriksson M, Bjerkner A, Tiveljung-Lindell A, Andersson B (2005) Cloning of a human parvovirus by molecular screening of respiratory tract samples. Proc Natl Acad Sci U S A 102:12891–12896

67. Longtin J, Bastien M, Gilca R, Leblanc E, de Serres G, Bergeron MG et al (2008) Human bocavirus infections in hospitalized children and adults. Emerg Infect Dis 14:217–221

68. Arnold JC, Singh KK, Spector SA, Sawyer MH (2006) Human bocavirus: prevalence and clinical spectrum at a children's hospital. Clin Infect Dis 43:283–288

69. van de Pol AC, Wolfs TF, Jansen NJ, Kimpen JL, van Loon AM, Rossen JW (2009) Human bocavirus and KI/WU polyomaviruses in pediatric intensive care patients. Emerg Infect Dis 15:454–457

70. Allander T, Jartti T, Gupta S, Niesters HG, Lehtinen P, Osterback R et al (2007) Human bocavirus and acute wheezing in children. Clin Infect Dis 44:904–910

71. Allander T (2008) Human bocavirus. J Clin Virol 41:29–33

72. Lu X, Chittaganpitch M, Olsen SJ, Mackay IM, Sloots TP, Fry AM et al (2006) Real-time PCR assays for detection of bocavirus in human specimens. J Clin Microbiol 44:3231–3235

73. Soderlund-Venermo M, Lahtinen A, Jartti T, Hedman L, Kemppainen K, Lehtinen P et al (2009) Clinical assessment and improved diagnosis of bocavirus-induced wheezing in children, Finland. Emerg Infect Dis 15:1423–1430

74. Dijkman R, Koekkoek SM, Molenkamp R, Schildgen O, van der Hoek L (2009) Human bocavirus can be cultured in differentiated human airway epithelial cells. J Virol 83:7739–7748

75. Wigand R, Sabin AB (1961) Properties of ECHO types 22, 23 and 24 viruses. Arch Gesamte Virusforsch 11:224–247

76. Baumgarte S, de Souza Luna LK, Grywna K, Panning M, Drexler JF, Karsten C et al (2008) Prevalence, types, and RNA concentrations of human parechoviruses, including a sixth parechovirus type, in stool samples from patients with acute enteritis. J Clin Microbiol 46:242–248

77. Benschop KS, Schinkel J, Luken ME, van den Broek PJ, Beersma MF, Menelik N et al (2006) Fourth human parechovirus serotype. Emerg Infect Dis 12:1572–1575

78. Al-Sunaidi M, Williams CH, Hughes PJ, Schnurr DP, Stanway G (2007) Analysis of a new human parechovirus allows the definition of parechovirus types and the identification of RNA structural domains. J Virol 81:1013–1021

79. Ito M, Yamashita T, Tsuzuki H, Takeda N, Sakae K (2004) Isolation and identification of a novel human parechovirus. J Gen Virol 85:391–398

80. Harvala H, Robertson I, McWilliam Leitch EC, Benschop K, Wolthers KC, Templeton K et al (2008) Epidemiology and clinical associations of human parechovirus respiratory infections. J Clin Microbiol 46:3446–3453

81. Schnurr D, Dondero M, Holland D, Connor J (1996) Characterization of echovirus 22 variants. Arch Virol 141:1749–1758

82. Watanabe K, Oie M, Higuchi M, Nishikawa M, Fujii M (2007) Isolation and characterization of novel human parechovirus from clinical samples. Emerg Infect Dis 13:889–895

83. Chieochansin T, Vichiwattana P, Korkong S, Theamboonlers A, Poovorawan Y (2011) Molecular epidemiology, genome characterization, and recombination event of human parechovirus. Virology 421:159–166

84. Pajkrt D, Benschop KS, Westerhuis B, Molenkamp R, Spanjerberg L, Wolthers KC (2009) Clinical characteristics of human parechoviruses 4–6

85. Harvala H, Robertson I, Chieochansin T, McWilliam Leitch EC, Templeton K, Simmonds P (2009) Specific association of human parechovirus type 3 with sepsis and fever in young infants, as identified by direct typing of cerebrospinal fluid samples. J Infect Dis 199:1753–1760

86. Harvala H, Wolthers KC, Simmonds P (2010) Parechoviruses in children: understanding a new infection. Curr Opin Infect Dis 23:224–230

87. Nix WA, Maher K, Johansson ES, Niklasson B, Lindberg AM, Pallansch MA et al (2008) Detection of all known parechoviruses by real-time PCR. J Clin Microbiol 46:2519–2524

88. Allander T, Andreasson K, Gupta S, Bjerkner A, Bogdanovic G, Persson MA et al (2007) Identification of a third human polyomavirus. J Virol 81:4130–4136

89. Gaynor AM, Nissen MD, Whiley DM, Mackay IM, Lambert SB, Wu G et al (2007) Identification of a novel polyomavirus from patients with acute respiratory tract infections. PLoS Pathog 3:e64

90. Nguyen NL, Le BM, Wang D (2009) Serologic evidence of frequent human infection with WU and KI polyomaviruses. Emerg Infect Dis 15:1199–1205

91. Abed Y, Wang D, Boivin G (2007) WU polyomavirus in children, Canada. Emerg Infect Dis 13:1939–1941

92. Norja P, Ubillos I, Templeton K, Simmonds P (2007) No evidence for an association between infections with WU and KI polyomaviruses and respiratory disease. J Clin Virol 40:307–311

93. Han TH, Chung JY, Koo JW, Kim SW, Hwang ES (2007) WU polyomavirus in children with acute lower respiratory tract infections, South Korea. Emerg Infect Dis 13:1766–1768

94. Babakir-Mina M, Ciccozzi M, Perno CF, Ciotti M (2011) The novel KI, WU, MC polyomaviruses: possible human pathogens? New Microbiol 34:1–8

95. Dawood FS, Jain S, Finelli L, Shaw MW, Lindstrom S, Garten RJ et al (2009) Emergence of a novel swine-origin influenza A (H1N1) virus in humans. N Engl J Med 360:2605–2615

96. Garten RJ, Davis CT, Russell CA, Shu B, Lindstrom S, Balish A et al (2009) Antigenic and genetic characteristics of swine-origin 2009 A(H1N1) influenza viruses circulating in humans. Science 325:197–201

97. Libster R, Bugna J, Coviello S, Hijano DR, Dunaiewsky M, Reynoso N et al (2010) Pediatric hospitalizations associated with 2009 pandemic influenza A (H1N1) in Argentina. N Engl J Med 362:45–55

98. Farias JA, Fernandez A, Monteverde E, Vidal N, Arias P, Montes MJ et al (2010) Critically ill infants and children with influenza A (H1N1) in pediatric intensive care units in Argentina. Intensive Care Med 36(6):1015–1022 (Epub 2010 Mar 18)

99. Bishop JF, Murnane MP, Owen R (2009) Australia's winter with the 2009 pandemic influenza A (H1N1) virus. N Engl J Med 361:2591–2594

100. Lister P, Reynolds F, Parslow R, Chan A, Cooper M, Plunkett A et al (2009) Swine-origin influenza virus H1N1, seasonal influenza virus, and critical illness in children. Lancet 374:605–607

101. Patel M, Dennis A, Flutter C, Khan Z (2010) Pandemic (H1N1) 2009 influenza. Br J Anaesth 104:128–142

102. Giannella M, Alonso M, Garcia de Viedma D, Lopez Roa P, Catalan P, Padilla B et al (2011) Prolonged viral shedding in pandemic influenza A (H1N1): clinical significance and viral load analysis in hospitalized patients. Clin Microbiol Infect 17:1160–1165

103. Bennett S, Gunson RN, MacLean A, Miller R, Carman WF (2011) The validation of a real- time RT-PCR assay which detects influenza A and types simultaneously for influenza A H1N1 (2009) and oseltamivir-resistant (H275Y) influenza A H1N1 (2009). J Virol Methods 171:86–90

104. Ginocchio CC, Zhang F, Manji R, Arora S, Bornfreund M, Falk L et al (2009) Evaluation of multiple test methods for the detection of the novel 2009 influenza A (H1N1) during the New York City outbreak. J Clin Virol 45:191–195

105. Faix DJ, Sherman SS, Waterman SH (2009) Rapid-test sensitivity for novel swine-origin influenza A (H1N1) virus in humans. N Engl J Med 361:728–729

106. Smith JR, Rayner CR, Donner B, Wollenhaupt M, Klumpp K, Dutkowski R (2011) Oseltamivir in seasonal, pandemic, and avian influenza: a comprehensive review of 10-years clinical experience. Adv Ther 28:927–959

107. Meijer A, Jonges M, Abbink F, Ang W, van Beek J, Beersma M et al (2011) Oseltamivir-resistant pandemic A(H1N1) 2009 influenza viruses detected through enhanced surveillance in the Netherlands, 2009–2010. Antiviral Res 92:81–89

108. Arguedas A, Soley C, Lindert K (2010) Responses to 2009 H1N1 vaccine in children 3 to 17 years of age. N Engl J Med 362:370–372

109. Lambert LC, Fauci AS (2010) Influenza vaccines for the future. N Engl J Med 363:2036–2044

110. Wu J, Xu F, Lu L, Lu M, Miao L, Gao T et al (2010) Safety and effectiveness of a 2009 H1N1 vaccine in Beijing. N Engl J Med 363:2416–2423

111. Gilca V, De Serres G, Hamelin ME, Boivin G, Ouakki M, Boulianne N et al (2011) Antibody persistence and response to 2010–2011 trivalent influenza vaccine one year after a single dose of 2009 AS03-adjuvanted pandemic H1N1 vaccine in children. Vaccine

112. (2011) Effectiveness and safety of the A-H1N1 vaccine in children: a hospital-based case- control study. BMJ Open 1:e000167

113. van den Hoogen BG, de Jong JC, Groen J, Kuiken T, de Groot R, Fouchier RA et al. (2001) A newly discovered human pneumovirus isolated from young children with respiratory tract disease. Nat Med 7:719–724

114. Sloots TP, Whiley DM, Lambert SB, Nissen MD (2008) Emerging respiratory agents: new viruses for old diseases? J Clin Virol 42:233–243

115. Bosis S, Esposito S, Osterhaus AD, Tremolati E, Begliatti E, Tagliabue C et al (2008) Association between high nasopharyngeal viral load and disease severity in children with human metapneumovirus infection. J Clin Virol 42:286–290

116. Semple MG, Dankert HM, Ebrahimi B, Correia JB, Booth JA, Stewart JP et al (2007) Severe respiratory syncytial virus bronchiolitis in infants is associated with reduced airway interferon gamma and substance P. PLoS One 2:e1038

117. Gerna G, Campanini G, Rognoni V, Marchi A, Rovida F, Piralla A et al (2008) Correlation of viral load as determined by real-time RT-PCR and clinical characteristics of respiratory syncytial virus lower respiratory tract infections in early infancy. J Clin Virol 41:45–48

118. McCullers JA (2006) Insights into the interaction between influenza virus and pneumococcus. Clin Microbiol Rev 19:571–582

119. McErlean P, Shackelton LA, Lambert SB, Nissen MD, Sloots TP, Mackay IM ( 2007) Characterisation of a newly identified human rhinovirus, HRV-QPM, discovered in infants with bronchiolitis. J Clin Virol 39:67–75

120. Peltola V, Heikkinen T, Ruuskanen O, Jartti T, Hovi T, Kilpi T et al (2011) Temporal association between rhinovirus circulation in the community and invasive pneumococcal disease in children. Pediatr Infect Dis J 30:456–461

121. Heikkinen T, Osterback R, Peltola V, Jartti T, Vainionpaa R (2008) Human metapneumovirus infections in children. Emerg Infect Dis 14:101–106

122. Aberle JH, Aberle SW, Redlberger-Fritz M, Sandhofer MJ, Popow-Kraupp T (2010) Human metapneumovirus subgroup changes and seasonality during epidemics. Pediatr Infect Dis J 29:1016–1018

123. Mullins JA, Erdman DD, Weinberg GA, Edwards K, Hall CB, Walker FJ et al (2004) Human metapneumovirus infection among children hospitalized with acute respiratory illness. Emerg Infect Dis 10:700–705

124. Chen X, Zhang ZY, Zhao Y, Liu EM, Zhao XD (2010) Acute lower respiratory tract infections by human metapneumovirus in children in Southwest China: a 2-year study. Pediatr Pulmonol 45:824–831

125. Manoha C, Bour JB, Pitoiset C, Darniot M, Aho S, Pothier P (2008) Rapid and sensitive detection of metapneumovirus in clinical specimens by indi-

rect fluorescence assay using a monoclonal anti-body. J Med Virol 80:154–158

126.   Williams J-V, Edwards K-M, Weinberg G-A, Griffin M-R, Hall C-B, Zhu Y et al (2010) Population Based Incidence of Human Metapneumovirus Infection among Hospitalized Children. J Infect Dis 201:1890–1898

127.   Bosis S, Esposito S, Niesters HG, Crovari P, Osterhaus AD, Principi N (2005) Impact of human metapneumovirus in childhood: comparison with respiratory syncytial virus and influenza viruses. J Med Virol 75:101–104

128.   Hackett S, Hill L, Patel J, Ratnaraja N, Ifeyinwa A, Farooqi M et al (2009) Clinical characteristics of paediatric H1N1 admissions in Birmingham, UK. Lancet 374:605

129.   Jain S, Kamimoto L, Bramley AM, Schmitz AM, Benoit SR, Louie J et al (2009) Hospitalized patients with 2009 H1N1 influenza in the United States, April-June 2009. N Engl J Med 361:1935–1944

# Urinary Tract Infections in Children: Microbial Virulence Versus Host Susceptibility

Catharina Svanborg

**Abstract**

Urinary tract infections (UTI) are common, dangerous and interesting. This review includes a general background on UTIs and molecular mechanisms of pathogenesis. In addition, we discuss UTI susceptibility and especially the effect of genetic variation on innate immunity.

The symptoms of acute pyelonephritis are caused by the innate immune response and inflammation in the urinary tract decreases renal tubular function and may give rise to renal scarring, especially in childhood. The disease severity is explained by pathogens and their virulence factors triggering signaling through Toll-like receptors (TLRs), interferon regulatory factors (IRFs) and type 1 interferons, and the activation of a host response mediating disease or pathology or clearance of infection. In children with asymptomatic bacteriuria (ABU), in contrast, bacteria persist without causing symptoms or pathology. ABU strains mostly lack virulence factors, and the lack of symptoms has largely been attributed to their lack of virulence. Recently, rapid progress has been made in the understanding of host susceptibility mechanisms. For example, genetic alterations that reduce TLR4 function are associated with ABU while polymorphisms reducing IRF3 or CXCR1 expression are associated with acute pyelonephritis and an increased risk for renal scarring.

Understanding bacterial virulence and host resistance promises new tools to improve the diagnostic accuracy in children with UTI. By combining information on bacterial virulence and the host response, it should be possible to start individualizing diagnosis and therapy. Finally, we propose that the prediction of future disease risk and decisions on prophylaxis and invasive diagnostic procedures might be improved by genetic analysis.

C. Svanborg (✉)
Department of Microbiology, Immunology and
Glycobiology, Institute of Laboratory Medicine,
Lund University, Lund, Sweden
e-mail: Catharina.Svanborg@med.lu.se

N. Curtis et al. (eds.), *Hot Topics in Infection and Immunity in Children IX,*
Advances in Experimental Medicine and Biology 764, DOI 10.1007/978-1-4614-4726-9_17,
© Springer Science+Business Media New York 2013

## 17.1   Introduction

Bacteria frequently invade the distal urinary tract but in most individuals, infection is not established, due to an efficient antibacterial host defense, reviewed below. However, in patients with voiding dysfunction, malformations and/or molecular immune defects, infection is facilitated, and even bacteria of reduced virulence may establish infection. Most children who develop UTI have a normal urinary tract, and to establish bacteriuria and symptomatic infection, specific virulence factors are required as well as a match with a susceptible host.

Urinary tract infection (UTI) is defined by significant bacteriuria, classically $>10^5$ CFU/ml of urine, but lower counts are considered relevant in a patient with acute symptoms and frequent voiding. Bacteriuria may lead to severe symptoms and even long-term sequelae such as renal scarring, or be asymptomatic and protect against symptomatic disease.

Acute pyelonephritis (APN) is the most severe form of UTI, accompanied by local and systemic inflammation and symptoms. Bacteria establish infection in the renal pelvis and may proceed to invade renal tissues. The patient experiences flank pain, fever $>38.5$ and general malaise and laboratory tests detect elevated levels of circulating acute phase reactants like CRP, ESR, cytokines and leucocytes. Increasingly used to also quantify local inflammation in the urinary tract are urine cytokine levels, adding much-needed information in addition to pyuria, the classical diagnostic tool to detect inflammation in the urinary tract. Adults with acute pyelonephritis develop bacteremia in about 30 % of cases and urosepsis remains an important cause of death. The frequency of bacteremia or uro-sepsis in children is lower, and, in the pediatric age group, it is not entirely clear to what extent bacteria invade the blood-stream.

Asymptomatic bacteriuria (ABU) is the most common form of UTI, occurring in about 1 % of girls, 2 % of pregnant women and 20 % of both men and women above 70 years of age. These patients carry $>10^5$ CFU/ml of urine, often for extended periods of time, without developing symptoms. The patients may develop a mild host response that can be quantified as low cytokine levels in urine (IL-8 but not IL-6) as well as variable, low level pyuria and after long-term carriage the patients may develop lymphoid follicles in the bladder mucosa, which resolve after antibiotic therapy. This host response is obviously not strong enough to cause symptoms and the relevance of variable host responses in ABU is unclear.

Acute cystitis is very common, but not well defined in childhood, as diagnosis largely relies on the patient being able to describe the symptoms. In adult women, acute cystitis episodes are exceedingly common, with frequency, dysuria and suprapubic pain and often require antibiotic therapy. In addition, lower tract symptoms may precede upper tract infection, creating a grey zone between patient groups with symptomatic UTI.

The level and severity of UTI is determined by the properties of the infecting strain; mostly *Escherichia coli*. Virulent uropathogenic *E. coli* that cause acute pyelonephritis are carried for extended periods of time in the fecal/perineal flora of highly UTI prone individuals. Invasion of the urinary tract is hindered by an intact urine flow and a highly efficient innate immune response, which senses the presence of invaders and destroys them, so that the urinary tract can regain its "sterility".

UTIs are also a fascinating model, to study the molecular features of host-pathogen interactions at mucosal surfaces. Unlike other mucosal sites that are populated by a complex bacterial flora, the urinary tract is usually infected by a single strain, making analyses of virulence factors and their contribution to human disease quite feasible. Host response variables may also be assessed and directly correlated to disease severity. This chapter focuses on determinants of bacterial virulence and on the host response, especially dysfunctions that tilt the balance towards inflammation, symptoms and disease.

## 17.2  Pathogenesis of UTI

*E. coli* strains cause 80–90 % of community-acquired acute pyelonephritis episodes and are especially common in children. Uropathogenic *E. coli* (UPEC) strains are defined by chromosomal virulence genes, which may be activated at different times during infection, to facilitate host tissue attack and bacterial survival. For example, bacterial adhesion to the mucosa is a crucial first step in disease induction, and fimbriae increase tissue exposure to secreted bacterial toxins, like hemolysin and LPS, that damage the tissues by interfering with cellular functions or by directly inducing cell death. UPEC strains can be identified, by typing for virulence factors like P fimbriae.

Critical mechanisms involved in antibacterial host defense have also been identified. Resistance to infection relies on innate immune mechanisms and defects in innate immunity drastically increase UTI susceptibility, both in mice carrying single gene deletions and in patients prone to acute pyelonephritis or ABU. In contrast, specific immunity executed by T lymphocytes and/or B lymphocytes is not involved in acute defense. These findings are consistent with clinical observations, showing that susceptibility to acute pyelonephritis is not associated with hypo-gammaglobulinemia or T lymphocyte deficiencies.

## 17.3  Bacterial Virulence Factors

### 17.3.1  Bacterial Adhesion

The ability to adhere to cells in the urinary tract is a critical virulence factor in UTI. UPEC trains express surface fimbriae that recognize specific receptors on host cells. The different fimbrial types are distinguished by their structural features and by the oligosaccharide receptor epitopes that they recognize. Type l fimbriae bind to mannosylated glycoproteins, such as the Tamm-Horsfall glycoprotein, secretory IgA or bladder cell uroplakin. S fimbriae bind sialic acid

epitopes present in sialylated glycoproteins and glycolipids. P-fimbriae recognize $Gal\alpha 1$-4$Gal\beta$-epitopes in the globoseries of glycolipids. The expression of P fimbriae is strongly linked to virulence, being present in about 80 % of strains causing uncomplicated acute pyelonephritis but in < 20 % of strains causing ABU. The frequency of P fimbriated strains is reduced in patients with impaired resistance, due to pregnancy, urodynamic defects or genetic abnormalities (see below).

### 17.3.2  Lipopolysaccharide (LPS)

LPS is the endotoxin of Gram-negative bacteria, containing the membrane anchor Lipid A which is responsible for the toxic effects. Injection of Lipid A mimics many of the symptoms associated with acute pyelonephritis, Gram negative septicemia or other serious Gramnegative infections, including fever, and activates the acute phase response. In addition, LPS carries an invariable oligosaccharide core and a repeating oligosaccharide that determines the O antigen. LPS activates Toll-like receptor 4 (TLR4) signaling, after binding to LBP, CD14 and MD2. This mechanism is essential for innate immune activation during systemic infection but is less important at mucosal surfaces, where the cell-bound CD14 receptor often is missing and MD2 may be weakly expressed.

### 17.3.3  Capsular Polysaccharide

Capsular polysaccharides are formed from highly repetitive, oligosaccharides and different capsular types are defined by variant antigenic oligosaccharide epitopes. The capsules surround bacteria in blood and tissues and contribute to bacterial survival in the presence of a functional host defense, by preventing lytic effects of complement and antibacterial peptides as well as phagocytosis. Mutant bacteria with altered capsule expression show reduced virulence in experimental UTI models.

### 17.3.4   Hemolysin

Hemolysins are cytotoxic, pore-forming proteins that permeate the cell membrane, commonly identified by their ability to lyse erythrocytes. Hemolysin production was observed in *E. coli* causing acute pyelonephritis in the 1940s, and hemolysin occurs more often in *E. coli* causing acute pyelonephritis than in fecal isolates even though the difference in frequency is less pronounced than for e.g. P fimbriae.

### 17.3.5   Metabolic Competition

Most *E. coli* strains causing acute pyelonephritis produce aerobactin, which is an iron binding protein with high affinity, as well as other iron sequestering proteins. By producing such proteins, bacteria compete with the host for nutrients such as iron. Host iron-binding proteins include lactoferrin in mucosal secretions and transferrin in the circulation, and their concentrations increase in response to infection.

### 17.4   The Innate Immune Response

The innate immune response is activated in a pathogen specific way by P fimbriae mediated adhesion to glycolipid receptors, followed by release of ceramide, activation of TLR4 signaling, and downstream transcription factors like IRF3, which trigger cytokine production and neutrophil recruitment, eventually killing the bacteria. Indeed, the presence of a mucosal cytokine response was first observed in UTI. IL-6 is secreted by uroepithelial cells and acts as an endogenous pyrogen, activates hepatocyte CRP production and stimulates mucosal B cells to produce IgA antibodies. The biological activities of IL-6 thus agree with the clinical presentation of acute pyelonephritis and IL-6 is detected in the blood and urine of patients with acute pyelonephritis.

Infected epithelial cells also secrete IL-8, which is chemotactic for neutrophils. IL-8 forms a gradient across the mucosa and neutrophils migrate towards the gradient, and cross the mucosal barrier and into the urine. The concentrations of IL-8 correlate with the number of leucocytes in urine, and the chemotactic activity of urine is inhibited by IL-8 antibodies, showing that IL-8 is an essential chemoattractant in UTI. In addition, infection increases the expression of IL-8 receptors, which are important targets for the IL-8 family of chemokines and support neutrophil migration and activation. These mechanisms explain "pyuria", i. e. the presence of leucocytes in urine, which has been used as a diagnostic tool in UTI since microscopy was first used in the clinic. More recently, a number of additional chemokines have been found in urine, probably reflecting a more complex immune response in the mucosa, especially during long-term bacterial carriage.

These mechanisms also determine the symptoms and signs of acute UTI and several studies have suggested that urine cytokine levels directly reflect the degree of renal involvement.

### 17.5   Host Factors that Predispose to UTI

The frequency of UTI varies with age, gender and socio-economic factors. UTI morbidity is higher in women than in men, except during the first year and at the end of life. Regardless of gender, certain individuals develop severe, acute infections, some recur and a subset develop a chronic state of tissue damage, leading to renal scarring. For decades, scientists have worked to better understand the factors that determine susceptibility and to develop diagnostic tools to discriminate these individuals from others, who only experience sporadic UTI episodes. Recent studies have started to identify genetic risk factors that distinguish pyelonephritis-prone children from those who consistently develop ABU. Even if the information is limited, it is clear that genetic variation affecting innate immunity influences UTI susceptibility.

Resistance to acute pyelonephritis is also influenced by dysfunctional voiding. Patients with reflux are susceptible to infection with bacteria of lower virulence than patients with anatomically normal urinary tracts. It is not clear if the urine flow, *per se*, is sufficient to maintain sterility, aided by antibacterial peptides that directly kill the bacteria.

### 17.5.1  Inflammation and Bacterial Clearance

The inflammatory response is crucial for the antibacterial defense of the urinary tract. Loss of functional Toll-like receptors impairs the innate immune response, and promotes long-term bacterial colonization and ABU. In contrast, defects in the antibacterial effector functions encoded by this pathway, result in severe acute infections, tissue damage and renal scarring.

### 17.5.2  Toll-like Receptor Signaling and ABU

Children with ABU express lower levels of the TLR4 receptor than age matched controls or children with acute pyelonephritis. They carry mutations in the TLR4 promoter that reduce the efficiency of TLR4 expression. Mice lacking the TLR4 gene also develop ABU, with $> 10^5$ CFU/ml of bacteria but no symptoms or tissue damage. Low TLR4 expression and the resulting absence of inflammation can thus be protective rather than destructive.

### 17.5.3  IRF3, CXCR1 and Acute Pelonephritis

The IRF3 transcription factor enhances the transcription of innate immune response genes including cytokines and chemokine receptors. Mice lacking *Irf3* develop severe acute pyelonephritis with urosepsis and renal abscesses. Mice lacking the IL-8 receptor (*mCxcr2*$^{-/-}$) also develop severe acute pyelonephritis with urosepsis and surviving animals have renal abscesses and renal scarring. These examples show that single gene deletions affect crucial functions in innate immune signaling, resulting in increased UTI susceptibility in mice. The pathology of the *mCxcr2*$^{-/-}$ was also due to poor neutrophil recruitment and neutropil activation. Like the *Irf3*$^{-/-}$ mice, the *mCxcr2*$^{-/-}$ mice showed acute renal abscesses and surviving mice developed renal scarring.

We also found *IRF3* and *CXCR1* polymorphisms in pyelonephritis prone patients. Children prone to acute pyelonephritis had reduced CXCR1 expression and heterozygous CXCR1 polymorphisms and low mRNA levels suggested that CXCR1 transcription is impaired in this patient group. We also found promoter polymorphisms that reduced promoter efficiency and IRF3 expression in pyelonephritis prone children.

### 17.5.4  Receptors, Blood Groups and UTI Susceptibility

Uropathogenic *E. coli* colonize mucosal surfaces through adhesion, and thus, resistance to infection can be modified by the repertoire of receptors that the patients express and by soluble, anti-adhesive molecules. Early studies by Stamey et al. proposed that women with recurrent UTI have increased vaginal carriage of uropathogenic bacteria and their uroepithelial cells have since been shown by several groups to have an increased tendency to bind uropathogenic *E. coli*, reflecting an increased density of receptors.

P fimbriated *E. coli* bind to the globoseries of glycolipids, which are antigens in the P blood group system. Through P blood group determination, it is possible to predict the mucosal receptor repertoire and thus, the P blood group can be used as a marker in epidemiologic studies. Individuals of blood group P lack a critical enzyme involved in glycolipid synthesis and lack functional receptors for P fimbriae predicting that they would be more resistant to infection with P fimbriated bacteria. Children of blood group P1 run an increased risk of acute pyelonephritis (about 11 fold) and carry P fimbriated *E. coli* in the intestinal flora, which is a UTI risk factor. Epithelial receptor expression is also influenced by the ABO blood group and individuals expressing globo A are preferentially infected by P fimbriated *E. coli* that recognize a combined Galα1-4Galβ and blood group A epitope.

### 17.5.5 Anti-adhesive Molecules in Urine

Urine contains soluble receptor oligosaccharides or glycoconjugates that bind fimbriae and prevent them from mediating adhesion. The Tamm-Horsfall protein, which is a normal urine constituent, carries terminal mannose residues, that bind type 1 fimbriated *E. coli* and sialic acid residues bind S fimbriae. P fimbriae, which show the strongest disease association, have no known soluble inhibitors in urine.

### 17.5.6 Specific Immunity and Resistance to UTI

Acute pyelonephritis gives rise to a specific immune response after three to seven days. The mucosal immune response has a relatively short memory, however, and thus most patients have low levels of specific antibodies in urine, when they develop recurrent infections. In experimental studies, mice lacking a specific immune system have surprisingly not shown a drastic increase in UTI susceptibility, and patients with different specific immuno-deficiencies do not primarily develop UTI, suggesting that the specific immune system is not essential for the defense against acute infection.

The main evidence that specific immunity may protect the urinary tract against infection has been generated through vaccination studies. Early studies showed that vaccination with formalin treated whole bacteria markedly increases specific antibody levels. Further studies have used purified antigens (LPS, kapsel, fimbriae, outer membrane proteins) and show that hyperimmunization can protect against infection, provided that the vaccine antigen is expressed by the bacterial strain used for infection and infectious challenge is given while the immune response remains active. The antigenic variation in uropathogenic *E. coli* is a problem for the implementation of vaccination in UTI, however, modern vaccine development gives rise to new optimism. The genetic studies also suggest that genetically susceptible individuals might be targeted for vaccination.

## 17.6 Summary

The symptoms of acute pyelonephritis are caused by the innate immune response and urinary tract inflammation decreases renal tubular function and may give rise to renal scarring, especially in children. Antibacterial defence relies on innate immunity and TLR4 signaling initiates this response. Reduced TLR4 function is associated with ABU, while reduced IRF3 or CXCR1 expression is associated with severe acute pyelonephritis and an increased risk of renal scarring.

Most cases of acute pyelonephritis are caused by *E. coli* strains expressing virulence factors that contribute to the establishment of bacteriuria. P fimbriae remain the single virulence factor that best describes the virulence of the infecting strain, but more extensive gene expression methods may also be used to get a more complete picture of the virulence repertoire.

The characterization of bacterial virulence factors and inflammatory profiles may provide new tools for the diagnosis of UTI. Virulent bacteria can, for example, be identified by the expression of P fimbriae. By combining information on bacterial virulence and the host response, it should be possible to start "individualizing" diagnosis and therapy in the large and diverse group of patients with UTI.

## References

1. Nielubowicz GR, Mobley HL (2010) *Host-pathogen interactions in urinary tract infection.* Nat Rev Urol 7(8):430–441
2. Ragnarsdottir B et al (2011) *Genetics of innate immunity and UTI susceptibility.* Nat Rev Urol 8(8):449–468
3. Bhat RG, Katy TA, Place FC (2011) *Pediatric urinary tract infections.* Emerg Med Clin North Am 29(3):637–653
4. Montini G, Tullus K, Hewitt I (2011) *Febrile urinary tract infections in children.* N Engl J Med 365(3):239–250

# Preventing Urinary Tract Infections in Early Childhood

# 18

Gabrielle J. Williams, Jonathan C. Craig
and Jonathan R. Carapetis

**Abstract**

Urinary tract infection (UTI) is common in children, causes them considerable discomfort, as well as distress to parents and has a tendency to recur. Approximately 20 % of those children who experience one infection will have a repeat episode. Since 1975, 11 trials of long-term antibiotics compared with placebo or no treatment in 1,550 children have been published. Results have been heterogeneous, but the largest trial demonstrated a small reduction (6 % absolute risk reduction, risk ratio 0.65) in the risk of repeat symptomatic UTI over 12 months of treatment. This effect was consistent across sub groups of children based upon age, gender, vesicoureteric reflux status and number of prior infections. Trials involving re-implantation surgery (and antibiotics compared with antibiotics alone) for the sub-group of children with vesicoureteric reflux have not shown a reduction in repeat UTI, with the possible exception of a very small benefit for febrile UTI. Systematic reviews have shown that circumcision reduces the risk of repeat infection but 111 circumcisions would need to be performed to prevent one UTI in unpredisposed boys. Given the need for anaesthesia and the risk of surgical complication, net clinical benefit is probably restricted to those who are predisposed (such as those with recurrent infection). Many small trials in complementary therapies have been published and many suggest some benefit, however inclusion of children is limited. Only three trials involving 394 children for cranberry products, two trials with a total of 252 children for probiotics and one trial with 24 children for vitamin A are published. Estimates of efficacy vary

J. R. Carapetis (✉)
Telethon Institute for Child Health Research, University
of Western Australia, Perth, Australia
email: jcarapetis@ichr.uwa.edu.au

G. J. Williams · J. C. Craig
School of Public Health, University of Sydney, Centre
for Kidney Research, Children's Hospital at Westmead,
Sydney, Australia
e-mail: gabrielle.williams1@health.nsw.gov.au

J. C. Craig
e-mail: Jonathan.craig@sydney.edu.au

N. Curtis et al. (eds.), *Hot Topics in Infection and Immunity in Children IX,*
Advances in Experimental Medicine and Biology 764, DOI 10.1007/978-1-4614-4726-9_18,
© Springer Science+Business Media New York 2013

widely and imprecision is evident. Multiple interventions to prevent UTI in children exist. Of those, long-term low dose antibiotics has the strongest evidence base, but the benefit is small. Circumcision in boys reduces the risk substantially, but should be restricted to those at risk. There is little evidence of benefit of re-implantation alone, and the benefit of this procedure over antibiotics alone is very small. Cranberry concentrate is probably effective.

## 18.1 Background

### 18.1.1 Frequency

Urinary tract infection (UTI) is a very common illness in children, affecting 2 % of boys and 8 % of girls by the age of 7 years [1]. It is also the most common serious bacterial infection in children with fever who present for assessment [2, 3] and causes an unpleasant acute illness with manifestations that include fever, lethargy, vomiting and cystitis symptoms.

### 18.1.2 Recurrence

Good evidence to quantify the risk of recurrence and identify factors that may predispose to repeat infections is quite scarce. Studies that follow children with UTI over time are required, and there have been only a few. Two studies of this design [4, 5] have demonstrated that about 12 % of children with first UTI experience a recurrence within one year. The placebo arm of a large, blinded trial [6] showed a recurrence rate of 19 % in 12 months, but eligibility criteria were not limited to the first infection, so this may be an over-estimate of the true risk.

### 18.1.3 Risk Factors for Recurrence

Some children are more at risk for future UTIs than others. Risk factors include an age less than six months at first UTI, grade III–IV vesicoureteric reflux and white race. Early observations

that UTI and vesicoureteric reflux were associated with renal damage [7–9] led to the standard practice of performing voiding cystourethrography to identify reflux in children with a history of UTI [10, 11].

## 18.2 Antibiotic Treatment

Children with reflux were routinely given daily low-dose antibiotics for many years [12] with the aim of preventing further UTI and renal damage. Until 1997, only four trials [13–16] with 171 children and conflicting findings provided the evidence base for this treatment (Table 18.1). Since then seven trials [6, 17–22] have been published with broader criteria, including children with reflux and designs which are less prone to bias. Six of the seven trials showed a reduced risk of repeat symptomatic UTI with prophylactic antibiotics but the magnitude of the effect was small and in most studies the difference did not reach statistical significance (Fig. 18.1). In the largest ($n = 576$) and importantly the only blinded study [6], the benefit was statistically significant, and showed a 6 % absolute risk reduction for repeat symptomatic UTI in children taking trimethoprim sulphamethoxazole for 12 months. Benefits did not vary according to baseline characteristics such as age, gender and reflux status in *a priori* sub-group analysis. Five trials [6, 18, 20–22] also reported rates of bacterial resistance to the prophylactic drug and all showed substantial increases. Overall, there appears to be a small benefit from prophylactic antibiotics (6 % absolute risk reduction, or an overall number needed to treat of 16 over 1 year) but this must be weighed against the proven risk of increased bacterial resistance to antibiotic and with consideration for suggested but uncertain effects such as susceptibility to asthma and inflammatory bowel disease [23, 24].

Five trials have compared one antibiotic with another [25–29], two of which compared cotrimoxazole with nitrofurantoin [26, 29]. These two trials were small (N of 120 and 132) but both demonstrated statistically significant superiority of nitrofurantoin (risk ratios (RR) 0.57 (95 % CI

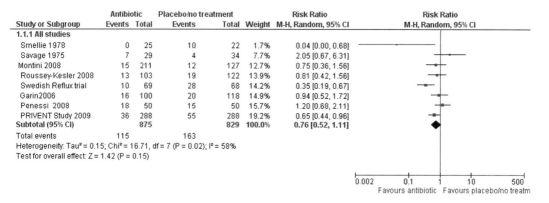

**Fig. 18.1** Randomised controlled trials of antibiotic compared with placebo/no treatment for the prevention of repeat symptomatic UTI in children

0.35–0.92) and 0.32 (95 % CI 0.19–0.56)). However, one study [26] reported 30 % of patients withdrawing from the study due to side effects of nitrofurantoin suggesting the acceptability of this treatment may be poor.

## 18.3   Surgical Treatment

Surgical correction of the physical abnormality of reflux can involve open surgery to reimplant the ureters or endoscopic injection of agents at the vesicoureteric junction. A recent systematic review of these treatments [30] showed no convincing evidence of a reduced risk of repeat symptomatic UTI at 1–2, 4–5 or 10 years after surgery (and antibiotics) compared with antibiotics alone (1–2 years RR 0.88 (95 % CI 0.26–3.01), 5–10 years RR 0.79 (95 % CI 0.49–1.26), 10 years RR 1.06 (95 % CI 0.78–1.44)). These trials and the numerous case series found in the literature usually demonstrate a very high rate of surgical correction of reflux in treated children but without matching reduction in risk of UTI, suggesting that reflux only has a modest attributable risk for further UTI.

### 18.3.1   Circumcision

A systematic review of trials and observational studies of circumcision [31] to prevent UTI showed that circumcision reduces the risk of UTI but that 111 circumcisions would need to be performed to prevent one UTI in normal boys with a baseline risk of 1 %. Major complications occur in approximately 2 %. Circumcision could be considered in boys predisposed to UTI to achieve net clinical benefit, such as those with recurrent infection and/or those with high grade reflux. In this scenario the number needed to treat would be about 5–10.

## 18.4   Complementary Treatments

Systematic reviews and trials of complementary treatments for prevention of UTI have been conducted but most do not include a substantial number of children. A systematic review of cranberry products [32] suggests some benefit in preventing recurrent UTI in women but insufficient data were available for conclusions about efficacy in children. Since that review three trials in children have been published [33–35], two only in abstract form [34, 35], and all demonstrated apparent benefit in cranberry product use but only one reached statistical significance [33]. The smallest study ($n=51$) compared cranberry with antibiotic treatment trimethoprim and suggested that cranberry may be more effective (relative risk of 0.65 (95 % CI 0.34–1.25) while the larger, blinded study ($n=263$) gave a less favourable relative risk of 0.74 (95 % CI 0.44–1.25). Neither

**Table 18.1** Randomised controlled trials for interventions to prevent urinary tract infection in children

| First author | Year | Number | Participants | | | Intervention | | | Comparator | Blinded | Event rate (%) | Outcomes | | | Results |
|---|---|---|---|---|---|---|---|---|---|---|---|---|---|---|---|
| | | | Age (years) | N VUR | VUR grades | NF | TMP-SMZ | Other | | | | Symptomatic UTI | Positive culture | Abn DMSA | Risk ratio (95 % CI) symptomatic UTI, (positive culture) |
| *Low dose, antibiotic therapy compared with placebo/no treatment* | | | | | | | | | | | | | | | |
| Savage | 1975 | 63 | 5–7.8 | 16 | I–III | • | • | | Placebo | x | ? | 7/63 (11.1) | • | • | | 2.05 (0.67, 6.31) 0.93 (0.74, 1.17) |
| Stansfield | 1975 | 45 | 0.5–14 | 13 | NS | | • | | ✓ | ✓ | | 12/45 (26.7) | | • | | 0.05 (0.00, 0.72) |
| Lohr | 1977 | 18# | 3–13 | 0 | NA | • | | | ✓ | ✓ | | 15/18 (83.3) | | • | | 0.13 (0.04, 0.50) |
| Smellie | 1978 | 45 | 2–12 | 0 | NA | • | • | | x | x | | 11/47 (23.4) | • | • | | 0.04 (0.0, 0.68) 0.04 (0.0, 0.52) |
| Reddy | 1997 | 43 | NS | 43 | ? | ? | ? | ? | x | ? | | 6/29 (20.7) | • | • | • | 0.25 (0.03, 1.85) |
| Garin | 2006 | 218 | 0.25–17 | 113 | I–III | • | • | | x | ? | | 36/218 (16.5) | • | • | • | 0.94 (0.52, 1.72) 0.74 (0.43, 1.28) |
| Montini | 2008 | 338 | 0.08–8.4 | 128 | I–III | • | • | • | x | x | | 27/338 (3.6) | • | • | • | 0.75 (0.36, 1.56) 0.50 (0.29, 0.86) |
| Roussey-Kesler | 2008 | 225 | 0.08–3 | 225 | I–III | | • | | x | ? | | 32/225 (14.2) | • | • | • | 0.81 (0.42, 1.56) 0.67 (0.40, 1.11) |
| Pennesi | 2008 | 100 | 0–2.5 | 100 | II–IV | | • | | x | x | | 33/100 (33.3) | • | • | • | 1.20 (0.68, 2.11) |
| PRIVENT | 2009 | 576 | 0–18 | 243 | I–V | | • | | ✓ | ✓ | | 91/576 (15.8) | • | • | • | 0.65 (0.44, 0.96) |
| Swedish Reflux Trial | 2010 | 203 | 1–2 | 203 | III–IV | • | • | • | x | x | | 49/203 (24.1) | • | • | • | 0.39 (0.21 0.76) |
| *Low dose, antibiotic A vs low dose antibiotic B* | | | | | | | | | | | | | | | |
| Carlsen | 1985 | 35# | 1–13 | 17 | I–III | • | | Pivmecilliam | Alternative antibiotic | ? | | 10/33 (30.3) | | • | | 0.68 (0.28, 1.65) |
| Brentrup | 1990 | 130 | 1–14 | 51 | NS | • | • | | Alternative antibiotic | ✓ | | 31/120 (25.8) | | • | | 0.32 (0.19, 0.56) |
| Lettgen | 2002 | 57 | 1–11 | NS | NS | • | | Cefixim | Alternative antibiotic | ? | | 5/57 (8.8) | • | | | 1.35 (0.24, 7.48) |

**Table 18.1** (continued)

| First author | Year | Number | Participants | | | Intervention | Comparator | Blinded | Event rate (%) | Outcomes | Results |
|---|---|---|---|---|---|---|---|---|---|---|---|
| Belet | 2004 | 80 | 0.5–15 | 0 | NA | • Cephadroxil/cefprozil | Alternative antibiotic | x | 12/80 (15.0) | • | 0.69 (0.20, 2.39) 1.35 (0.71, 2.56) TMP_SMX vs cefprozil; 1.79 (0.33, 9.70) 5.95 (1.46, 24.21) TMP_SMX vs cephadroxil; 0.39 (0.09 to 1.71) 0.23 (0.06, 0.92) cephadroxil vs cefprozil |
| Falaflaki | 2007 | 132 | 0.25–12 | 57 | I–IV | • • | Alternative antibiotic | ? | 47/132 (35.6) | • | 0.57 (0.35 to 0.92) |
| *Alternative dose* | | | | | | | | | | | |
| Baciulus | 2003 | 33 | 0–16 | 8 | ? | Cefdroxil | Every vs alternate night dose | x | 7/33 (21.2) | • | 1.11 (0.29, 4.21) |
| *Complementary therapies* | | | | | | | | | | | |
| Salo^ | 2010 | 263 | NS | 35 | ? | Cranberry juice | Placebo juice | ✓ | 48/263 (18.3) | • | 0.74 (0.44, 1.25) |
| Uberos^ | 2010 | 51 | NS | NS | ? | Cranberry syrup | Trim-ethoprim | ✓ | 23/51 (45.1) | • | 0.65 (0.34, 1.25) (cranberry vs trim) |
| Ferrara | 2009 | 80 | 3–14 | 0 | NA | Cranberry or probiotic | No treatment | x | 34/80 (42.5) | • | 0.28 (0.12, 0.64) cranberry vs no treatment; 0.63 (0.38, 1.07) probiotic vs no treatment; 0.44 (0.18, 1.09) cranberry vs probiotic |
| Lee^ | 2010 | 132 | ~0.08–0.8 | 132 | ? | Probiotic | TMP-SMX | ? | 47/128 (36.7) | • | 0.81 (0.51, 1.28) |
| Lee | 2007 | 120 | 1.08–3 | 120 | I–V | Probiotic | TMP-SMX | ? | 24/120 (20.0) | • | 0.85 (0.41 to 1.74) |
| Yilmaz | 2007 | 24 | >1–12 | 0 | NA | Vitamin A (single dose) | Placebo | ✓ | ? | • | RR not calculable, *Vit A 0.29/mo vs 0.44/mo 6–12 mo* |

# Cross over trial

^ Published only in abstract form

study demonstrated statistical significance. This evidence suggests cranberry products may reduce the risk of repeat UTI in children but there is considerable uncertainty. None of these trials reported adverse events, however trials in adults suggest most adverse events are minor gastrointestinal issues.

A systematic review of methenamine hippurate [36] for preventing UTI concluded that the intervention may be effective in patients without renal tract abnormalities but no children were included in the trials. No new trials have been published to help resolve the uncertainty. A blinded, randomised placebo controlled trial of herbal products *Tropaeoli majoris* and *Armoraciae rusticanae* [37] in adults showed no difference in average number of recurrences between the two groups using intention to treat analysis. Two randomised trials of probiotics compared with cotrimoxazole treatment to prevent UTI in children with vesicoureteric reflux have been published [38, 39]. Neither study demonstrated a statistically significant difference but the point estimates favoured probiotics 0.85 (95 % CI 0.41–1.74) and 0.81 (95 % CI 0.51–1.28) [36, 37]. There remains uncertainty and imprecision about the efficacy of this intervention to prevent recurrent UTI in children.

A meta-analysis of trials [40] of the immune active agent, Uro-Vaxom showed it to be an effective treatment for preventing recurring UTIs, however none of the five included trials were large or optimally designed and each included only adults. Several treatment studies in children have been published [41, 42] but without randomisation nor a comparator these are not a firm basis for decision making.

A randomised, placebo controlled trial of vitamin A to prevent recurrent UTI in children [43] showed a reduced rate of UTI in the follow-up period, but only 24 children participated so the estimate of efficacy is imprecise and there are also concerns over selection bias.

## 18.5 Conclusion

Good quality evidence demonstrates that low dose antibiotics reduce the risk of repeat UTI in children by approximately 6 %, and is consistent across all groups of children. Clinicians now have a clear estimate of risk reduction along with details on adverse effects with which to discuss treatment options with parents. Given the relatively small absolute benefit, clinicians and families may opt for prophylactic antibiotics when the risk of recurrence is relatively high (e.g. in those with recurrent infection) or when the potential seriousness of an additional event is very significant (e.g. in a very young infant). Circumcision reduces the risk of repeat UTI but is best limited to those at higher risk of recurrence. Complementary therapies have been explored using randomised controlled trials and usually found to be effective but study design is generally poor and studies are small, leading to potential bias and imprecision. However, most of these interventions are usually free of adverse events and parents may elect to try them. They should be reminded to be alert to further infections and seek treatment when appropriate.

## References

1. Hellström A, Hanson E, Hansson S, Hjälmås K, Jodal U (1991) Association between urinary symptoms at 7 years old and previous urinary tract infection. Arch Dis Child 66:232–234
2. Shaikh N, Morone NE (2008) Prevalence of urinary tract infection in childhood: a meta-analysis. Pediatr Infect Dis 27:302–308
3. Craig JC, Williams GJ, Jones M, Codarini M, Macaskill P, Hayen A, Irwig L, Fitzgerald DA, Isaacs D, McCaskill M (2010) The accuracy of clinical symptoms and signs for the diagnosis of serious bacterial infection in young febrile children: a prospective cohort study of 15,781 febrile illnesses. British Medical Journal 340:c1594
4. Panaretto K, Craig JC, Knight JF et al (1999) Risk factors for recurrent urinary tract infection in preschool children. J Paediatr Child Health 35:454–459
5. Conway PH, Cnaan A, Zaoutis T et al (2007) Recurrent urinary tract infections in children: risk factors and association with prophylactic antimicrobials. JAMA 298:179–186

6. Craig JC, Simpson JM, Williams GJ, Lowe A, Reynolds GJ, McTaggart SJ, Hodson EM, Carapetis JR, Cranswick NE, Smith G, Irwig LM, Caldwell PHY, Hamilton S, Roy LP, PRIVENT Investigators (2009) Antibiotic prophylaxis and recurrent urinary tract infection in children. N Engl J Med 361:1748–1759

7. Bailey RR (1973) The relationship of vesicoureteric reflux to urinary tract infection and chronic pyelonephritis-reflux nephropathy. Clin Nephrol 1:132–141

8. Arant BS Jr (1991) Vesicoureteric reflux and renal injury. Am J Kidney Dis 17:491–511

9. Disney APS (1991) Reflux nephropathy in Australia and New Zealand: prevalence, incidence and management 1975–1989. In: Bailey RR (ed) Proceedings of the Second CJ Hodson Symposium on Reflux Nephropathy. Design Printing Services, Christchurch, p 53–56

10. American Academy of Pediatrics (1999) Committee on Quality Improvement, Subcommittee on Urinary Tract Infection. Practice parameter: the diagnosis, treatment, and evaluation of the initial urinary tract infection in febrile infants and young children. Pediatrics 103:843–852 (Errata, Pediatrics 1999;103:1052, 1999;104:118, 2000;105:141)

11. Guidelines for the management of acute urinary tract infection in childhood (1991) Report of a working group of the Research Unit, Royal College of Physicians. J R Coll Physicians Lond 25:36–42

12. Elder JS, Peters CA, Arant BS Jr et al (1997) Pediatric Vesicoureteral Reflux Guidelines Panel summary report on the management of primary vesicoureteral reflux in children. J Urol 157:1846–1851

13. Savage DC, Howie G, Adler K, Wilson MI (1975) Controlled trial of therapy in covert bacteriuria of childhood. Lancet 1:358–361

14. Stansfeld JM (1975) Duration of treatment for urinary tract infections in children. Br Med J 3:65–66

15. Smellie JM, Katz G, Gruneberg RN (1978) Controlled trial of prophylactic treatment in childhood urinary-tract infection. Lancet 2:175–178

16. Lohr JA, Nunley DH, Howards SS, Ford RF (1977) Prevention of recurrent urinary tract infections in girls. Pediatrics 59(4):562–565

17. Reddy PP, Evans MT, Hughes PA et al (1997) Antimicrobial prophylaxis in children with vesicoureteral reflux: a randomised prospective study of continuous therapy vs intermittent therapy vs surveillance. Pediatrics 100(Suppl):555–556

18. Montini G, Rigon L, Zucchetta P et al (2008) Prophylaxis after first febrile urinary tract infection in children? A multicenter, randomised controlled non-inferiority trial. Pediatrics 122:1064–1071

19. Garin EH, Olavarria F, Garcia NV, Valenciano B, Campos A, Young L (2006) Clinical significance of primary vesicoureteral reflux and urinary antibiotic prophylaxis after acute pyelonephritis: a multicenter, randomized, controlled study. Pediatrics 117:626–632

20. Roussey-Kesler G, Gadjos V, Idres N et al (2008) Antibiotic prophylaxis for the prevention of recurrent urinary tract infection in children with low grade vesicoureteral reflux: results from a prospective randomized study. J Urol 179:674–679

21. Pennesi M, Travan L, Peratoner L et al (2008) Is antibiotic prophylaxis in children with vesicoureteral reflux effective in preventing pyelonephritis and renal scars? A randomized, controlled trial. Pediatrics 121:e1489–1494

22. Brandstrom P, Esbjorner E, Herthelius M, Swerersson S, Jodal U, Hansson S (2010) The Swedish reflux trial in children: III. Urinary tract infection pattern. J Urol 184(4):286–291

23. Hviid A, Svanstrom H, Frisch M (2011) Antibiotic use and inflammatory bowel diseases in childhood. Gut 60:49–54

24. Murk W, Risnes KR, Bracken MB (2011) Prenatal or early-life exposure to antibiotics and risk of childhood asthma: a systematic review. Pediatrics 127(6):1125–1138

25. Carlsen NLT, Hesselbjerg U, Glenting P (1985) Comparison of long-term, low dose pivmecillinam and nitrofurantoin in the control of recurrent urinary tract infection in children. J Antimicrob Chemother 19:509–517

26. Brendstrup L, Hjelt K, Petersen KE, Petersen S, Andersen EA, Daugbjerg PS, Stagegaard BR Nielsen OH, Vejlsgaard R, Schou G (1990) Nitrofurantoin versus trimethoprim prophylaxis in recurrent urinary tract infection in children. Acta Paediatr Scand 79:1225–1234

27. Lettgen B, Troster K (2002) Prophylaxis of recurrent urinary tract infections in chdilren, results of an open, controlled randomised study about the efficacy and tolerance of cefixime compared to nitrofurantoin. Klin Padiatria 214(6):353–358

28. Belet N, Islek I, Belet U, Sunter AT, Kucukoduk S (2004) Comparison of trimethoprim-sulfamethoxazole, cephadroxil and cefprozil as prophylaxis for recurrent urinary tract infections in children. J Chemother 16(1):77–81

29. Falakaflaki B, Fallah R, Jamshidi MR, Moezi F, Torabi Z (2007) Comparison of nitrofurantoin and trimethoprim-sulphamethoxazole for long-term prophylaxis in chdilren with recurrent urinary tract infections. Int J Pharmacol 3(2):179–182

30. Nagler EVT, Williams G, Hodson EM, Craig JC (2011) Interventions for primary vesicoureteric reflux. Cochrane Database Syst Rev 6:CD001532. doi: 10.1002/14651858.CD001532.pub4

31. Singh-Grewal D, Macdessi J, Craig JC (2005) Circumcision for the prevention of urinary tract infection in boys: a systematic review of randomised trials and observational studies. Arch Dis Child 90:853–858

32. Jepson RG, Mihaljevic L, Craig J (2007) Cranberries for preventing urinary tract infections. Cochrane Database Syst Rev 2007:CD001321. doi: 10.1002/14651858.CD001321.pub4

33. Ferrara P, Romaniello L, Vitelli O, Gatto A, Serva M, Cataldi L (2009) Cranberry juice for the prevention

of recurrent urinary tract infections: a randomized controlled trial in children. Scand J Urol Nephrol 43(5):369–372

34. Salo J, Kontiokari T, Helminen M, Korppi M, Nieminen T, Pokka T et al (2010) Randomized trial of cranberry juice for the prevention of recurrences of urinary tract infections in children. Clin Microbiol Infect 16(Suppl 2):386

35. Uberos J, Rodriguez-Belmonte R, Fernandez-Puentes V, Narbona-Lopez E, Molina-Carballo A, munozhoyos A (2010) Cranberry sydrup vs trimethoprim in the prophylaxis of recurrent urinary infection: a double blind randomized clinical trial. Acta Paediatr 99(Suppl 462):48

36. Lee BB, Simpson JM, Craig JC, Bhuta T (2007) Methenamine hippurate for preventing urinary tract infections. Cochrane Database Syst Rev 4:CD003265. doi:10.1002/14651858.CD003265.pub2

37. Albrecht U, Goos KH (2007) A randomised, double-blind, placebo-controlled trial of a herbal medicinal product containing Tropaeoli majoris herba (Nasturtium) and Armoraciae rusticanae radix (Horseradish) for the prophylactic treatment of patients with chronically recurrent lower urinary tract infections. Curr Med Res Opin 23:22

38. Lee SJ, Shim YH, Cho SJ, Lee JW (2007) Probiotics prophylaxis in children with persistent primary vesicoureteral reflux. Pediatr Nephrol 22:1315–1320

39. Lee SJ, Lee M (2010) Probiotic versus antibiotic prophylaxis in infants with primary vesicoureteral reflux. Pediatr Nephrol 25(9):72

40. Bauer HW, Rahlfs VW, Lauener PA, Blessmann GS (2002) Prevention of recurrent urinary tract infections with immuno-active E. coli fractions: a meta-analysis of five placebo-controlled double-blind studies. Int J Antimicrob Agents 19:451–456

41. Golabek B, Nowakowska K, Slowik M, Paruskiewicz G (2002) Usefulness of Uro-Vaxom in complex treatment of recurrent urinary tract infections in girls. Polksi Merkuriusz 12(70):269–272

42. Goszczyk A, Bochniewska V, Jung A (2000) Clinical assessment of Uro_vaxom in the treatment and prophylaxis of recurrent urinary tract infection in chdilren: preliminary results. Pol merkur Lekarski 8(46):242–230

43. Yilmaz A, Bahat E, Yilmaz GG, Hasanoglu A, Akman S, Guven AG (2007) Adjuvant effect of vitamin A on recurrent lower urinary tract infections. Pediatr Int 49:310–313

# How to Get and Get Rid of Gonorrhea

**19**

Jennifer C. Smith, Tim Mailman and Noni E. MacDonald

**Abstract**

Gonorrhea remains as a significant public health concern with an estimated 88 million new cases per year globally. Gonorrhea is a disease of sexual networks and is most prevalent in youth, men who have sex with men, and the socioeconomically disadvantaged. Highly adaptive through years of co-evolution, gonorrhea has developed multiple ways of evading the human immune system. Although new molecular-based strategies have opened avenues for less invasive testing, education and accessibility issues persist. Novel strategies, including use of the internet and social media, are required to better target high risk groups for education, testing, and treatment. Increasing the availability of youth-friendly health services will also help foster earlier gonorrhea diagnosis and management. The inappropriate and overuse of antibiotics and propensity of gonococcus for mutation has led to growing microbe resistance. Treatment failures now include both oral and intravenous formulations of third generation cephalosporins; key front line recommended gonococcal treatment in many countries. With treatment options dwindling, the need for better preventative strategies has never been more important. This overview highlights some of the major aspects of gonococcal infection, including the epidemiology of the disease with an emphasis on sexual networks, new diagnostic techniques, treatment options in the face of evolving gonococcal resistance, and notes potential new preventative strategies.

N. E. MacDonald (✉)
Dalhousie University, Canadian Center for Vaccinology,
IWK Health Center, Halifax, Canada
e-mail: noni.macdonald@dal.ca

J. C. Smith
Pediatric Infectious Diseases, Dalhousie University,
Canadian Center for Vaccinology, IWK Health Center,
Halifax, Canada
e-mail: jaurora777@yahoo.ca

T. Mailman
Dalhousie University, IWK Health Center, Halifax,
Canada
e-mail: tim.mailman@iwk.nshealth.ca

N. Curtis et al. (eds.), *Hot Topics in Infection and Immunity in Children IX,*
Advances in Experimental Medicine and Biology 764, DOI 10.1007/978-1-4614-4726-9_19,
© Springer Science+Business Media New York 2013

## 19.1 Introduction

*Neisseria gonorrhoeae* has been a sexually transmitted scourge of humans for eons. The relationship has been so long and so enduring that a piece of human DNA became entwined with the microbe's genome many millennia ago [1]. Unfortunately, gonorrhea continues to cause significant morbidity and mortality worldwide. A disease largely of the disadvantaged, gonorrhea disproportionately affects inhabitants of developing countries and those of lower social economic status, globally. Worldwide, the gonorrhea epidemic is primarily focused among young people due to a variety of social and behavioural factors shared by this group.

While little has changed in the clinical presentation of gonorrhea, our understanding of organism pathogenesis and the availability of diagnostic tools has grown. Additionally, multidrug resistant gonococcal strains have now emerged. Without alternative antibiotics or a candidate vaccine in the pipeline, gonorrhea may soon become untreatable [2, 3]. With dwindling therapeutic options, new prevention strategies based on an understanding of sexual networks and utilizing youth-friendly information modalities, need to be pursued.

Rather than an exhaustive review, the purpose of this article is to supply a snapshot of gonorrhea in the world today with an emphasis on young people living in developed countries. Key topics explored include current epidemiology, risk factors for transmission, benefits and limitations of new diagnostic modalities, treatment options in the context of growing antimicrobial resistance, and novel preventative strategies.

## 19.2 Gonococcal Pathogenesis

Through the co-evolution of *N. gonorrhea* and humans, gonococcus has become highly adept at evading host immune defenses. These pathogenic mechanisms have important clinical correlates in terms of infectivity, disease manifestations, development of antibiotic resistance, and the potential for vaccine development.

A number of characteristics are known to contribute to the virulence of *N. gonorrhoeae* including pili, porin protein, opacity proteins, lipooligosaccharides, reduction modifiable proteins, IgA proteases, and iron- or oxygen-repressible proteins. A detailed discussion of these gene products and molecular systems are available elsewhere [4, 5]. While some gonococcal pathogenic mechanisms are well described, others remain unclear. For example, the recently discovered horizontal gene transfer between humans and *N. gonorrhea* is currently of unknown significance, beyond its affirmation of the extended length of time the two have co-existed [1].

Acquired through sexual contact, *N. gonorrhea* establishes infection of the urogenital tract by interacting with non-ciliated epithelial cells (Fig. 19.1). The molecular mechanisms responsible for gonococcal infection appear to differ between males and females. For example, invasion of female but not male urogenital cells is facilitated by cervix-specific CR3 complement receptors. Male urethral epithelial cells do not express these receptors. CR3 mediated invasion, in turn, is associated with asymptomatic cervicitis, perhaps explaining the higher incidence of asymptomatic disease in women than men [6]. The formation of biofilms on cervical cells may also play a role in the pathogenesis of *N. gonorrhea* in women [7].

After adhering to epithelial cell surfaces, gonococcus is engulfed into cells where it replicates intracellularly within phagocytes. The microbe later exits from the basal surface via exocytosis and enters the submucosal space. Mucosal epithelial damage and submucosal invasion lead to the influx of neutrophils with the resultant production of purulent material (e.g., "drip") into the lumen of infected tissues [4]. Both organism-specific factors, including peptidoglycan and lipopolysaccharide, alongside host-related responses, such as phagocyte production of tumor necrosis factor, contribute to tissue damage [5].

While the majority of invading gonococci are ingested and destroyed by neutrophils, a minority escape innate and early adaptive immune responses, thereby, allowing for persistent infection. Infrequently, disseminated infection occurs [8]. Normally, human serum is capable of killing

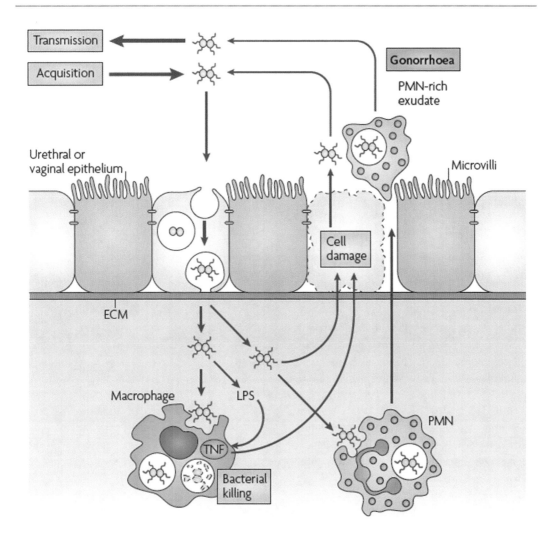

**Fig. 19.1** Pathogenesis of gonococcal infection [5]. (With kind permission from Springer Science+Busness Media B.V. and M Virji)

circulating and mucosal gonococci via the activation of complement. However, some gonococcal strains are serum resistant, inducing defective terminal complement deposition. For example, gonococcus is able to bind C4BP, a key inhibitor of the classical pathway, thereby eluding complement-mediated lysis [9].

Recurrent gonococcal infection is the rule rather than exception. Although natural infection prompts antibody production against gonococcal cell wall components, the impact of the humoral immune response tends to be minimal. Antigenic and phase variation of outer gonococcal components, alongside mimicry of host glycolipids, fur-

ther limits an effective immune response [4, 5, 10] and is a major reason why vaccines developed using traditional techniques have failed. Additionally, interspecies and as well as interstrain genetic exchange within the *Neisseria* genus continually introduces novel antigenic components [4] including antibiotic resistance genes.

## 19.3 Epidemiology

Gonorrhea remains a significant public health concern worldwide. An estimated 88 million people are newly infected on a yearly basis with

**Table 19.1** Global gonorrhea epidemiology: estimated new cases of adult gonorrhea infection (in millions) in 1995 and 1999. (Adapted from [27])

| Region | 1995 (in millions) | 1999 (in millions) |
|---|---|---|
| North America | 1.75 | 1.56 |
| Western Europe | 1.23 | 1.11 |
| North Africa and Middle East | 1.54 | 1.47 |
| Eastern Europe and Central Asia | 2.32 | 3.31 |
| Sub Saharan Africa | 15.67 | 17.03 |
| South and South East Asia | 29.11 | 27.2 |
| East Asia and Pacific | 3.27 | 3.27 |
| Australia and New Zealand | 0.13 | 0.12 |
| Latin America and Caribbean | 7.12 | 7.27 |
| Total | 62.15 | 62.35 |

the majority of cases occurring in the developing world [11]. The reported rate in the developing world is about 20 times the rate in industrialized countries (Table 19.1). Gonorrhea is the second most commonly reported bacterial sexually transmitted infection in many industrialized countries including the United States (US), Canada, and the United Kingdom (UK) [12–14]. While prevalence and incidence statistics are available for many countries, the current methods of data collection have limitations and likely substantially underestimate infection rates [15–17].

In general, surveillance data tends to be more robust from countries with national reporting systems such as those found in many industrialized countries including the US, Canada, the UK, Australia, and the Netherlands [18]. For example, the Gonococcal Isolate Surveillance Project (GISP) serves as a sentinel surveillance system in the US, monitoring gonococcal antimicrobial susceptibilities though ongoing testing of a select segment of the US population. Additionally, a number of European countries share a surveillance program called the European Gonococcal Antimicrobial Surveillance Programme (EURO-GASP), part of a broader European STI surveillance network known as the European Surveillance of Sexually Transmitted Infections (ESSTI). A number of these national and multinational systems collect detailed epidemiological and behavioural information in addition to antimicrobial resistance data [19], aiding in informing preventative

**Table 19.2** Global gonorrhea epidemiology incidence per 100,000 population in selected countries over time. (Adapted from [12, 14, 107, 140–142])

| Country | 1980 | 1995 | 1999 | 2008 |
|---|---|---|---|---|
| Austria | 92.62 | 11.29 | 5.44 | 9.85 |
| Azerbaijan | 32.64 | 26.2 | 12.08 | 13.38 |
| Belarus | 116.94 | 165.18 | 97.7 | 56.5 (2007) |
| Canada | 223 | 19 | 14.9 (1997) | 33.1 (2009) |
| Croatia | 58.54 | 1.11 | 0.99 | 0.25 |
| Czech Republic | 86.46 | 19.73 | 10.28 | 7.63 |
| Denmark | 215.5 | 5.55 | 6.28 | 7.26 |
| Estonia | 237.36 | 200.22 | 83.09 | 10.88 |
| Finland | 200.5 | 7.4 | 4.94 | 3.79 |
| France | – | 0.35 | 0.57 | 1.7 |
| Georgia | 46.64 | 23.69 | 12.43 | 15.88 |
| Germany | 129.07 | 4.99 | 2.69 | – |
| Israel | 22.53 | 0.76 | 3.98 | 4 |
| Italy | 3.57 | 0.55 | 0.5 | 0.29 |
| Norway | 257.02 | 3.99 | 4.26 | 6.31 |
| Poland | 104.03 | 4.04 | 2.08 | 0.87 (2007) |
| Romania | 91.1 | 24.71 | 17.76 | 2.95 |
| Russian Federation | 196.34 | 172.81 | 118.5 | 56.38 |
| Switzerland | 23.98 | 3.77 | 4.61 | 12 |
| UK and Northern Ireland | 98.73 | 17.95 | 28.62 | 31.33 (2007) |
| United States | 442.1 | 147.5 | 129.2 | 111.6 |

as well as treatment strategies [20]. Although an important step in the right direction, surveillance systems remain prone to inaccuracies and the paucity of data from developing countries hinders global evaluation of gonococcal rates and assessment of intervention programs [17, 21, 22]. Strengthening of local country data collecting systems and international collaboration are needed to improve surveillance of gonorrhea on a global scale. Table 19.2 compares rates of gonorrhea in selected countries in Europe with Canada and the United States. The variation in reported rates per 100,000 population is quite wide ranging from a low of 0.25 and 1.7 in Croatia and France, respectively, to much higher rates of 56.38, 56.5 and 111.6 in the Russian Federation, Belarus and the US, respectively.

**Fig. 19.2** Example of an extensive social network among young people in Denver, US [38]. (With permission from A.A. Al-Tayyib)

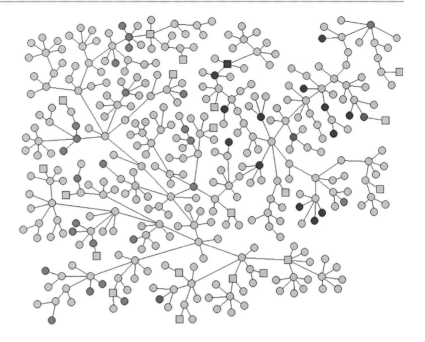

Comprehensive reviews of the historical epidemiology of gonorrhea are available elsewhere [17, 23]. In brief, gonorrhea rates have fluctuated worldwide over the last hundred years. Peaking around the Second World War in the 1940s and during the "sexual liberation" of the late 1960s and early 1970s, declining gonococcal rates were subsequently observed in the US, Canada and many European countries [12, 24, 25]. Following a transient rise in the 1990s and early 2000s, infection rates have since stabilized or declined in some countries but now show resurgence in others. For example, while gonorrhea rates have fallen in the US, they have risen in Canada [12, 14]. High infection rates also persist in many lower-income countries including Eastern Europe, the Western Pacific, Asia, Africa and Latin America [26–28].

## 19.4   Risk Groups: Emphasis on Young People & Sexual Networks

The factors needed to sustain a sexually transmitted infection epidemic are summarized by the case-reproduction ratio (Ro), an equation used to determine the number of new cases generated by an infected person. This ratio is calculated by the following equation: $Ro=\beta CD$, where "$\beta$" indicates the efficiency of transmission, "C" the number and/or rate of sexual partner change, and "D" the average duration of infectivity [29]. When Ro equals one, the infection is endemic and stable. Ro values greater than or less than one reflect increasing or decreasing infection rates, respectively.

Each component of the case-reproduction ratio reflects a variety of sexually transmitted infection risk factors. For example, the efficacy of transmission ("$\beta$") is determined, in part, by gonococcal strain subtype and infectious inoculum [30]. To illustrate the latter, male to female genital-to-genital transmission is more efficient than the reverse (e.g. male to female: 50–70 %; female to male: 20 %), likely due to a higher infectious inoculum in male urethral discharge than vaginal secretions [5, 31, 32]. Host behaviours, such as condom use, also play a role in transmission efficiency ("$\beta$") [33–35].

The variable duration of infectivity ("D") reflects the time between sexually transmitted infection acquisition and successful treatment or clearance of infection. Gonorrhea is mild or asymptomatic in up to 50 % of infected females and in less than 10 % of infected males [36], delaying or preventing diagnosis, especially in women. Fear of stigmatization, inaccessible diagnostic and treatment services, and therapy noncompliance are other potential reasons for longer periods of infectivity and, hence, increased risk of transmission [17].

The last variable in the case-reproduction ratio, the number and/or rate of sexual partner change ("C"), is self-explanatory. However, this variable is worth exploring further from the standpoint of sexual networks. Long term monogamous sexual relationships are not part of a sexual network. Rather, sexual networks are characterized by core groups of people with frequent partner exchange and high degrees of connectivity [37]. These networks can be remarkably extensive as recently demonstrated by Al-Tayyib & Rietmeijer [38] (Fig. 19.2). Although some sexually transmitted infection cases are detected and treated, infection within the network remains making reinfection probable. Behavioural factors, such as sporadic condom use within the group, promote rapid spread and persistence of sexually transmitted infections [39]. More prevalent in core groups than the general population, gonorrhea is an archetype of sexually transmitted infection transmission and persistence within sexual networks.

Teenagers and young adults, due to a combination of social, behavioural and biological factors, are disproportionately affected by sexually transmitted infections [40]. Along with men who have sex with men, street youth, intravenous drug users, and persons in correctional facilities, young people often fit the "core group criteria" of sexual networks [23, 41]. Contributing teen behaviours include multiple concurrent sexual partners [39, 42–47], sequential sexual partnerships of limited duration [40, 48, 49], failure to use barrier protection and the mixing of alcohol, drugs and risky sexual activity [47]. Additionally, the smaller introitus and cervical immaturity of young females confers a biological risk of sexually transmitted infection acquisition [50].

The 2009 National Youth Risk Behavior Survey in the United States reported that 46 % of grade 9 (ages 14–15 years) to grade 12 (ages 17–18 years) students had engaged in sexual intercourse at least once, 34.2 % were currently sexually active (i.e., had sexual intercourse with at least one person during the 3 months prior to the study), and 13.8 % had had sex with ≥4 people. Of the 34.2 % of currently sexually active teens, 61.1 % reported condom use by themselves or their partner and 21.6 %, alcohol or drugs before the last sexual intercourse. Of the teens surveyed, 7.4 % reported being forced into sex [47].

Early sexual debut further puts teens at risk [51, 52] with 25 % of young people acquiring their first sexually transmitted infection within a year of first sexual intercourse [49, 53]. Although the 2009 US National Youth Risk Behavior Survey showed improvement in some measures as compared to a decade ago (e.g., 2000—49.9 % reported ever having had sexual intercourse; 16.2 %, having ≥4 lifetime sexual partners) [54], the sexual behaviours of young people continue to put them at increased risk for sexually transmitted infections including gonorrhea.

Gonorrhea rates also vary by racial or ethnic group and by geographic region most likely reflecting background differences in rates within sexual networks. For example, in the US and UK, black ethnic population groups have disproportionately higher gonococcal infection rates [12, 19] than whites. In the US, the gonorrhea rate is 20 times higher in blacks than whites with young black women bearing the heaviest gonorrhea burden [12]. Geographically, gonorrhea rates vary by population density and region. For example, gonorrhea is disproportionately concentrated in London in the UK, and in the South in the US [12, 13].

## 19.5   Clinical Presentation

The clinical presentation of gonorrhea is well-described elsewhere [55]. Rather than provide a comprehensive overview, only a brief description of infectious manifestations in teenagers and young adults is provided below. A discussion of gonorrhea in neonates and young children is available elsewhere [56].

## 19.5.1 Genitourinary Infection

As noted above, gonococcal infection in males is asymptomatic in 5–10 % of cases. Symptomatic males most commonly suffer from urethritis, which begins 2–7 days after exposure. Complications are rare in men. Gonorrhea in females, on the other hand, is mild or asymptomatic in up to 50 % of cases. Symptomatic females usually develop cervicitis, typically 3–5 days after exposure. Complications are more common in women than men with the most feared, pelvic inflammatory disease, occurring in 10–20 % of untreated females [57]. Additionally, gonococcal infection during pregnancy poses a risk to both the mother and unborn child [58, 59].

## 19.5.2 Pharyngeal & Anorectal Gonorrhea

Pharyngeal gonorrhea results from orogenital contact. Although historically considered to only affect men who have sex with men and commercial sex workers, the increasing rates of orogenital sex amongst heterosexual couples has escalated their risk [60]. Pharyngeal gonorrhea is usually clinically silent, tends to resolve spontaneously within 10–12 weeks, and rarely causes complications [59]. However, pharyngeal infection, if not treated, has been linked to disseminated gonorrhea and serves as a potential reservoir for transmission to sexual partners [60]. Additionally, through the genetic exchange between commensal and gonococcal *Neisseria* species, gonococcal infection of the pharynx facilitates antimicrobial resistance transfer [61].

In males, anorectal gonorrhea occurs almost exclusively from receptive rectal intercourse amongst men who have sex with men. In women, transmural inoculation from vaginal secretions is likely the most common source [62]. Like pharyngeal gonorrhea, anorectal infection is asymptomatic in the majority of cases but serves as an infectious reservoir for transmission to sexual partners [63], a site for microbial genomic exchange and for persistence within sexual networks.

## 19.5.3 Disseminated Gonococcal Infection (DGI)

Disseminated disease requires bloodstream invasion from local sites of gonococcal infection. However, signs and symptoms of mucous membrane involvement rarely precede disseminated infection. Disseminated gonococcal infection is a rare entity, estimated to occur in only 1–3 % of infected adult patients [59]. Although usually clinically silent, disseminated gonococcal infection can be symptomatic, typically causing mild to moderate illness. Even without treatment, symptoms usually resolve in a matter of days and long-term sequelae are atypical [64]. Infrequently, disseminated gonococcal infection can cause fulminant sepsis and death [65]. Disseminated disease is also associated with infection at other sites, most commonly septic arthritis. Less frequently, disseminated gonococcal infection presents as a triad of tenosynovitis, dermatitis and polyarthralgias without purulent arthritis [64]. Given an association between disseminated gonococcal infection and terminal complement deficiency, patients with recurrent disseminated disease should be tested for immune deficiency [66].

Other forms of disseminated gonococcal disease, including osteomyelitis, meningitis, endocarditis and myocarditis, are infrequently reported [59, 61].

## 19.6 Diagnosis

## 19.6.1 Screening

Screening decisions should be based on a solid understanding of epidemiology of gonorrhea. Given that gonorrhea rates vary considerably within and between populations, mass screening for this pathogen has largely been abandoned [23]. While some countries limit testing to symptomatic individuals and their sexual contacts, most recognize the need to extend gonorrhea testing to subpopulations with high prevalence rates and/or risk factors. For example, numerous organizations in the US support

**Table 19.3** Comparison of Gonorrhea Diagnostic Tests

|  | Culture based | Molecular testing |
|---|---|---|
| Specimen transport | Requires rapid transport and stringent control of environmental conditions | Less stringent requirements. Testing able to detect nonviable organisms. More amenable to self-collected specimens. Allows batched, centralized testing |
| Minimum turn-around time | 1–3 Days | Hours |
| Equipment | Agar plates, incubator, basic confirmatory tests | Commercial molecular detection platform +/− nucleic acid extraction system. |
| Enables antibiotic susceptibility testing | Yes | No |
| Cost per test | Inexpensive | Variable |
| Training | General clinical microbiology training | Molecular microbiology training |
| Simultaneous Chlamydia detection | No | Available on several platforms |

routine testing of pregnant women with risk factors, men who have sex with men, and adolescent patients [67]. Less agreement has been reached over screening groups with low prevalence rates but high risk of asymptomatic, complicated infections (e.g., low risk pregnant women). Additionally, optimal screening intervals for at risk populations and the need for post-treatment testing requires further research.

Certainly, both the potential benefits and harms of screening must be considered prior to implementing a testing program. In favor of screening is the identification and subsequent treatment of asymptomatic patients, thereby, preventing disease complications and infectious spread to others (i.e., decreasing the prevalence in a sexual network). However, screening tools must be carefully selected, considering both the sensitivity and specificity of the test, alongside its applicability to the population under investigation [68]. The inverse relationship between disease prevalence and false positivity rate is a well recognized statistical phenomenon. In the context of sexually transmitted infection testing, false positive results are expected to cause needless patient anxiety, potential partner discord, and increase unnecessary antibiotic use. Hence, the limitations of a screening test must always be considered prior to its application.

Given the rise of multi-resistant gonorrhea (see below), the role of screening as a secondary prevention strategy may need to be reevaluated.

## 19.6.2 Microbiology/Diagnosis

Table 19.3 compares two current microbiological tests, culture and nucleic acid amplification tests (molecular testing), for gonorrhea, noting specimen transport requirements, potential for self collection, rapidity of test, necessary equipment, availability of antibiotic sensitivity testing, relative cost and expertise needed, and ease of coincident chlamydia detection.

### 19.6.2.1 Diagnosis of Genitourinary Gonorrhea

**Gram Stain**

Gram stain is a rapid tool for diagnosis of gonococcal urethritis in symptomatic men. *N. gonorrhoeae* characteristically appear as Gram-negative intracellular diplococci within polymorphonuclear cells. The coupling of opposing flattened bacterial surfaces produces the organism's distinctive "kidney bean" shape. In men with symptomatic urethritis, Gram stain has comparable specificity to culture [69]. While a presumptive diagnosis of gonorrhea in this demographic can be made on the basis of Gram stain alone, a negative result does not rule out gonococcal infection [59]. Furthermore, nonurethral specimens from nonsterile sites are not appropriate for Gram stain due to poor specificity.

## Culture

Culture remains an option for gonorrhea testing from genital sites. The appropriate site(s) for specimen collection depend on patient age, gender, and sexual practices as well as clinical manifestations of infection. While the primary site of culture collection from women has been the endocervix, alternative sample sites include the urethra and vagina. Vaginal or urine specimens are recommended for women post-hysterectomy and for prepubertal girls. The primary collection site from men is the urethra. Rectal and pharyngeal swabs should be collected in patients who have engaged in oral and anal intercourse, respectively. Conjunctival cultures should be sent in the cases of suspected gonorrhea ophthalmia. Lastly, blood, synovial, and cerebrospinal fluid specimens should be considered in suspected disseminated disease.

Culture is performed on selective media and incubated in carbon dioxide at 35–37 °C for up to 72 h. Colonies are generally heterogeneous in appearance reflecting variations in piliation and are both catalase and oxidase positive. A variety of options exist for confirmation including carbohydrate utilization, chromogenic enzyme substrate tests, immunologic methods and deoxyribonucleic acid (DNA) probe confirmation tests. Culture has several advantages including acceptable sensitivity, high specificity, relative low cost, and suitability for a variety of clinical specimens. It does not require the costly instrumentation associated with molecular detection and provides a viable organism for antibiotic susceptibility testing and tracking epidemiological trends. There are, however, multiple disadvantages to culture. It requires more invasive sampling techniques; in particular, urethral swabs for men and endocervical swabs for women. Culture also necessitates strict storage and transportation conditions given that N. gonorrhoeae has specialized growth requirements and is highly susceptible to environmental variations. Lastly, culture takes 24–72 h to produce a definite result.

## Nucleic Acid Amplification Diagnostic Techniques (NAATs)

The first non-culture diagnostic tests were based on the detection of specific gonococcal antigens. Suboptimal performance characteristics and high cost favoured the continued use of culture [69]. Antigen-based tests were subsequently supplanted by nucleic acid amplification tests (NAATs). Although more expensive than culture, nucleic acid detection methods offer a number of advantages over culture including rapid turnaround time and the ability to detect organisms that are nonviable for culture. NAATs are based on detection of species-specific DNA or ribonucleic acid (RNA) sequences. The enhanced sensitivity of NAATs permits testing on a wider variety of specimen types including urethral swabs, endocervical and vaginal swabs as well as urine. A variety of commercial NAATs, involving different amplification techniques and nucleic acid targets, have been approved for the detection of gonorrhea in most developed countries. While NAATs have been approved for endocervical, urethral, vaginal, and urine specimens, culture has been the standard for pharyngeal and anorectal specimens [70]. However, this situation is in transition as increasing data validates the use of NAATs for non-urogenital sites [71, 72].

NAATs sensitivity and specificity depends upon the amplification of organism-specific nucleic acid sequences. The success of a particular NAAT relates to the choice of target gene. Commercial and in-house NAATs have used a variety of genetic targets including the cryptic plasmid (cppB gene), opa genes, cytosine DNA methyltransferase and the 16SrRNA gene. With the potential exception of urine specimens, NAATs generally offer superior sensitivity to culture, attributable to their ability to produce a positive signal from as little as a single copy of target DNA or RNA [69].

NAATs have replaced culture in many countries as the preferred method of diagnosing gonorrhea from genital sites. Many assays offer the advantage of testing for both gonococcal and chlamydia from a single swab or urine specimen—a benefit for both patients and clinicians. The ability to test less invasive specimens such as urine samples and vaginal swabs has a significant advantage over culture as it improves patient acceptance of testing [73] and provides the additional option of self-collected specimens—which many patients find preferable [74, 75].

Theoretical limitations to NAAT sensitivity are assay-specific. Interstrain genetic differences may result in false negatives due to variable or absent nucleic acid sequences. Given that gonococcal strain distribution varies geographically and temporally, the performance of NAATs may differ both among and within groups over time. However, in general, commercial assays have targeted very stable genes to minimize this issue. Competitive amplification, whereby targets of lesser concentration fail to amplify in multiplex assays, and the presence of inhibitors may also reduce assay sensitivity [70]. Positive internal controls identify some of these technical limitations. Notably, commercial NAATs do not routinely detect antibiotic resistance genes although these may be future targets.

NAATs typically exhibit high specificity. Interspecies homology within the *Neisseria* genus [76] and frequent horizontal genetic exchange among *Neisseria* species [70] are theoretical causes of false positive results. Cross-contamination during processing is a risk that must be controlled for with any NAAT to maintain test specificity. Maintaining high specificity is essential in order to maximize the positive predictive value of the test, particularly in low prevalence populations. New multiplex NAATs, which target multiple genes in a single assay, may provide improved target specificity. Additionally, limiting screening programs to high prevalence populations will maximize the positive predictive value of the utilized test.

### 19.6.2.2   Diagnosis of Pharyngeal and Rectal Gonorrhea

Cultured pharyngeal and rectal swabs generally have poor sensitivity [60]. Several studies have demonstrated the superior sensitivity of NAATs over culture in detecting pharyngeal and rectal gonococcal infection [71, 73, 77, 78]. In addi-

tion, although concerns have been raised regarding cross-reactivity with commensal *Neisseria* species at extragenital sites, many NAATs have been found to have superior specificity to culture. Ongoing validation of assays for the detection of pharyngeal and rectal gonorrhea will help to minimize potential reservoirs of infection and, secondarily, reduce the development of antimicrobial resistance (pharyngeal gonorrhea).

### 19.6.2.3   Diagnosis of Disseminated Gonococcal Infection

The yield from cultures is limited in disseminated gonococcal infections. However, it is prudent to culture all potentially infected areas, including genitourinary, rectal, and pharyngeal sites. The presence of a septic joint should prompt arthrocentesis. Mean synovial fluid leukocyte count in disseminated gonococcal infections is typically around 50,000 cells/mm$^3$ although it can be less than 10,000 cells/mm$^3$ [64]. Synovial fluid gram stain and culture are often negative. However, molecular testing of aspirated joint fluid with 16S rDNA PCR can augment diagnostic yield [79, 80].

Although blood cultures are negative in 50–75 % of cases, at least two blood specimens should be sent for culture in patients presenting with suspected disseminated disease [64]. The positivity rate may be higher in patients with tenosynovitis, dermatitis, and polyarthralgias than those presenting with purulent arthritis. Associated cutaneous lesions can be swabbed for culture but are also of low diagnostic value [81].

## 19.7   Gonococcal Resistance & Treatment

### 19.7.1   Resistance History

Since the introduction of antibiotics in 1937 [82], *N. gonorrhea* has evolved multiple resistance mechanisms, both chromosomal- and plasmid-mediated. The rise in gonococcal resistance can be attributed to both the adaptability of the organism (see pathogenesis above) as well as antibiotic prescribing habits. With regards to the latter, the increased use of over-the-counter and broad spectrum antibiotics combined with inappropriate prescribing policies for many clinical situations have facilitated the emergence of gonococcal resistance.

Introduced in the 1930s, sulfanilamide was the first curative treatment for gonorrhea. However, antibiotic resistance developed rapidly [82] such that penicillins replaced sulfanilamides as first-line gonococcal therapy in the 1940s [83]. Although penicillin remained the mainstay of therapy for several decades, low-level resistance slowly developed with progressive increases in penicillin minimal inhibitory concentrations necessitating higher treatment doses [84]. Tetracycline, used initially to treat penicillin allergic patients with gonorrhea, also became less effective over time [85]. In the late 1970s, the emergence of highly-resistant, penicillinase-producing *N. gonorrhoeae* marked the end of penicillin as a therapeutic agent for gonorrhea [86].

Starting in the mid-1980s, quinolones replaced penicillins as the first-line treatment for gonorrhea [87]. Effective single oral dosing, a favorable side effect profile, and excellent efficacy at all anatomical sites made quinolones an attractive therapeutic option. However, isolates resistant to quinolones began to appear by the 1990s, initially in Asia and then quickly spreading—likely initially via travelers and then endemically within high risk groups—to Australia, Hawaii and later to North America and Europe [61, 88]. By the mid- to late-1990s, quinolones were no longer the drug of choice for gonorrhea in most Asia-Pacific countries and by the mid to late 2000s, England, Europe and North America followed suit [73, 89].

Of note, despite a decrease in the use of quinolones for treatment of gonorrhea in the UK, the rate of resistance has continued to climb in England [90] highlighting the resistance problem with gonorrhea. Because gonococcal infection is relatively common and the organism readily shares resistance genes, stopping particular antibiotic treatment of gonorrhea does not result in a diminution of resistance. The gonococcus continues to share resistance genes and still experiences ongoing resistance selection pressure because these antibiotics are widely used to treat other infections.

Extended-spectrum cephalosporins are currently considered first-line therapy for gonorrhea in most countries. Although cephalosporins remain largely effective, growing resistance has been noted throughout the Asia-Pacific, North America and Europe [21, 61, 91–98]. Both the UK and US based surveillance programmes have reported a minimum inhibitory concentration creep to both ceftriaxone and cefixime in recent years [21, 91, 99]. For example, in the US, from 2000 to 2010, the minimum inhibitory concentrations to cephalosporins increased from 0.2 to 1.5 % for cefixime and from 0.1 to 0.4 % for ceftriaxone in 2000 and 2011, respectively [91, 92]. Cases of treatment failure have been reported after oral cephalosporin use in Asia, the UK and Norway [94, 99–101]. Although suspected treatment failure with ceftriaxone is rarely reported [102, 103], a highly multi-resistant *N. gonorrhea* strain has recently been identified in a sex worker in Japan [104].

Considering the inability to prevent spread of resistant gonococci in the past, the emergence and spread of multi-resistant cephalosporin strains appears probable. Given global travel and sexual tourism this likely will not take long. With no effective single or combined antibiotic treatment to replace cephalosporins, gonorrhea may soon become untreatable. The obvious need for new antibiotic regimens, as the microbe becomes more resistant, is an area that has not received the research attention it deserves. Although the need has clearly been recognized for a number of years [105], novel antibiotic regimens for treating gonococcal infection remain inadequately addressed.

Potential strategies to combat cephalosporin resistance include the promotion of more appropriate antimicrobial use, the development of novel treatment agents and regimens, improved surveillance and reporting networks to facilitate the identification and rapid containment of resistant isolates, as well as more effective sexually transmitted infection control and prevention programs [2, 3, 21, 61, 106].

**Table 19.4** Example of Gonorrhea Treatment Guidelines for Selected Infection Sites: Centers for Disease Control and Prevention 2010 [79, 92]

| Clinical indication | Recommended regimens |
| --- | --- |
| Uncomplicated genitourinary or anorectal gonorrhea | Ceftriaxone 250 mg IM in single dose OR, IF NOT AN OPTION Cefixime 400 mg PO in a single dose[a] PLUS Azithromycin 1 gm PO in sgine dose OR, IF NOT AN OPTION Doxycycline 100 mg PO × 7 days |
| Uncomplicated pharyngeal gonorrhea | Ceftriaxone 250 mg IM in single dose[a] |
| Adult gonococcal conjunctivitis | Ceftriaxone 250 mg IM in single dose |
| Disseminated gonococcal infection | Ceftriaxone 1 g IV q. 24 h × 7 days[b] |

[a] Testing and treatment for chlamydia is recommended. Treatment regimens for chlamydia, Azithromycin 1 gm orally in single dose OR Doxycycline 100 mg a day for 7 days, is part of dual therapy recommended for gonorrhea
[b] Consider changing to oral therapy (Cefixime 400 mg PO BID) 24–48 h after clinical improvement

### 19.7.2 Uncomplicated Genitourinary and Anorectal Gonorrhea Treatment Recommendations

Given the differences among national guidelines and the variability of resistance rates between countries, the details on recommended therapeutic regimens have been kept to a minimum. Additionally, the discussion is limited to therapy for adolescent and adult gonorrhea. Unless indicated otherwise, treatment recommendations provided here reflect the recently updated 2010 Centers for Disease Control and Prevention Sexually Transmitted Disease Guidelines (outlined in Table 19.4) [79, 92].

As of 2012, the CDC no longer recommends cefixime as a first-line agent for the treatment of uncomplicated genitourinary and anorectal gonococcal infections. This decision is based on declining cefixime susceptibility among urethral *N. gonorrhea* isolates from the Gonococcal Isolate Surveillance Program. Dual therapy with intramuscular (IM) ceftriaxone and either a single dose of oral azithromycin or 7 day course of oral doxycycline is now the recommended first-

line treatment regimen for genitourinary and anorectal gonorrhea.

In addition to treating coincident *Chlamydia trachomatis* infection, ceftriaxone combined with azithromycin or doxycycline may serve to improve treatment efficacy and delay the development and spread of cephalosporin-resistant gonococcal isolates. Azithromycin is the preferred agent over doxycycline given the ease and compliance of single-dose treatment regimens and lower prevalence of gonococcal resistance to azithromycin than tetracycline [91].

Patients should avoid sexual activity until their antimicrobial therapy is completed and their symptoms have resolved. In the case of persistent gonococcal infection following first-line therapy, further treatment advice should be sought from an infectious diseases expert.

### 19.7.3 Treatment of Pharyngeal Gonorrhea

Although usually self-limited, pharyngeal gonorrhea serves as a potential reservoir for infectious transmission and antimicrobial resistance [60, 61]. Hence, pharyngeal gonococcal infection should be treated. More difficult to eradicate than urogenital and anorectal gonococcal infections, current therapy options are limited to a single dose of IM ceftriaxone. Alternative injectable and oral antibiotics, including cefixime, have unreliable efficacy in this treatment context and should be avoided. As with genitourinary and anorectal gonorrhea, co-treatment with azithromycin and/or doxycycline is recommended for pharyngeal gonorrhea.

### 19.7.4 Treatment of Disseminated Gonococcal Infection

Treatment of DGI requires parenteral therapy with ceftriaxone or an alternative third-generation cephalosporin (e.g., cefotaxime or ceftizoxime). Oral therapy can be considered in patients whose illness is sufficiently mild to permit close outpatient follow-up. However, a patient with uncertain diagnosis, complicated disease, and/or difficulty tolerating oral medications should be hospitalized for the initiation of therapy. Transition to oral cefixime can be considered for susceptible organisms following 24–48 h of clinical improvement on intravenous therapy.

There have been no controlled clinical trials with respect to the duration of treatment for DGI. Antimicrobial therapy should likely be for a minimum of 7 days or continued at least until clinical signs of infection have resolved. Patients with purulent arthritis require joint drainage and 7–14 days of antibiotic therapy [64]. Treatment of meningitis or endocarditis consists of ceftriaxone for 10–14 days or a minimum of 4 weeks, respectively.

### 19.7.5 Special Treatment Circumstances & Alternative Agents

Ceftriaxone is also first-line treatment for gonococcal infections in pregnant women. Patients with human immunodeficiency virus (HIV) should receive the same treatment for gonorrhea as persons without HIV infection.

A non-beta lactam antibiotic may be needed in the case of severe penicillin or cephalosporin allergy [108]. Potential alternative options to cephalosporins include quinolones, macrolides, and spectinomycin. However, treatment in these cases should be performed in consultation with infectious diseases and, if possible, directed by isolate susceptibility testing.

Alternative safe and effective single-dose injectable cephalosporin regimens are available for urogenital and anorectal gonorrhea and include ceftizoxime, cefoxitin (in combination with probenecid), and cefotaxime. None of these injectable alternatives offers any advantage over ceftriaxone for urogenital infection. Oral substitutes include cefpodoxime and cefuroxime axetil. However, alternative oral and injectable cephalosporins may provide suboptimal efficacy for pharyngeal gonorrhea. Furthermore, caution is necessary when using oral cephalosporins to treat urogenital and anorectal gonorrhea due to increasing rates of gonococcal resistance to cefixime.

Developed and marketed specifically for the treatment of gonorrhea, spectinomycin is an effective alternative for genitourinary but not pharyngeal infection. Resistance tends to be rare and sporadic. However, spectinomycin is an expensive, injectable drug that is not readily available in many countries including the US.

Uncomplicated gonococcal infection can be effectively treated with high-dose azithromycin (2 gm PO once). However, significant gastrointestinal side effects and rapid development of gonococcal resistance necessitate the use of high-dose azithromycin only under special circumstances [79, 91, 109]. As noted above, a single does of azithromycin (1 gm PO) is now recommended *in combination* with ceftriaxone (250 mg IM) to treat gonorrhea. However, due to growing microbe resistance, treatment failures have been reported even on this combined antibiotic regimen [110].

Quinolones can be considered as second-line treatment agents. However, given high global resistance rates to this antibiotic class, treatment with a quinolone should be delayed until susceptibility results are available. Furthermore, quinolones are contraindicated in pregnant women and persons younger than 18 years of age.

In some parts of the world, aminoglycosides are used to treat gonorrhea. For example, gentamicin is the national first-line treatment for gonococcal infection in Malawi. It has also been used to treat gonorrhea in Mongolia [111–113]. The precise mode of action of aminoglycosides against gonococcus remains unclear. Despite uncertain microbiological resistance breakpoints and limited clinical resistance data, treatment failures with aminoglycosides appear rare [114, 115].

### 19.7.6 Treatment: Other Recommendations

Reassessment following treatment bestows a number of benefits (e.g., to confirm compliance, ensure resolution of symptoms, enquire about adverse reactions, provide further education and partner notification). However, follow-up visits may not be practical in all settings [36].

As per the 2010 CDC guidelines and 2012 update, repeat testing is not recommended upon completion of gonorrhea treatment with first-line therapy [91, 92]. However, a test of cure should be considered in the following circumstances: if second-line agents are utilized, following treatment of pharyngeal gonorrhea, and/or if symptoms or signs of infection persist. Some countries, such as the UK, recommend a test of cure in all cases, recognizing the emerging problem of resistance to extended-spectrum cephalosporins [36].

Evidence pertaining to the method and timing of test of cure is limited. In cases of persistent symptoms or signs, a repeat culture should be performed at least 72 h after completion of antimicrobial therapy [116]. NAATs are expected to remain positive for weeks following successful treatment and do not allow for the identification of resistant organisms. Thus, a positive NAAT should be followed by a confirmatory culture whenever possible.

Given high rates of recurrence, the identification of *N. gonorrhea* following treatment is more likely to reflect reinfection rather than treatment failure [117–120]. With this in mind, retesting for re-infection 3 months after treatment is recommended, especially in high risk patients. Additionally, in the case of disease recurrence, further patient education and partner treatment should be pursued.

*Chlamydia trachomatis* co-infection is common in patients with gonorrhea. With this in mind, patients treated for *N. gonorrhea* should also be tested and treated for chlamydia. Other sexually transmitted infections should also be considered, including syphilis, HIV, and hepatitis B. The therapy of choice for chlamydia, azithromycin (1 gm PO once) or doxycycline (100 mg BID×7 days), corresponds to the dual therapy now suggested by the CDC for gonorrhea.

Positive gonorrhea cases should be reported to the local public health agency, as per national guidelines, and contact tracing pursued. Contact tracing reduces the risk of re-infection and facilitates the identification and treatment of asymptomatic infections [121]. All sexual partners

exposed to an index case within 60 days of symptom onset or diagnosis should be evaluated and treated. If greater than 60 days has passed since the index case has had sexual contact, the most recent sex partner should be contacted. Contact tracing may be performed by patients, physicians or public health departments depending on local and national strategies. To achieve compliance with notification, creative maneuvers have been used. The sexually transmitted infection notification program in British Columbia, Canada (www.gender-focus.com/2011/08/08/std-notification-by-e-card-starts-in) uses humour to garner attention and interest. Patient-delivered therapy to sexual partners has been shown to decrease the rate of gonococcal reinfection [122]. However, this approach may be poorly received by some patients and/or partners. Patient-delivered therapy also leads to missed opportunities to educate sexual partners and to diagnose coexisting sexually transmitted infections. This is particularly problematic for men who have sex with men where the risk of coexisting sexually transmitted infections is high. Given ethical and medicolegal issues around partner delivered therapy, local laws should be reviewed before utilizing partner delivered therapy [123].

Even with well funded and supported public health and/or patient contact tracing, all infected contacts in a sexual network are unlikely to be tracked down as no one member of the network knows all members. Leaving an infected sexual network member untreated keeps the organism ricocheting around the network.

## 19.8  Prevention

Most, if not all, countries recognize gonorrhea as a public-health priority. However, the strategies employed to address this sexually transmitted infection vary widely. While early diagnosis and treatment is uniformly pursued, emphasis on prevention has varied among countries [23].

The prevention of sexually transmitted infections in young people has been reviewed elsewhere [124]. To summarize, few trials on the efficacy of sexually transmitted infection prevention programs have been published within the last decade. Amongst published studies, considerable variability in program type and efficacy has been identified. While some prevention strategies have been associated with diminished infection-related risk behaviors, few report corresponding reductions in confirmed sexually transmitted infections. Given that rates of gonorrhea appear to be rising in some industrialized countries as noted earlier, this is especially concerning.

One well-studied preventative measure is condom use. Condoms, as well as other barrier contraceptive methods, provide a high degree of protection against sexually transmitted infections [33–35]. Post-exposure antibiotic prophylaxis also reduces risk of infection but is unlikely to be cost effective [125]. Another strategy, gonococcal mass treatment programs for at risk groups, has been trialed with limited success. Both antibiotic prophylaxis and mass treatment strategies carry the risk of fostering antimicrobial resistance, a serious concern given the emergence of multi-resistant gonococci [126, 127].

New and innovative approaches to prevention are required. In general, recent decades have seen a movement towards multidisciplinary and multileveled approaches with emphasis placed on the integration of gonorrhea prevention programs into already existing community and health-care structures [128]. Additionally, the empowerment of individuals and communities to develop and foster preventative strategies relevant to their particular social contexts is essential. Social networks influence the opinions and decisions of members [129]. Therefore, targeted core group counseling may provide the opportunity for better information dissemination within groups. For example, peer-led rather than outsider-initiated sexual health interventions (i.e., on condom use, contact tracing) are more likely to succeed, as demonstrated in adolescent and other populations [38, 130, 131].

Another important prevention strategy is the use of appropriate communication tools. Young people increasingly rely on the internet and social media as sources of health information [132–134]. With this in mind, sexual health educators need to develop creative ways of getting their message across to teens and young adults. For example, novel Internet-based strategies are being trialed in Alberta, Canada, to tackle the ongoing syphilis epidemic within that province. One approach utilizes a spoof on online dating (http://www.plentyofsyph.com) that has been associated with a significant increase in young people coming forward for testing [2]. Similar strategies could prove useful for other sexually transmitted infections including gonorrhea.

Lastly, more adolescent-friendly health care services are required. Teens tend to live in the "now" and are less likely to seek treatment at appointment-only clinics, especially if wait times are long [135]. Online triage systems and options for testing-only express visits may enhance the efficiency of busy sexually transmitted infection clinics [136, 137]. Drop in clinics, especially if located where young people frequent (e.g., malls and schools) are another option, allowing teens to follow through with "spur of the moment" decisions to get sexually transmitted infection testing.

Learning about sex, sexuality and the prevention of sexually transmitted infections is a basic human right of adolescents [138]. The strategy of "abstinence" or "abstinence until marriage" lacks scientific evidence and is flawed from a medical-ethical standpoint [139]. Given that sexually transmitted infections, including gonorrhea, disproportionately affect young adults, we have a duty to provide adolescents with meaningful knowledge and the skills to protect themselves from acquiring and transmitting sexually transmitted infections. This will require collaboration on many levels, ranging from the family unit and the community to local and national governmental levels.

## 19.9  Conclusion

Moving into the second decade of the 21st century, *N. gonorrhoeae* continues to represent a significant public health challenge. Gonococcal rates persist above national and international targets, resulting in needless morbidity and mortality. The emergence of multi-resistant gonorrhea is only going to make this situation worse.

Although the development of NAATs has improved our ability to diagnose gonorrhea, at least in industrialized countries, further refinement of diagnostic techniques is required. In particular, the role of pharyngeal gonorrhea as a significant reservoir of infection and antimicrobial resistance necessitates the development of more effective easy to use diagnostic tools. Furthermore, because resistance is such an important issue, traditional culture and sensitivity testing remains essential for tracking evolving gonococcal resistance both nationally and globally. Sentinel sites where cultures are done routinely will be key.

*N. gonorrhea* has proved to be a highly adaptable organism. Increasing gonococcal resistance to cephalosporins, with no single dose alternative antibiotics in the pipeline, threatens to make this organism untreatable. Effective and prudent antimicrobial strategies alongside robust national and international gonococcal surveillance programs are very much in need.

Lastly, sexually transmitted infection prevention strategies have never been more important. Public health services should aim to address the behavioral and social factors that put people at risk, including youth-friendly education and health care services. As these move forward, rigorous evaluation and collaboration of prevention strategies will be required at local, national and global levels.

# References

1. Anderson MT, Seifert HS (2011) Opportunity and means: horizontal gene transfer from the human host to a bacterial pathogen. mBio. doi: 10.1128/mBio.00005-11
2. MacDonald NE, Stanbrook MB, Flegel K, Hebert PC (2011) Gonorrhea: what goes around comes around. CMAJ. doi:10.1503/cmaj.111393. http://www.cmaj.ca/content/early/2011/09/19/cmaj.111393.long
3. Fisman D, Laupland KB (2011) Sexually transmitted infections in Canada: a sticky situation. Can J Infect Dis Med Microbiol 22(3):80–82
4. Sparling PF (2008) Biology of Neisseria gonorrhoeae. In: Holmes K, Sparling P, Stamm W, Piot P, Wasserheit J, Corey L, Cohen M (eds) Sexually Transmitted Diseases, 4th edn. McGraw-Hill, China, p 607–626
5. Virji M (2009) Pathogenic neisseriae: surface modulation, pathogenesis and infection control. Nat Rev Microbiol 7(4):274–286
6. Edwards JL (2008) The role of complement in gonococcal infection of cervical epithelia. Vaccine 26(Suppl 8):156–161
7. Steichen CT, Shao JQ, Ketterer MR, Apicella MA (2008) Gonococcal cervicitis: a role for biofilm in pathogenesis. J Infect Dis 198(12):1856–1861
8. Goldenberg DL, Sexton DJ (2011) Disseminated gonococcal infection (Internet). Waltham, MA: UpToDate; 2011. http://www.uptodate.com/contents/disseminated-gonococcal-infection?source=see_link. Accessed 19 Sept 2011
9. Blom AM, Ram S (2008) Contribution of interactions between complement inhibitor C4b-binding protein and pathogens to their ability to establish infection with particularly emphasis on Neisseria gonorrhoeae. Vaccine 26(Suppl 8):149–155
10. Jarvis GA (1995) Recognition and control of neisserial infection by antibody and complement. Trends Microbiol 3(5):198–201
11. Department of Reproductive Health and Research/World Health Organization (2011) Emergence of multi-drug resistant Neisseria gonorrhoeae—Threat of global rise in untreatable sexually transmitted infections 2011. http://whqlibdoc.who.int/hq/2011/WHO_RHR_11.14_eng.pdf
12. Centers for Disease Control and Prevention (2011) Sexually Transmitted Disease Surveillance 2009. Atlanta: US Department of Health and Human Services; 2010a http://www.cdc.gov/std/stats09/surv2009-Complete.pdf. Accessed 19 Sept 2011
13. Health Protection Agency (2011) Healthy Protection Report weekly report 2011. http://www.hpa.org.uk/hpr/archives/2011/hpr2411.pdf. Accessed 19 Sept 2011
14. Public Health Agency of Canada (2011) Reported cases of notifiable STI from January 1 to June 30, 2009 and January 1 to June 30, 2010 and corresponding annual rates for the years 2009 and 2010, 2010. http://www.phac-aspc.gc.ca/std-mts/stdcases-casmts/index-eng.php Accessed 19 Sept 2011
15. Catchpole MA (1996) The role of epidemiology and surveillance systems in the control of sexually transmitted diseases. Genitourin Med 72(5):321–329
16. Kent C (2007) STD Surveillance: Critical and Costly, but Do We Know if It Works? Sex Trans Dis 34(2):81–82
17. Tapsall J (2011) Antimicrobial resistance in Neisseria gonorrhoeae. WHO; 2001. http://www.who.int/drugresistance/Antimicrobial_resistance_in_

Neisseria_gonorrhoeae.pdf. Accessed 19 Sept 2011

18. Paine TC, Fenton KA, Herring A, Turner A, Ison C, Martin A et al (2001) GRASP: a new national sentinel surveillance initiative for monitoring gonococcal antimicrobial resistance in England and Wales. Sex Transm Infect 77(6):398–401

19. Delpech V, Marin IMC, Hughes G, Nichols T, James L, Ison CA, Gonococcal Resistance to Antibiotics Surveillance Program steering group (2009) Epidemiology and clinical presentation of gonorrhoea in England and Wales: findings from the Gonococcal Resistance to Antimicrobials Surveillance Programme 2001–2006. Sex Transm Infect 85(5):317–321

20. Dubois-Arber F, Jeannin A, Spencer B, Gervasoni JP, Graz B, Elford J et al (2010) Mapping HIV/STI behavioural surveillance in Europe. BMC Infect Dis 4(10):290–300

21. Tapsall JW, Ndowa F, Lewis DA, Unemo M (2009) Meeting the public health challenge of multidrug- and extensively drug-resistant Neisseria gonorrhoeae. Expert Rev Anti Infect Ther 7(7):821–834

22. Unemo M, Fasth O, Fredlung H, Limnios A, Tapsall J (2009) Phenotypic and genetic characterization of the 2008 WHO Neisseria gonorrhoeae reference strain panel intended for global quality assurance and quality control of gonococcal antimicrobial resistance surveillance for public health purposes. J Antimicrob Chemother 63(6):1142–1151

23. Barnes RC, Holmes KK (1984) Epidemiology of gonorrhea: current perspectives. Epidemiol Rev 6:1–30

24. Alary M (1997) Gonorrhea: epidemiology and control strategies. Can J Hum Sex 6(2):151–159

25. Fenton KA, Lowndes CM (2004) Recent Trends in the epidemiology of sexually transmitted infections in the European Union. Sex Transm Infect 80(4):255–263

26. Uuskula A. Puur A, Toompere K, DeHovitz J (2010) Trends in the epidemiology of bacterial sexually transmitted infections in eastern Europe. Sex Transm Infect 86(1):6–14

27. World Health Organization (2011) Global prevalence and incidence of selected curable sexually transmitted infections: overview and estimates. World Health Organization; 2001. http://www.who.int/hiv/pub/sti/who_hiv_aids_2001.02.pdf. Accessed 19 Sept 2011

28. Centralized Information System for Infectious Diseases (homepage on the Internet) (2011) Europe: World Health Organization. http://data.euro.who.int/cisid. Accessed 19 Sept 2011

29. May RM, Anderson RM (1987) Transmission dynamics of HIV infection. Nature 326(6):137–142

30. Brooks GF (1985) Pathogenesis and immunology of gonococcal infection. In: Brooks GF, Donegan EA (eds) Gonococcal infection. Edward Arnold, London, p 51–82

31. Holmes KK, Johnson DW, Trostle HJ (1970) An estimate of the risk of men acquiring gonorrhea by sexual contact with infected females. Am J Epidemiol 91(2):170–174

32. Hooper RR, Reynolds GH, Jones OG, Zaidi A, Wiesner PJ, Latimer KP et al (1978) Cohort study of venereal disease.1: the risk of gonorrhea transmission from infected women to men. Am J Epidemiol 108(2):136–144

33. Kigbu JH, Nyango DD (2009) A critical look on condoms. Niger J Med 18(4):354–359

34. Steiner MJ, Cates W Jr (2006) Condoms and sexually-transmiited infections. N Engl J Med 354(25):2642–2643

35. Warner L, Stone KM, Macaluso M, Buehler JW, Austin HD (2006) Condom use and risk of gonorrhea and Chlamydia: a systemic review of design and measurement factors assessed in epidemiologic studies. Sex Transm Dis 33(1):35–51

36. Bignell C, FitzGerald M (2011) UK National guideline for the management of gonorrhoea in adults. British Association for Sexual Health and HIV, 2011. http://www.bashh.org/documents/3611 Accessed 19 Sept 2011

37. Brunham RC (1997) Core group theory: a central concept in STD epidemiology. Venereology 10:34–39

38. Al-Tayyib AA, Rietmeijer CA (2011) Detecting Chlamydia and gonococcal infections through social and sexual networks. Sex Transm Infect. doi:10.1136/sextrans-2011-050102.59

39. Ford K, Sohn W, Lepkowski J (2002) American adolescents: sexual mixing patterns, bridge partners, and concurrency. Sex Transm Dis 29(1):13–19

40. Berlan ED, Holland-Hall C (2010) Sexually transmitted infections in adolescents: advances in epidemiology, screening, and diagnosis. Adolesc Med State Art Rev 21(2):332–346

41. Stoner BP, Whittington WL, Hughes JP, Aral SO, Homes KK (2000) Comparative epidemiology of heterosexual gonococcal and chlamydial networks: implications for transmission patterns. Sex Transm Dis 27(4):215–223

42. MacDonald NE, Wells GA, Fisher WA, Warren WK, King MA, Doherty JA et al (1990) High-risk STD/HIV behavior among college students. JAMA 263(23):3155–3159

43. MacDonald NE, Fisher WA, Wells GA, Doherty JA, Bowie WR (1994) Canadian street youth: correlates of sexually risk-taking activity. Pediatr Infect Dis J 13(8):690–697

44. Shafti T, Burnstein GR (2004) An overview of sexually transmitted infections in adolescents. Adolesc Med Clin 15(2):201–214

45. Robertson AA, Thomas CB, St Lawrence JS, Pack R (2005) Predictors of infection with chlamydia

or gonorrhea in incarcerated adolescents. Sex Transm Dis 32(2):115–122

46. Public Health Agency of Canada (2011) Street youth in Canada. Findings from enhanced surveillance of canadian street youth, 1999–2003 (Internet), March 2006. http://www.phac-aspc.gc.ca/std-mts/reports_06/youth-eng.php. Accessed 19 Sept 2011

47. Eaton DK, Kann L, Kinchen S, Shanklin S, Ross J, Hawkins J et al (2010) Youth risk behaviour surveillance-United States, 2009. MMWR Surveill Summ 59(5):1–142

48. Ford K, Lepkowski JM (2004) Characteristics of sexual partners and STD infection among American adolescents. Int J STD AIDS 15(4):260–265

49. Forhan SE, Gottlieb SL, Sternberg MR, Xu F, Datta SD, McQuillan GM et al (2009) Prevalence of sexually transmitted infections among female adolescents aged 14 to 19 in the United States. Pediatrics 124(6):1505–1512

50. Kahn JA, Rosenthal SL, Succop PA, Ho GYF, Burk RD (2002) Mediators of the association between age at first sexual intercourse and subsequent human papillomavirus infection. Pediatrics 109(1):5

51. Kaestle CE, Halpern CT, Miller WC, Ford CA (2005) Young age at first sexual intercourse and sexually transmitted infections in adolescents and young adults. Am J Epidemiol 161(8):774–780

52. Jayakody A, Sinha S, Tyler K, Khadr SN, Clark C, Klineberg E et al (2001) Early sexual risk among black and minority ethnicity teenagers: a mixed methods study. J Adolesc Health 48(5):499–506

53. Tu W, Batteiger BE, Wiche S, Ofner S, Van Der Pol B, Katz BP et al (2009) Time from first intercourse to first sexually transmitted infection diagnosis among adolescent women. Arch Pediatr Adolesc Med 163(12):1106–1111

54. Kann L, Kichen SA, Williams BI, Ross JG, Lowry R, Grunbaum JA et al (2000) Youth risk behaviour surveillance-United States, 1999. MMWR CDC Surveill Summ 49(5):1–32

55. Hook EW, Handsfield HH (2008) Gonococcal Infections in the Adult. In: Holmes K, Sparling P, Stamm W, Piot P et al (eds) Sexually transmitted infections, 4th edn. McGraw-Hill, China, p 627–646

56. Woods WR (2005) Gonococcal infections in neonates and young children. Semin Pediatr Infect Dis 16(4):258–270

57. Platt R, Rice PA, McCormack WM (1983) Risk of acquiring gonorrhea and prevalence of abnormal adnexal findings among women recently exposed to gonorrhea. JAMA 250(23):3205–3209

58. Edwards LE, Barrada MT, Hamman AA, Hakanson EY (1978) Gonorrhea in pregnancy. Am J Obstet Gynecol 132(6):637–641

59. Woods CR, Jr (2009) Gonococcal infections. In: Feigin RD, Cherry JD, Demmler-Harrison GJ, Kaplan SL (eds) Feigin & Cherry's textbook of pediatric infectious diseases, 6th edn. Elsevier, Philadelphia, p 1366–1393

60. Kinghorn G (2010) Pharyngeal gonorrhea: a silent cause for concern. Sex transm infect 86(6):413–414

61. Barry PM, Klausner JD (2009) The use of cephalosporins for gonorrhea: the impending problem of resistance. Expert Opin Pharmacother 10(4):555–577

62. Kinghorn GR, Rashid S (1979) Prevalence of rectal and pharyngeal infection in women with gonorrhoea in Sheffield. Br J Vener Dis 55(6):408–410

63. Hunte T, Alcaide M, Castro J (2010) Rectal infections with chlamydia and gonorrhoea in women attending a multiethnic sexually transmitted diseases urban clinic. Int J STD AIDS 21(12):819–822

64. Goldenberg, DL, Sexton DJ (2011) z gonococcal infection (Internet). Waltham, MA: UptoDate; 2011 (cited 2011 Sept 11). http://www.uptodate.com/contents/disseminated-gonococcal-infection

65. Pasquariello CA, Plotkin SA, Rice RJ, Hackney JR (1985) Fatal gonococcal septicemia. Pediatr Infect Dis 4(2):204–206

66. Rice PA (2005) Gonococcal arthritis (disseminated gonococcal infection). Infect Dis Clin North Am 19(4):853–861

67. US Preventative Services Task Force (2005) Screening for Gonorrhea: Recommendation Statement. Ann Fam Med 3(3):263–267

68. Ross JDC (2010) Gonorrhoea: to screen or not screen? Sex Transm Infect 86(6):411–412

69. Johnson RE, Newhall WJ, Papp JR, Knapp JS, Black CM, Gift TL et al (2002) Screening tests to detect Chlamydia trachomatis and Neisseria gonorrhoeae infections—2002. MMWR Recomm Rep 51(RR-15):1–38

70. Whiley DM, Tapsall JW, Sloots TP (2006) Nucleic acid amplification testing for Neisseria gonorrhoeae: an ongoing challenge. J Mol Diagn 8(1):3–15

71. Bachmann LH, Johnson RE, Cheng H, Markowitz L, Papp JR, Palella FJ Jr et al (2010) Nucleic acid amplification tests for diagnosis of Neisseria gonorrhoeae and Chlamydia trachomatis rectal infections. J Clin Microbiol 48(5):1827–1832

72. Fairley CK, Chen MY, Bradshaw CS, Tabrizi SN (2011) Is it time to move to nucleic acid amplification tests for pharyngeal and rectal gonorrhoea in men who have sex with men to improve gonorrhoea control? Sex Health 8(1):9–11

73. Centers for Disease Control and Prevention (2007) Update to CDC's sexually transmitted diseases treatment guidelines, 2006: fluoroquinolones no longer recommended for treatment of gonococcal infections. MMWR Morb Mortal Wkly Rep 56(14):332–336

74. Graseck AS, Shih SL, Peipert JF (2011) Home versus clinic-based specimen collection for Chla-

mydia trachomatis and Neisseria gonorrhoeae. Expert Rev Anti Infect Ther 9(2):183–194

75. Howard EJ, Xu F, Taylor SN, Stoner BP, Mena L, Nsuami MJ et al (2011) Screening methods for Chlamydia trachomatis and Neisseria gonorrhoeae infections in sexually transmitted infection clinics: what do patients prefer? Sex Transm Infect 87(2):149–151

76. Ison C (2006) GC NAATs: is the time right? Sex Transm Infect 82(6): 515

77. Bachmann LH, Johnson RE, Cheng H, Markowitz LE, Papp JR, Hook EW 3rd (2009) Nucleic acid amplification tests for diagnosis of Neisseria gonorrhoeae oropharyngeal infections. J Clin Microbiol 47(4):902–907

78. Razali MF, Fairley CK, Hocking J, Bradshaw CS, Chen MY (2010) Sampling technique and detection rates for pharyngeal gonorrhea using culture. Sex Transm Dis 37(8):522–524

79. Centers for Disease Control and Prevention (2010) Sexually transmitted diseases treatment guidelines, 2010. MMWR Recomm Rep 59(RR-12):1–110

80. Bonilla H, Kepley R, Pawlak J, Belian B, Raynor A, Saravolatz LD (2011) Rapid diagnosis of septic arthritis using 16S rDNA PCR: a comparison of 3 methods. Diagn Microbiol Infect Dis 69(4):390–395

81. Burstein GR, Murray PJ (2003) Diagnosis and management of sexually transmitted disease pathogens among adolescents. Pediatr Rev 24(3):75–82

82. Dees JE, Colston JAC (1937) The Use of Sulfanilamide in Gonococcic Infections. JAMA 108:1855–1858

83. Herrell WE, Cook EN, Thomspon L (1943) Use of Penicillin in Sulfonamide-resistant Gonorrhea Infections. JAMA 122(5):289–292

84. Whittington WL, Knapp JS (1988) Trends in resistance of Neisseria gonorrhoeae to antimicrobial agents in the United States. Sex Transm Dis 15(4):202–210

85. Reyn A, Korner B, Bentzon MW (1958) Effects of penicillin, streptomycin and tetracycline on N. gonorrhoeae isolated in 1944 and in 1957. Br J Vener Dis 34:227–239

86. Phillips I (1976) Beta lactamase-producing, penicillin-resistant gonococcus. Lancet 2(7987): 656–657

87. Scott GR, McMillan A, Young H (1987) Ciprofloxacin versus ampicillin and probenecid in the treatment of uncomplicated gonorrhoeae in men. J Antimicrob Chemother 20(1):177–121

88. Dan M (2004) The use of fluoroquinolones in gonorrhoea: the increasing problem of resistance. Expert Opin Pharmacother 5(4):829–854

89. Health Protection Agency (2011) Susceptibility testing of N. gonorrhoeae (Internet) 2009a. http://www. hpa.org.uk/web/HPAweb&HPAwebStandard/ HPAweb_C/1195733778434. Accessed 19 Sept 2011

90. Health Protection Agency (2011) STIs Annual Slide Set 2000–2009 (Internet). 2009b. http:// www.hpa.org.uk/Topics/InfectiousDiseases/InfectionsAZ/STIs/STIsAnnualDataTables/Annual STISlideset/. Accessed 19 Sept 2011

91. Centers for Disease Control and Prevention (2011) Cephalosporin susceptibility among Neisseria gonorrhoeae Isolates—United States, 2000–2010. MMWR Morb Mortal Wkly Rep 60(26):873–877

92. Centers for Disease Control and Prevention (2012) Update to CDC's sexually transmitted diseases treatment guidelines, 2010: oral cephalosporins no longer a recommended treatment for gonococcal infections. MMWR 61(31):590–594

93. Chisholm SA, Mouton JW, Lewis DA, Nicols T, Ison CA, Livermore DM (2010) Cephalosporin MIC creep among gonococci: time for a pharmacodynamic rethink? J Antimicrob Chemother 65(10):2141–2148

94. Cole MJ, Chisholm SA, Unemo M, Hoffmann S, van de Laar MJW, Ison CA (2011) European gonococcal antimicrobial surveillance programme (Euro-GASP): towards timelier monitoring. Sex Transm Infect. doi:10.1136/ sextrans-2011-050102.63

95. Kirkcaldy RD, Ballard RC, Dowell D (2011) Gonococcal resistance: are cephalosporins next? Curr Infect Dis Rep 13(2):196–204

96. Martin IM, Hoffmann S, Ison CA (2006) European surveillance of sexually transmitted infections (ESSTI): the first combined antimicrobial susceptibility data for Neisseria gonorrhoeae in Western Europe. J Antimicrob Chemother 58(3):587–593

97. WHO Western Pacific Gonococcal Antimicrobial Surveillance Programme (2008) Surveillance of antibiotic resistance in Neisseria gonorrhoeae in the WHO Western Pacific Region. Commun Dis Intell 32(1):48–51

98. Wang SA, Lee MV, O'Connor N, Iverson CJ, Ohye RG, Whitican RM et al (2003) Multidrugresistant Neisseria gonorrhoeae with decreased susceptibility to cefixime-Hawaii. Clin Infect Dis 37(6):849–852

99. Tapsall JW (2009) Neisseria gonorrhoeae and emerging resistant to extended spectrum cephalosporins. Curr Opin Infect Dis 22(1):87–91

100. Deguchi T, Yasuda M, Yokoi S, Ishida K, Ito M, Ishihara S et al (2003) Treatment of uncomplicated gonococcal urethritis by double-dosing of 200 mg cefixime at a 6 h interval. J Infect Chemother 9(1):35–39

101. Ito M, Yasuda M, Yokoi S, Ito S, Takahashi Y, Ishihara S et al (2004) Remarkable increase in central Japan in 2001–2002 of Neisseria gonorrhoeae isolates with decreased susceptibility to penicillin, tetracycline, oral cephalosporins, and fluoroquinolones. Antimicrob Agents Chemother 48(8):3185–3187

102. Tapsall J, Read P, Carmody C, Bourne C, Ray S, Limnios A et al (2009) Two cases of failed ceftri

axone treatment in pharyngeal gonorrhoeae verified by molecular microbiological methods. J Med Microbiol 58(Pt 5):683–687

103. Unemo M, Golparian D, Hestner A (2011) Ceftriaxone treatment failure of pharyngeal gonorrhea verified by international recommendations, Sweden, July 2010. Euro Surveill 16(6): pii:19792. http://www.eurosurveillance.org/ViewArticle.aspx?ArticleId=19792

104. Ohnishi M, Saika T, Hoshina S, Iwasaku K, Nakayama S, Watanabe H et al (2011) Ceftriaxone-resistant *Neisseria gonorrhoeae,* Japan. Emerg Infect Dis 17(1):148–149

105. Newman LM, Moran JS, Workowski KA (2007) Update on the management of Gonorrhea in adults in the United States. Clin Infect Dis 44(Suppl 3):84–101

106. Deguchi T, Nakane K, Yasuda M, Maeda S (2010) Emergence and spread of drug resistant Neisseria gonorrhoeae. J Urol 184(3):851–858

107. Public Health Agency of Canada (2011) Canadian guidelines on sexually transmitted infections, 2008 Edition (Internet). Ottawa, ON: Public Health Agency of Canada; 2008. http://www.phac-aspc.gc.ca/std-mts/sti-its/guide-lignesdir-eng.php. Accessed 19 Sept 2011

108. Pichichero ME (2005) A review of evidence supporting the American Academy of Pediatrics recommendation for prescribing cephalosporin antibiotics for penicillin-allergic patients. Pediatrics 115(4):1048–1057

109. Bignell C, Garley J (2010) Azithromycin in the treatment of infection with Neisseria gonorrhoeae. Sex Transm Infect 86(6):422–426

110. Ison CA, Hussey J, Sankar KN, Evans J, Alexander S (2011) Gonorrhoea treatment failures to cefixime and azithromycin in England. Euro Surveill 16(14):pii: 19833. http://www.eurosurveillance.org/images/dynamic/EE/V16N14/art19833.pdf

111. Daly CC, Hoffman I, Hobbs M, Maida M, Zimba D, Davis R et al (1997) Development of an antimicrobial susceptibility surveillance system for Neisseria gonorrhoeae in Malawi: comparison of methods. J Clin Microbiol 35(11):2985–2989

112. Ieven M, Van Looveren M, Sudigdoadi S, Rosana Y, Goossens W, Lammens C et al (2003) Antimicrobial susceptibilities of Neisseria gonorrhoeae strains isolated in Java, Indonesia. Sex Trans Dis 30(1):25–29

113. Lkhamsuren E, Shultz TR, Limnios EA, Tapsall JW (2001) The antibiotic susceptibility of Neisseria gonorrhoeae isolated in Ulaanbaatar, Mongolia. Sex Transm Infect 77(3):218–219

114. Brown LB, Krysiak R, Kamanga G, Mapanie C, Kanyamula H, Banda B et al (2010) Neisseria gonorrhoeae antimicrobial susceptibility in Lilongwe Malawi, 2007. Sex Transm Dis 37(3):169–172

115. Vakulenko SB, Mobashery S (2003) Versatility of aminoglycosides and prospects for their future. Clin Microbiol Rev 16(3):430–450

116. Jephcott AE (1997) Microbiological diagnosis of gonorrhoea. Genitourin Med 73(4):245–252

117. Fung M, Scott KC, Kent CK, Klausner JD (2007) Chlamydial and gonococcal reinfection among men: a systematic review of data to evaluate the need for retesting. Sex Transm Infect 83(4):304–309

118. Hosenfeld CB, Workowski KA, Berman S, Zaidi A, Dyson J, Mosure D et al (2009) Repeat infection with chlamydia and gonorrhea among females: a systematic review of the literature. Sex Transm Dis 36(8):478–489

119. Kissinger PJ, Reilly K, Taylor SN, Leichliter JS, Rosenthal S, Martin DH (2009) Early repeat Chlamydia trachomatis and Neisseria gonorrhoeae infections among heterosexual men. Sex Transm Dis 36(8):498–500

120. Peterman TA, Tian LH, Metcalf CA, Satterwhite CL, Malotte CK, DeAugustine N et al (2006) High incidence of new sexually transmitted infections in the year following a sexually transmitted infection: a case for rescreening. Ann Intern Med 145(8):564–572

121. Du P, Coles B, Gerber T, McNutt LA (2007) Effects of partner notification on reducing Gonorrhea incidence rate. Sex Transm Dis 34(4):189–194

122. Kissinger P, Schmidt N, Mohammed H, Leichliter JS, Gift TL, Meadors B et al (2006) Patient-delivered partner treatment for Trichomonas vaginalis infection: a randomized controlled trial. Sex Transm Dis 33(7):445–450

123. Saperstein AK, Firnhaber GC (2010) Should you test or treat partners of patients with gonorrhea, chlamydia or trichomoniasis? J Fam Practice 59(1):46–48

124. DiClemente RJ, Crosby RA (2006) Preventing sexually transmitted infections among adolescents: the glass is half full. Curr Opin Infect Dis 19(1):39–43

125. Harrison WO, Hooper RR, Weisner PJ, Campbell AF, Karney WW, Reynolds GH et al (1979) A trial of minocycline given after exposure to prevent gonorrhea. NEJM 300(19):1074–1078

126. Holmes KK, Johnson DW, Kvale PA, Halverson CW, Keys TF, Martin DH (1996) Impact of a Gonorrhea control program, including selective mass treatment in female sex workers. J Infect Dis 174(Suppl 2):230–9

127. Manhart LE, Holmes KK (2005) Randomized controlled trials of individual-level, population-level, and multileveled interventions for preventing sexually transmitted infections: what has worked? J Infect Dis 191(Suppl 1):7–24

128. Jourden J, Etkind P (2004) Enhancing HIV/AIDS and STD prevention through program integration. Public Health Rep 119(1):4–11

129. Yee L, Simon M (2010) The role of the social network in contraception decision-making among young, african american and latina women. J Adolesc Health 47(4):374–380

130. Kim CR, Free C (2008) Recent evaluations of the peer-led approach in adolescent sexual health education: a systemic review. Int Fam Plann Perspect 34(2):89–90

131. Berenson AB, Wu ZH, Breitkopf CR, Newman J (2006) The relationship between source of sexual information and sexual behavior among female adolescents. Contraception 73(3):274–278

132. Borzekowski DLG, Rickert VI (2001) Adolescent cybersurfing for health information: a new resource that crosses barriers. Arch Pediatr Adolesc Med 155(7):813–817

133. Gray NJ, Klein JD, Noyce PR, Sesselberg TS, Cantrill JA (2005) Health information-seeking behaviour in adolescence: the place of the internet. Soc Sci Med 60(7):1467–1478

134. Vance K, Howe W, Dellavalle RP (2009) Social internet sites as a source of public health information. Dermatol Clin 27(2):133–136

135. Ward H, Robinson AJ (2006) Still waiting: poor access to sexual health services in the UK. Sex Transm Infect 82(1):3

136. Jones R, Menton-Johansson JR, Waters AM, Sullivan AK (2010) eTriage—a novel, web-based triage and booking service: enabling timely access to sexual health clinics. Int J STD AIDS 21(2):30–33

137. Shamos SJ, Mettenbrink CJ, Subiadur JA, Mitchell BL, Rietmeijer CA (2008) Evaluation of a testing-only "express" visit option to enhance efficiency in a busy STI clinic. Sex Transm Dis 35(4):336–340

138. Ruiz MS, Gable AR, Kaplan EH, Stoto MA, Fineberg HV, Trussell J (eds) (2001) No time to lose: getting more from HIV prevention. National Academy Press, Washington DC

139. Santelli J, Ott MA, Lyon M, Rogers J, Summers D (2006) Abstinence-only education policies and program: a position paper of the Society for Adolescent Medicine. J Adolesc Health 38(1):83–87

140. World Health Organization (2011) Regional Office for Europe. Sexually transmitted infections. http://data.euro.who.int/cisid/Default.aspx?TabID=272714. Accessed 19 Sept 2011

141. European Centre for Disease Control and Prevention (2011) Annual epidemiological report on communicable diseases in Europe, 2010. http://ecdc.europa.eu/en/publications/Publications/1011_SUR_Annual_Epidemiological_Report_on_Communicable_Diseases_in_Europe.pdf. Accessed 19 Sept 2011

142. Centers for Disease Control and Prevention (2011) Sexually Transmitted Diseases Surveillance, 2008. http://www.cdc.gov/std/stats08/tables/1.htm Accessed 19 Sept 2011

# Management of Severe Malaria: Results from Recent Trials

**20**

Peter Olupot-Olupot and Kathryn Maitland

**Abstract**

Globally, malaria remains a substantial public health burden with an estimated 349–552 million clinical cases of *P. falciparum* malaria each year—leading to 780,000 deaths directly attributable to the disease. Whilst the outcome from severe malaria in Africa children remains poor, recent developments in the management of malaria have come from two key sources—the introduction of new, safe and rapidly-effective anti-malarials and high quality evidence from two of the largest clinical trials ever conducted in African children with severe malaria. As a result, the time-honoured anti-malarial treatment for severe malaria, quinine, will now be replaced by artesunate, a water-soluble artemisinin derivative. Supportive care, specifically the management of shock, has been informed by a large late phase clinical trial which concluded that bolus resuscitation is harmful and therefore should be avoided in children with severe malaria, including the high risk group with severe metabolic acidosis and advanced shock.

## 20.1 Background

Of the five species of malaria that infect humans, *Plasmodium falciparum* causes most severe and fatal disease in the tropics. In recent years, *Plasmodium vivax* has been also been described as causing severe and complicated malaria in Asia [1].

K. Maitland (✉)
KEMRI-Wellcome Trust Programme, Kilifi, Kenya
e-mail: kathryn.maitland@gmail.com

Department of Paediatrics, Imperial College, London

P. Olupot-Olupot
Mbale Regional Referral Hospital, Mbale, Uganda

Globally, the most reliable recent estimate indicates that there were 349–552 million clinical cases of *P. falciparum* malaria in 2007 [2], with up to 780,000 deaths directly attributable to the disease [3]. The burden and spectrum of malaria varies from one geographical region to another, with intensity of transmission and between adults and children. There have been some encouraging reports of declining malaria incidence, but these relate to selected areas and may not reflect the general situation [4]. Better malaria control, early diagnosis and improved drug treatment may each contribute to changing malaria epidemiology but overall trends are not yet comprehensively understood. In areas with high malaria transmission,

N. Curtis et al. (eds.), *Hot Topics in Infection and Immunity in Children IX,*
Advances in Experimental Medicine and Biology 764, DOI 10.1007/978-1-4614-4726-9_20,
© Springer Science+Business Media New York 2013

**Table 20.1** Clinical signs of severe malaria

| High priority: emergency management | | |
|---|---|---|
| Clinical | Laboratory | Intermediate risk: need for high dependency care |
| Hypoxia (oxygen saturations <92 %) | Severe anaemia haemoglobin <6 g/dl | Haemoglobin 6–10 g/dl |
| Respiratory distress | Hypoglycaemia <3 mmol | History of convulsions during this illness |
| Hypotension (systolic blood pressure <50, <60 or <70 mmHg if <1 year, 1–5 years or >5 years) | Metabolic acidosis (base deficit >8) | Hyperparasitaemia >5 % |
| Depressed conscious state (V, P or U on AVPU scale) | Lactate ≥5 mmol/l | Visible jaundice |
| Abnormal posturing | Creatinine >80 mmol/l | *P. falciparum* in a child with sickle cell disease |
| Inter-current seizures | Severe hyperkalaemia (potassium >5.5 mmol/l) | |

incidence has not changed and in some areas increases in disease burden have been reported [5, 6].

Prompt diagnosis and treatment is crucial to prevent mortality especially in high-risk groups including young children and some visitors, who have little or no immunity. Travellers, of any age, from non-endemic areas are particularly at risk, including those formerly from endemic areas and returning for short or long visits to their home or other endemic areas, because naturally-acquired immunity wanes rapidly. In the past decade there have been substantial therapeutic advances in the management of both mild and severe disease, with newer drugs, in particular artemisinin based combinations replacing time-honoured chloroquine and quinine. These developments are welcome as parasite resistance to malaria treatments continues to threaten malaria control and effective case management globally. This chapter focuses on the management of severe malaria with specific reference to the latest developments in treatment.

In 2010 and 2011, two of the largest clinical trials ever conducted in African children with severe malaria were published, both with groundbreaking results [7, 8]. Most of the latter part of this chapter is dedicated to discussing these trials. Both manuscripts are openly accessible on-line and we recommend that readers refer to these manuscripts as necessary. By way of introduction and to put into to context the relevance of the trials to the clinical management of paediatric malaria in Africa and, by extrapolation, to the management of children in the UK, we first describe the clinical spectrum of severe malaria.

## 20.2 What Defines Severe Malaria?

For decades severe malaria in African children had been rationalised into two non-over-lapping major syndromic presentations, those of cerebral malaria and severe malarial anaemia [9, 10]. Research over the years has established that severe acidosis, clinically manifesting as respiratory distress (Kausmaul's breathing), is a common complication and the leading independent factor predicting fatal outcome [11, 12]. Severe falciparum malaria has more recently been described as a complex syndrome, affecting many organ systems, which has many features in common with sepsis syndrome [13]. Indeed, clinical distinction between severe malaria and other common causes of life-threatening febrile illnesses including bacterial sepsis, pneumonia and CNS infection is often impossible at the time of presentation [14] (see Table 20.1 for summary of clinical characteristics). The most comprehensive description of the clinical spectrum of severe malaria in African children is the AQUAMAT trial [7]. The trial involved 11 centres in 9 countries including a mixture of urban and rural hospitals and populations with differing malaria endemicities. All 5,425 participants in the trial had clinical evidence of severe malaria and the diagnosis was reinforced by quality controlled laboratory evidence of *Plasmodium falciparum* infection. Overall, at baseline, 37 % had coma, 32 % had convulsions, 9 % had compensated shock (capillary refilling time greater or equal to 3 s) and 3 % decompensated shock (systolic blood pressure < 70 mmHg and cool peripheries). Only 2 % had clinical jaundice. Among admission laboratory parameters, 30 % had severe anaemia (haemoglobin <5 g/dl), 10 % were hypoglycaemic (blood sugar level <3 mmol/l) and 43 % had severe acidosis (base deficit >8). Co-morbidities included 13 % with (clinician diagnosed) sepsis (including 10 % with culture proven sepsis); 13 % with radiological evidence of pneumonia and 2 % with HIV, malnutrition or meningitis. Whilst dark urine (or haemoglobinuria also known as 'blackwater fever') and clinical jaundice were infrequent in these centres, there are other areas in Africa where these complications are more common (Olupot-Olupot, unpublished data). Key independent predictors of fatal outcome included acidosis, cerebral involvement, renal impairment and comorbidity [15].

We have recently published a review of the features of severe malaria and proposed UK management guidelines [16], which are based on information obtained from personal archives of references, the current World Health Organization (WHO) guidelines [17] and the Advanced Paediatric Life Support (APLS) guidelines [18]. Our recommendations for diagnosis are largely unchanged with the exception of shock—which we suggest should be defined by low systolic blood pressure, the rationale for which is discussed below as are the implications of recent results for therapeutic and supportive management recommendations.

### 20.2.1   Bacterial Infection

Comorbidity in children presenting to hospital is common in malaria-endemic Africa. Invasive bacterial disease is reported in 5–12 % of children with recent or current malaria, commonly Gram-

negative (GN) infections [19, 20]. However, the sensitivity of blood cultures to detect septicaemia is low, so it is likely that the figures are underestimates [21]. Non-typhoidal salmonellae (NTS) are the most common isolate, particularly in younger children suffering from moderate or severe anaemia [22, 23]. The mechanisms underlying this apparent susceptibility remain uncertain. Since children in malaria-endemic areas often present to hospital with malaria parasitaemia—there is a possibility that the association with bacteraemia is largely observer bias. However, the most persuasive data that there is a significant biological association come from the study of Nadjm and colleagues in Tanzania [20], describing the spectrum of bacteraemia in unselected children admitted to hospital. Three subgroups were described, children who were malaria-negative, cases with recent malaria (malaria antigen positive but slide negative) and children with positive malaria slides. Among malaria-negative children with bacteraemia, 51 % had Gram-positive organisms isolated and 49 % Gram-negative (GN). Among children with recent or intercurrent malaria 76 % of bacteraemias were with non-*Haemophilus influenzae* GN organisms (of which 90 % were NTS). There was also a trend towards increasing proportion of non-*Haemophilus influenzae* (and non-NTS) Gram-negative enteric ('other' GN) organisms with increasing density of malaria parasitaemia. However, invasive bacterial disease is much less commonly reported in imported malaria in non-endemic settings [20].

## 20.3   Recent Changes in the Management of Severe Malaria

### 20.3.1   The AQUAMAT Trial

#### 20.3.1.1   Background
Quinine has been used to treat malaria for over three centuries and until recently was the mainstay of treatment in African children with severe disease. In South East Asia the problem of quinine resistance, the substantial case fatality of severe malaria in adults and the availability of artemisinin derivatives prompted the performance of a

large multicentre randomised trial (SEAQUA-MAT) comparing parenteral artesunate with quinine in four countries [24]. The trial included 1,461 patients, most of whom were adults, and was stopped early by the data monitoring and safety committee following a planned interim analysis, which showed a substantial benefit in favour of artesunate. Mortality in patients receiving artesunate was 17 % (107/730) compared to 22 % (164/731) in patients receiving quinine. This equates to a relative reduction in mortality following artesunate compared to quinine of 34.7 % (95 % CI 18.5–47.6 %; $p = 0.0002$). Since this trial, artesunate has been the drug recommended for treatment of severe malaria in adults by the WHO.

A separate trial in African children was justified on several grounds. First, there were important differences between the clinical manifestations, the tempo of the disease and the response to treatment in African children with severe malaria compared with adults in South East Asia. In severe malaria in children most deaths occur within the first 24 h—far sooner than in most adults in the SEAQUAMAT trial. This provides a much narrower time-window for any advantage of artesunate in killing young ring stage parasites more rapidly to translate into clinical benefit. Second, malaria parasites in Africa are more quinine sensitive, with no reported quinine resistance at the time of the trial. Third, drug cost and availability were also potential issues. Finally, previous trials in children in Africa comparing artemether with quinine found little evidence of superiority. However, artemether is an oil-based formulation of artemesinin, which needs to be given by intramuscular injection. Pharmacokinetic studies showed that the drug is only slowly and erratically released from the intramuscular injection site [25, 26] suggesting a possible reason for the previous failure to show a significant difference between quinine and artemisinin derivatives in African children. The advantages of intravenous (IV) artesunate include its potency compared to artemether [27] and instant bioavailability by both IV and intramuscular routes leading to rapid peak concentrations within 1 h [26, 28]. All these considerations

supported the need for a conclusive trial to establish best practice.

### 20.3.1.2 Study Design

The AQUAMAT was conducted in nine African countries, and ran between October 2005 and July 2010 when the recruitment target was reached. Participating hospitals included university hospitals, research institutes and rural hospitals without prior research experience. None of the facilities had intensive care units—so the large majority of children were managed pragmatically on general paediatric wards. Children were eligible if they were under 15 years old with clinical evidence of severe malaria and a positive rapid malaria test (Optimal) for *Plasmodium falciparum* malaria (indicating current malaria infection). Treatment, at recommended dosages, with parenteral quinine or an artemisinin derivative for more than 24 h before admission were the only exclusion criteria. The trial was an open label trial of both antimalarials given parenterally (either by intramuscular or intravenous administration). Randomisation was masked as study numbers were kept inside opaque sealed paper envelopes and opened in numerical order. The card inside directed the clinician to a separate, sealed trial pack which contained the assigned intervention was on the first page. Blinding was not possible owing to the different methods of drug dilution and administration—artesunate requiring a two-step procedure for preparation and intravenous quinine needs to be given as an infusion, whereas artesunate is given as an injection. The interventions were given until the child was able to take an oral combination treatment, Coartem® (Artemether-Lumafantrine)—which was used to complete the treatment course. The primary outcome was in-hospital mortality and secondary outcomes included the incidence of severe complications and neurological sequelae.

### 20.3.1.3 Results

By intention to treat 572 (9.7 %) of the 5,425 children enrolled in the trial died. Whilst baseline characteristics of the children in the two arms were similar, mortality was substantially lower in the artesunate group 230/2,712 (8.5 %) compared to 297/2,713 (10.9 %) in the quinine arm. Comparing quinine to artesunate-treated children, the relative risk of death was 22.5 % (95 % CI 8.1–36.9) greater in the quinine arm ($p=0.002$). There was no difference in outcome whether the drugs were administered intravenously or intramuscularly, nor were there any differences in outcome across any of the major clinical spectra defining severe malaria.

The development of coma and/or convulsion was more common during hospital admission in the quinine arm ($p=0.02$); however there were no other differences between the arms in number of complications, additional treatments prescribed, time to recovery or numbers with neurological sequelae—which included 61 persistent mild, moderate or severe such sequelae among survivors.

### 20.3.1.4   Outcome: Change of Policy

Based upon the results of the SEAQUAMAT trial and now the AQUAMAT trial, parenteral artesunate is now recommended by the WHO and being adopted by individual countries as policy as the first line treatment for severe malaria in both adults and children. Artesunate should be given at admission then at 12 and 24 h, and once a day thereafter until the children can take and retain oral medication. Treatment must then be completed with a full course of artemisinin based combination therapy (ACT). If parenteral artesunate is not available, artemether or quinine are acceptable alternatives.

### 20.3.2   FEAST Trial

#### 20.3.2.1   Background

The FEAST trial (Fluid Expansion As Supportive Treatment) was the first large controlled trial of fluid resuscitation comparing the use of early bolus resuscitation with local standard of care (no bolus). At the time this trial was conducted, the recommendations for fluid resuscitation in Africa were substantially different from those in other parts of the world [29] and use of fluid resuscitation was highly controversial in children with severe malaria [30–32]. Although there was

evidence to suggest that hypovolaemia contributes to the pathophysiology of severe malaria [33, 34], there were conflicting recommendations as to how it should be treated. Published WHO guidelines suggested that children might respond favourably to extra fluid in the presence of concomitant dehydration. However, specific recommendations regarding the type, rate or volume of fluid administration were not provided [35]. Use of invasive central venous pressure monitoring was recommended to monitor fluid replacement [35]. However, for the vast majority of children in sub-Saharan African hospitals this was neither available nor feasible. Therefore most such children with severe malaria were receiving no specific fluid management apart from blood transfusion in some cases, as this is recommended for the management of severe anaemia [36].

FEAST was designed to address this controversy [8]. However, severe malaria and bacteraemic sepsis syndrome are similar and simple bedside and laboratory assessment cannot distinguish between them [14, 37–40]. The FEAST trial was designed as a pragmatic trial incorporating both populations in order to provide practical information for hospitals with few diagnostic facilities.

#### 20.3.2.2   Choice of Study Fluids

The trial was an open label randomised controlled trial of fluid resuscitation. Comparisons were at three levels; firstly, 0.9 % saline (a crystalloid) was compared with 5 % human albumin solution (colloid). Secondly, each of the fluids was compared to no bolus (control arm) and thirdly, both fluids were compared against no bolus (control). The inclusion of albumin (HAS) was based upon the results of three Phase II trials in Kenya comparing different fluids in children with severe malaria and signs of hypovolaemia [41–43]. The first trial had shown that up to 40 ml/kg of 0.9 % saline or HAS was safe and corrected haemodynamic indices of hypovolaemia [41]. The second trial was an open label, randomised comparison of saline and HAS in 117 children with severe malaria and acidosis. Mortality in children treated with HAS was 3.6 % (2/56), compared to 18 % (11/61) in those treated with

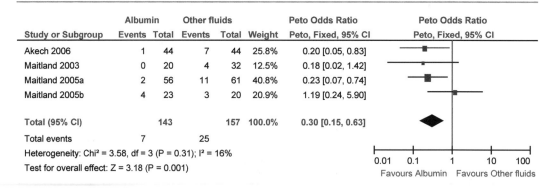

**Fig. 20.1** Forrest plots comparing effect of albumin versus other resuscitation fluids

saline (p=0.01) [42]. In the subgroup of children with coma, only 1/21 (5 %) of children receiving HAS died, compared with 11/24 (46 %) who received saline (p=0.002). Possibly owing to the greater survival rate in the HAS arm there was a non-significantly higher rate of neurological deficit (11 %; 6/54) compared to the saline arm (4 %, 2/50) [42]. In a further trial, mortality among children treated with HAS was lower (2.3 %; 1/44) than among those treated with the synthetic colloid Gelofusine (16 %; 7/44) [43]. In the sub group with coma, 1/ 25 (4 %) of HAS recipients died compared to 6/23 (26 %) in the Gelofusine arm (p=0.04). Neurological sequelae complicated 3/43 (7 %) survivors receiving HAS and 1/37 (3 %) survivors receiving Gelofusine [43]. A meta-analysis of the combined data from the four trials (Fig. 20.1) demonstrated a reduced mortality with HAS compared to saline or other colloids (Peto odds ratio 0.30; 95 % confidence interval 0.15, 0.63; p=0.001 and test for heterogeneity-$\chi^2$=3.58, p=0.31) but also noted the potential with such small sample sizes to overestimate effect. On the basis of the evidence from these previous trials, it was decided to include randomisation to HAS in the FEAST trial.

### 20.3.2.3  FEAST Trial Design

Children aged >60 days presenting with severe febrile illness and signs of impaired perfusion in six centres in three East African countries (Uganda, Kenya and Tanzania) were eligible for inclusion. Cases of gastroenteritis, severe malnutrition and non-medical causes (burns and trauma) were excluded from the trial, since the design and interventions were not relevant to these easily identifiable clinical presentations. Children were enrolled in two strata according to systolic blood pressure at presentation: 3,141 children without severe hypotension were enrolled in stratum A and received either an immediate bolus of 20 ml/kg given over 1 h (later increased to 40 ml/kg after protocol amendment) of 5 % HAS (1,050 children) or 0.9 % saline solution (1,047 children) or no bolus as controls (1,044 children); stratum B enrolled 29 children with severe hypotension (systolic blood pressure <50 mmHg if <12 m; <60 mmHg if 1–5years; <70 mmHg if >5 years) were randomized to receive either 40 ml/kg (later increased to 60 ml/kg after protocol amendment) HAS or saline boluses—there were no control arm in stratum B. After 1 h saline-bolus and albumin-bolus arms, but not the control arm, received an additional 20 ml/kg bolus if signs of impaired perfusion persisted. Beyond this time point further fluid boluses in all three arms were only prescribed for severe hypotension (see above definition)—where an additional 40 ml/kg boluses of study fluid (saline for the control arm) was given over 1 h. No crossover between the saline and albumin was permitted [8].

### 20.3.2.4  Supportive Care

All children were managed on general paediatric wards. Ventilation facilities other than short-term 'bag and mask' support were unavailable. Training in triage and emergency paediatric life support was given throughout the trial to optimize case recognition, supportive management and ensure protocol adherence. Basic infrastructural support for emergency care, oxygen saturation and automatic blood pressure monitoring was provided. All children received maintenance fluids, anti-malarials and/or antibiotics as required. Throughout the hospital admission time, episodes of hypotensive shock within 48 h and Adverse Events (AE) potentially related to fluid resuscitation Pulmonary Edema (PE), Raised Intracranial Pressure (RICP) and severe allergic reaction were actively solicited. AEs were sent to the Clinical Trials Facility, Kilifi, Kenya within 2 days where they were reviewed by an independent clinician. All AEs were subsequently monitored on site against source documents by visiting monitors. Twenty-eight-days reviews for neurological deficit were performed by an independent clinician, blind to treatment allocation. Children with neurological impairment at day 28 were re-assessed at 6 months.

The primary endpoint was mortality at 48-h after randomisation. Secondary endpoints included mortality at 28-days, neurological sequelae at 28-days and 24 weeks, episodes of hypotensive shock within 48 h of randomisation and adverse events related to fluid resuscitation (pulmonary oedema, intracranial hypertension or severe allergic reaction among those receiving HAS).

### 20.3.2.5  Results

Among the 3,141 children in stratum A, there were no major differences in baseline characteristics across the three arms. Median age was 24 months (IQR 13–38); 62 % had prostration, 15 % were comatose and 83 % had increased work of breathing. Moderate-severe acidosis was present in 1,070 (52 %) and severe lactic acidosis in 1,159 (39 %). Severely anaemic children were 987 (32 %) and 187 (6 %) had hypoglycaemia. Malaria was confirmed in 1,793 (57 %); 126/1,070 (12 %) had bacteraemia and 4.4 % were HIV seropositive.

In stratum A, 48-h mortality was 111/1,050 (10.6 %), 110/1,047 (10.5 %) and 76/1,044 (7.3 %) in HAS-bolus, saline-bolus and control arms, respectively. Risk ratios (95 % confidence intervals): saline-bolus versus control was 1.44 (1.09–1.90, $p=0.01$); HAS-bolus versus saline-bolus was 1.01 (0.78–1.29, $p=0.96$); bolus (HAS

or saline) versus control was 1.45 (1.13–1.86, $p=0.003$). Most fatalities occurred before 24 h (259; 87 %). The small number of fatalities occurring after 48 h provided no evidence that children in the control arm had excess delayed mortality [8].

In stratum B (children with severe hypotension and impaired perfusion) a total of 29 children, were enrolled with a median systolic blood pressure of 57 mmHg (interquartile range, 51–59). Thirteen were randomised to HAS of whom 9 (69 %) died and 16 randomised to saline and 9 died (56 %) died (relative risk of 48 h mortality with HAS bolus, 1.23; 95 % CI, 0.70–2.16; $p=0.45$).

Excess mortality associated with bolus arms compared to control was consistent across all pre-specified subgroups. Moreover, there was no difference in outcome between the two bolus arms (for all saline versus albumin comparisons), including the children in coma among whom we had hypothesised that albumin infusion might be neuroprotective [13]. At 28 days, HAS-saline-control mortality and neurological abnormality rates were 12 %, 12 % and 8.7 % ($p=0.004$ for bolus versus control) and 2.1 %, 1.3 %, 1.8 % ($p=0.85$ for bolus versus control); pulmonary oedema or raised intracranial pressure were reported in 30, 24 and 17 children, respectively.

### 20.3.2.6  Interpretation

Since the paper was published, there has been considerable debate through commentary papers, letters, at journal clubs and web-based discussion groups about the interpretation of the FEAST trial results [44–49]. The FEAST trial was even the subject of a pro/con debate at the IIC Meeting, Oxford July 2011. Many commentators on the FEAST trial have speculated on the reason for this surprising result, some raising concerns about the broad entry criteria. One of the major causes for concern was the definition of shock. In a subsequent letter to the NEJM, responding to nine of the numerous letters selected for publication, the investigating group presented new data on the relevance of the FEAST trial cohort to other definitions of shock [50]. These data are summarised in Table 20.2.

**Table 20.2** Application of various shock definitions to FEAST admission data

| | Shock definition | Mortality (%) | | | Absolute risk difference (95 % CI) |
|---|---|---|---|---|---|
| | | Overall (all arms) | Bolus (saline or albumin) | No bolus (control) | |
| FEAST inclusion criteria | Impaired consciousness (prostration or coma) and/or respiratory distress plus ≥ 1 of: CRT > 2 s; lower limb temperature gradient; weak pulse; HR > 180 (< 12 months), > 160 (12 months to 5 years), > 140 (>5 years) | 297/3,141 (9.5 %) | 221/2,097 (10.6 %) | 76/1,044 (7.3 %) | 3.3 % (1.2–5.3) |
| *Paediatric shock scores* | | | | | |
| ACCM-PALS | Pyrexia or hypothermia (core temperature > 37.40 C or < 36.0 °C) with either: hypotension (SBP < 70 mmHg (< 12 m); < 80 mmHg (≥ 12 m) Or ≥ 2 of: altered mental status (agitation, prostration or inability to localize a painful stimulus); CRT > 2 s; temperature gradient; bounding or weak pulse; HR > 180 (< 12 months), > 160 (12 months to 5 years), > 140 (>5 years) | 194/2,030 (10 %) | 150/1,389 (11 %) | 44/641 (7 %) | 3.9 % (1.4–6.5) |
| Surviving sepsis campaign | Lactate > 5 mmol/l | 216/1,159 (19 %) | 157/764 (21 %) | 59/395 (15 %) | 5.6 % (1.1–10.1) |
| WHO | Presence of cold hands or feet with both CRT 4 or more seconds and a weak, fast pulse | 27/65 (42 %) | 24/50 (48 %) | 3/15 (20 %) | 28 % (3.4–52.5) |

## 20.3.4  Implications for Policy

The results of the FEAST trial challenge the primacy of bolus resuscitation as a life-saving intervention in paediatric emergencies. The trial results showed that 89.4 % of those given boluses survived the first 48 h in hospital. But those given only maintenance fluids did better: 92.7 % of them survived. Put another way, this means that compared to maintenance fluids, boluses cause more than three children (3.3 %) to die out of every hundred treated.

These results are applicable to Africa, under the conditions that prevailed in the FEAST trial. There may be reasons why fluid boluses are harmful in such African children including specific range of pathogens and nutritional status. However, there should be further research to re-examine the physiological changes in shock and host response. For severe malaria, which was present in 57 % of the study population, the results are clear and should inform future policy. Hypovolaemic shock in severe malaria should not be resuscitated with boluses of isotonic crystalloid or colloids as outcome is worse than in children receiving maintenance fluids only. Children with signs of dehydration can safely be managed with dextrose/isotonic saline (0.9 %) maintenance fluids (4 ml/kg/h) since outcome is better than more aggressive rehydration or resuscitation with isotonic fluid boluses.

## References

1. Anstey NM, Russell B, Yeo TW, Price RN (2009) The pathophysiology of vivax malaria. Trends Parasitol 25(5):220–227
2. Hay SI, Guerra CA, Gething PW, Patil AP, Tatem AJ, Noor AM et al (2009) A world malaria map: Plasmodium falciparum endemicity in 2007. PLoS Med 6(3):e1000048

3.  Hay SI, Okiro EA, Gething PW, Patil AP, Tatem AJ, Guerra CA et al (2010) Estimating the global clinical burden of Plasmodium falciparum malaria in 2007. PLoS Med 7(6):e1000290

4.  Okiro EA, Alegana VA, Noor AM, Mutheu JJ, Juma E, Snow RW (2009) Malaria paediatric hospitalization between 1999 and 2008 across Kenya. BMC Med 7(1):75

5.  Okiro EA, Alegana VA, Noor AM, Snow RW (2010) Changing malaria intervention coverage, transmission and hospitalization in Kenya. Malar J 9:285

6.  Okiro EA, Bitira D, Mbabazi G, Mpimbaza A, Alegana VA, Talisuna AO et al (2011) Increasing malaria hospital admissions in Uganda between 1999 and 2009. BMC Med 9:37

7.  Dondorp AM, Fanello CI, Hendriksen IC, Gomes E, Seni A, Chhaganlal KD et al (2010) Artesunate versus quinine in the treatment of severe falciparum malaria in African children (AQUA-MAT): an open-label, randomised trial. Lancet 376(9753):1647–1657

8.  Maitland K, Kiguli S, Opoka RO, Engoru C, Olupot-Olupot P, Akech SO et al (2011) Mortality after fluid bolus in African children with severe infection. N Engl J Med 364(26):2483–2495

9.  Greenwood BM, Bradley AK, Greenwood AM, Byass P, Jammeh K, Marsh K et al (1987) Mortality and morbidity from malaria among children in a rural area of The Gambia, West Africa. Trans R Soc Trop Med Hyg 81(3):478–486

10. Newton CR, Krishna S (1998) Severe falciparum malaria in children: current understanding of pathophysiology and supportive treatment. Pharmacol Ther 79(1):1–53

11. English M, Waruiru C, Amukoye E, Murphy S, Crawley J, Mwangi I et al (1996) Deep breathing in children with severe malaria: indicator of metabolic acidosis and poor outcome. Am J Trop Med Hyg 55(5):521–524

12. Marsh K, Forster D, Waruiru C, Mwangi I, Winstanley M, Marsh V et al (1995) Indicators of life-threatening malaria in African children. N Engl J Med 332(21):1399–1404

13. Maitland K (2006) Severe malaria: lessons learned from the management of critical illness in children. Trends Parasitol 22(10):457–462

14. English M, Punt J, Mwangi I, McHugh K, Marsh K (1996) Clinical overlap between malaria and severe pneumonia in Africa children in hospital. Trans R Soc Trop Med Hyg 90(6):658–662

15. von Seidlein L, Olaosebikan R, Hendriksen IC, Lee SJ, Adedoyin OT, Agbenyega T et al (2012) Predicting the clinical outcome of severe falciparum malaria in african children: findings from a large randomized trial. Clin Infect Dis 54(8):1080–1090

16. Maitland K, Nadel S, Pollard AJ, Williams TN, Newton CR, Levin M (2005) Management of severe malaria in children: proposed guidelines for the United Kingdom. BMJ 331(7512):337–343

17. World Health Organization, Communicable Diseases Cluster (2000) Severe falciparum malaria. Trans R Soc Trop Med Hyg 94(Suppl 1):S1–90

18. Paediatric basic and advanced life support guidelines: An update. Paediatrics & child health. 2007; 12(6):495–500

19. Berkley J, Mwarumba S, Bramham K, Lowe B, Marsh K (1999) Bacteraemia complicating severe malaria in children. Trans R Soc Trop Med Hyg 93(3):283–286

20. Nadjm B, Amos B, Mtove G, Ostermann J, Chonya S, Wangai H et al (2010) WHO guidelines for antimicrobial treatment in children admitted to hospital in an area of intense Plasmodium falciparum transmission: prospective study. BMJ 340:c1350

21. Mancini N, Carletti S, Ghidoli N, Cichero P, Burioni R, Clementi M (2010) The era of molecular and other non-culture-based methods in diagnosis of sepsis. Clin Microbiol Rev 23(1):235–251

22. Graham SM, Walsh AL, Molyneux EM, Phiri AJ, Molyneux ME (2000) Clinical presentation of non-typhoidal Salmonella bacteraemia in Malawian children. Trans R Soc Trop Med Hyg 94(3):310–314

23. Mtove G, Amos B, von Seidlein L, Hendriksen I, Mwambuli A, Kimera J et al (2010) Invasive salmonellosis among children admitted to a rural Tanzanian hospital and a comparison with previous studies. PLoS One 5(2):e9244

24. Dondorp A, Nosten F, Stepniewska K, Day N, White N (2005) Artesunate versus quinine for treatment of severe falciparum malaria: a randomised trial. Lancet 366(9487):717–725

25. Murphy SA, Mberu E, Muhia D, English M, Crawley J, Waruiru C et al (1997) The disposition of intramuscular artemether in children with cerebral malaria; a preliminary study. Trans R Soc Trop Med Hyg 91(3):331–334

26. Hien T, Davis T, Chuong L, Ilett K, Sinh D, Phu N et al (2004) Comparative pharmacokinetics of intramuscular artesunate and artemether in patients with severe falciparum malaria. Antimicrob agents chemother 48:4234–4239

27. Brockman A, Price R, van Vugt M, Heppner D, Walsh D, Sooklo P et al (2000) Plasmodium falciparum antimalarial drug susceptibility on the northwestern border of Thailand during five years of extensive artesunate-mefloquine use. Trans R Soc Trop Med Hyg 94:537–544

28. Nealon C, Dzeing A, Muller-Romer U, Planche T, Sinou V, Kombila M et al (2002) Intramuscular bioavailability and clinical efficacy of artesunate in gabonese children with severe malaria. Antimicrob Agents Chemother 46(12):3933–3939

29. World Health Organization (2005) Hospital care for children: guidelines for the management of common illnesses with limited resources. Geneva, Switzerland. Report No.: ISBN 92 4 154670 0

30. Planche T, Onanga M, Schwenk A, Dzeing A, Borrmann S, Faucher JF et al (2004) Assessment of vol-

ume depletion in children with malaria. Plos Med 1(1):e18

31. Planche T (2006) Malaria and fluids—balancing acts. Trends Parasitol 21:562–567

32. Crawley J, Chu C, Mtove G, Nosten F (2010) Malaria in children. Lancet 375(9724):1468–1481

33. Maitland K, Levin M, English M, Mithwani S, Peshu N, Marsh K, Severe P et al (2003) Falciparum malaria in Kenyan children: evidence for hypovolaemia. QJM 96(6):427–434

34. Maitland K, Pamba A, Newton CR, Levin M (2003) Response to volume resuscitation in children with severe malaria. Pediatr Crit Care Med 4(4):426–431

35. Beare NA, Riva CE, Taylor TE, Molyneux ME, Kayira K, White VA et al (2006) Changes in optic nerve head blood flow in children with cerebral malaria and acute papilloedema. J Neurol Neurosurg Psychiatry 77(11):1288–1290

36. English M (2000) Life-threatening severe malarial anaemia. Trans R Soc Trop Med Hyg 94(6):585–588

37. Reyburn H, Mbatia R, Drakeley C, Carneiro I, Mwakasungula E, Mwerinde O et al (2004) Over-diagnosis of malaria in patients with severe febrile illness in Tanzania: a prospective study. BMJ 329(7476):1212

38. Evans JA, Adusei A, Timmann C, May J, Mack D, Agbenyega T et al (2004) High mortality of infant bacteraemia clinically indistinguishable from severe malaria. QJM 97(9):591–597

39. Taylor TE, Fu WJ, Carr RA, Whitten RO, Mueller JG, Fosiko NG et al (2004) Differentiating the pathologies of cerebral malaria by postmortem parasite counts. Nat Med 10(2):143–145

40. Bronzan RN, Taylor TE, Mwenechanya J, Tembo M, Kayira K, Bwanaisa L et al (2007) Bacteremia in Malawian children with severe malaria: prevalence, etiology, HIV coinfection, and outcome. J Infect Dis 195(6):895–904

41. Maitland K, Pamba A, Newton CR, Levin M (2003) Response to volume resuscitation in children with severe malaria. Pediatr Crit Care Med 4(4):426–431

42. Maitland K, Pamba A, English M, Peshu N, Marsh K, Newton CRJC et al (2005) Randomized trial of volume expansion with albumin or saline in children with severe malaria: preliminary evidence of albumin benefit. Clin Infect Dis 40:538–545

43. Akech S, Gwer S, Idro R, Fegan G, Eziefula AC, Newton CR et al (2006) Volume expansion with albumin compared to gelofusine in children with severe malaria: results of a controlled trial. PLoS Clin Trials 1(5):e21

44. Southall DP, Samuels MP (2011) Treating the wrong children with fluids will cause harm: response to 'mortality after fluid bolus in African children with severe infection'. Arch Dis Child 96(10):905–906

45. Myburgh JA (2011) Fluid resuscitation in acute illness–time to reappraise the basics. N Engl J Med 364(26):2543–2544

46. Hilton AK, Bellomo R (2011) Totem and taboo: fluids in sepsis. Critical Care 15(3):164

47. Duke T (2011) What the African fluid-bolus trial means. Lancet 378(9804):1685–1687

48. Bates I (2011) Detrimental effect of fluid resuscitation in the initial management of severely ill children in Africa. Transfus Med 21(5):289–290

49. Maitland K, Babiker A, Kiguli S, Molyneux E (2012) The FEAST trial of fluid bolus in African children with severe infection. Lancet 379(9816):613; (author reply − 4)

50. Maitland K, Akech S, Russell E (2011) Mortality after fluid bolus in African children with sepsis reply. N Engl J Med 365(14):1351–1353

# Index

N. Curtis et al. (eds.), *Hot Topics in Infection and Immunity in Children IX*,
Advances in Experimental Medicine and Biology 764, DOI 10.1007/978-1-4614-4726-9,
© Springer Science+Business Media New York 2013

Printed by Printforce, the Netherlands